DATA STRUCTURES WITH MODULA-2

MICHAEL B. FELDMAN

George Washington University

Prentice Hall, Englewood Cliffs, New Jersey 07632

Library of Congress Cataloging-in-Publication Data

FELDMAN, MICHAEL B.
   Data structures with Modula-2.

   Includes index.
   1. Modula-2 (Computer program language) 2. Data structures (Computer science) I. Title.
QA76.73.M63F45 1987   005.7′3   87-11368
ISBN 0-13-197344-4

Editorial/production supervision and
  interior design: **Ellen B. Greenberg**
Cover design: **20/20 Services, Inc.**
Manufacturing buyer: **Barbara Kelly Kittle**

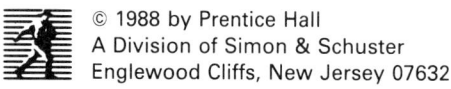
© 1988 by Prentice Hall
A Division of Simon & Schuster
Englewood Cliffs, New Jersey 07632

All rights reserved. No part of this book may be
reproduced, in any form or by any means,
without permission in writing from the publisher.

Printed in the United States of America

10 9 8 7 6 5 4 3 2 1

ISBN 0-13-197344-4 025

PRENTICE-HALL INTERNATIONAL (UK) LIMITED, *London*
PRENTICE-HALL OF AUSTRALIA PTY. LIMITED, *Sydney*
PRENTICE-HALL CANADA INC., *Toronto*
PRENTICE-HALL HISPANOAMERICANA, S.A., *Mexico*
PRENTICE-HALL OF INDIA PRIVATE LIMITED, *New Delhi*
PRENTICE-HALL OF JAPAN, INC., *Tokyo*
SIMON & SCHUSTER ASIA PTE. LTD., *Singapore*
EDITORA PRENTICE-HALL DO BRASIL, LTDA., *Rio de Janeiro*

TO MY THREE BEST FRIENDS:

MY WIFE RUTH AND MY BOYS BEN AND KEITH

CONTENTS

Preface xiii

1 Abstraction and Abstract Data Types 1

- 1.1 **GOAL OF THE CHAPTER** 1
- 1.2 **ABSTRACTION VS. IMPLEMENTATION** 1
  - 1.2.1 Numbers and Variables, 2
  - 1.2.2 Two-dimensional Arrays, 3
  - 1.2.3 Blocked Files, 4
- 1.3 **THE ABSTRACT DATA TYPE (ADT)** 6
  - 1.3.1 Data Types, Abstract Data Types, and Data Structures, 6
  - 1.3.2 ADTs and Complexity Reduction, 7
- 1.4 **THE LIBRARY MODULE: MODULA-2's MECHANISM FOR ADTs** 8
  - 1.4.1 Libraries, Clients, and Import/Export, 8
  - 1.4.2 Modula-2 Libraries: Definition and Implementation Modules, 9
- 1.5 **CONSTRUCTING AN ADT IN MODULA-2: FRACTIONS** 9
  - 1.5.1 Fractions as Mathematical Entities, 10
  - 1.5.2 Implementing Fractions in Modula-2, 11
  - 1.5.3 Classifying Operations: Constructors, Selectors, and Predicates, 11

1.5.4 Putting It All Together, 13
1.5.5 Using the Library Module in a Client Program, 15

1.6 FUNCTION/EXPRESSION NOTATION AND POINTERS 17
1.6.1 The Pointer in Modula-2, 18
1.6.2 Encapsulating All the Details: Opaque Export, 19
1.6.3 What Happens to the Allocated Memory?, 20

1.7 DESIGN: AN ABSTRACT DATA TYPE FOR TEXT STRINGS 22

1.8 STYLE GUIDE: USING ADTs in OLDER LANGUAGES 25

1.9 SUMMARY 27

1.10 EXERCISES 27

2 Algorithms, Recursion, and Performance Prediction   29

2.1 GOAL OF THE CHAPTER 29

2.2 ALGORITHMS AND ALGORITHM DESIGN 30

2.3 RECURSIVE ALGORITHMS 31
2.3.1 Reverse of a String, 32
2.3.2 Permutations of a Set, 33
2.3.3 Recursive Binary Search, 36
2.3.4 Recursive Merge/Sort, 38

2.4 PERFORMANCE PREDICTION AND THE "BIG O" NOTATION 39
2.4.1 Algorithm Growth Rates, 40
2.4.2 Estimating the Growth Rate of an Algorithm, 40
2.4.3 Some Examples Of Performance Prediction, 48

2.5 DESIGN: MAINTAINING A DYNAMIC TABLE 49
2.5.1 First Implementation: Unordered Array, 50
2.5.2 Second Implementation: Array Ordered by Key, 51

2.6 SUMMARY 53

2.7 EXERCISES 54

3 Arrays, Vectors, Matrices, and Lists 55

- 3.1 GOAL OF THE CHAPTER 55
- 3.2 CLASSICAL REPRESENTATION OF VECTORS AND MATRICES 55
  - 3.2.1 Vectors and One-Dimensional Arrays, 56
  - 3.2.2 Matrices and Two-Dimensional Arrays, 58
  - 3.2.3 Higher-Dimensional Structures, 61
- 3.3 VECTORS, MATRICES, AND MODULA-2 ARRAYS 63
  - 3.3.1 Vector Arithmetic and Open Array Parameters, 63
  - 3.3.2 Matrix Arithmetic, 68
- 3.4 DENSELY PACKED STRUCTURES WITH MANY "ZERO" ELEMENTS 71
  - 3.4.1 Lower Triangular Matrices, 71
  - 3.4.2 Symmetric Matrices, 73
  - 3.4.3 Band Matrices, 73
- 3.5 SPARSE VECTORS AND MATRICES 74
  - 3.5.1 Sparse Vectors, 75
  - 3.5.2 Linking the Elements Together, 79
- 3.6 DESIGN: VECTORS, MATRICES, AND FUNCTIONAL NOTATION 83
  - 3.6.1 A Dynamic Storage Scheme for Vectors and Matrices, 84
  - 3.6.2 A General Module for Matrices, 85
- 3.7 STYLE GUIDE: LANGUAGES WITHOUT RECORD TYPES 86
- 3.8 SUMMARY 87
- 3.9 EXERCISES 91

4 Linked Lists, Pointers, and Cursors 93

- 4.1 GOAL OF THE CHAPTER 93
- 4.2 LINKED STRUCTURES 94
- 4.3 POINTERS AND DYNAMIC MEMORY MANAGEMENT 94

- **4.4 USING DYNAMIC ALLOCATION FOR LINKED STRUCTURES** 97
  - 4.4.1 Creating a One-Way Linked List, 97
  - 4.4.2 An ADT for One-Way Linked Lists, 99
  - 4.4.3 One-Way Lists with Head and Tail Pointers, 102
  - 4.4.4 Building An Ordered List, 104
  - 4.4.5 Two-Way Linked Lists, 109
- **4.5 DESIGN ONE: SPARSE VECTORS AND MATRICES REVISITED** 111
  - 4.5.1 Vector Arithmetic, 111
  - 4.5.2 Sparse Matrices, 116
- **4.6 DESIGN TWO: TEXT HANDLING REVISITED** 117
- **4.7 STYLE GUIDE ONE: SIMULATING DYNAMIC ALLOCATION** 119
- **4.8 STYLE GUIDE TWO: FRACTIONS REVISITED** 127
- **4.9 SUMMARY** 131
- **4.10 EXERCISES** 131

5 Queues and Stacks 133

- **5.1 GOAL OF THE CHAPTER** 133
- **5.2 QUEUES AND STACKS** 133
- **5.3 QUEUES** 134
  - 5.3.1 Array Implementation of a Queue, 135
  - 5.3.2 Circular Array Implementation of a Queue, 135
  - 5.3.3 Linked-List Implementation of a Queue, 141
- **5.4 STACKS** 141
  - 5.4.1 Array Implementation of a Stack, 142
  - 5.4.2 Linked-List Implementation of a Stack, 144
- **5.5 STACKS, EVALUATION OF EXPRESSIONS, AND POLISH NOTATION** 144
  - 5.5.1 Evaluating RPN Expressions, 145
  - 5.5.2 Converting Manually from Infix to RPN Form, 147
- **5.6 DESIGN ONE: AN INFIX-TO-RPN TRANSLATOR** 150

Contents

- 5.7 DESIGN TWO: AN EVENT-DRIVEN SIMULATION 155
- 5.8 SUMMARY 156
- 5.9 EXERCISES 161

6 Directed Graphs 163

- 6.1 GOAL OF THE CHAPTER 163
- 6.2 INTRODUCTION 163
- 6.3 PROPERTIES OF DIGRAPHS 164
- 6.4 IMPLEMENTATIONS OF DIRECTED GRAPHS 169
  - 6.4.1 Adjacency Matrix, 169
  - 6.4.2 Adjacency List, 170
  - 6.4.3 Weighted Adjacency Matrix, 172
  - 6.4.4 State Table, 172
- 6.5 GRAPH TRAVERSALS 173
  - 6.5.1 Depth-First Search, 174
  - 6.5.2 Breadth-First Search, 174
- 6.6 DESIGN: A SIMPLE LEXICAL SCANNER 177
- 6.7 SUMMARY 179
- 6.8 EXERCISES 180

7 Tree Structures 182

- 7.1 GOAL OF THE CHAPTER 182
- 7.2 TREES 183
- 7.3 BINARY TREES 188
  - 7.3.1 Properties of Binary Trees, 188
  - 7.3.2 Implementing Binary Trees, 191
  - 7.3.3 Traversals of Binary Trees, 193
- 7.4 EXPRESSION TREES 195
  - 7.4.1 Constructing Expression Trees, 195
  - 7.4.2 Traversing Expression Trees, 199
- 7.5 BINARY SEARCH TREES (BSTs) 200
  - 7.5.1 The Report Operation: Traversing a BST, 200
  - 7.5.2 The Update Operation: Inserting a Record in a BST, 201

- 7.5.3 The Search Operation: Finding a Record in a BST, 204
- 7.5.4 The Delete Operation: Deleting a Record from a BST, 205

7.6 GENERAL TREES 208
- 7.6.1 Digital Search Trees, 208
- 7.6.2 B-Trees, 211

7.7 DESIGN ONE: BUILDING AN EXPRESSION TREE 215

7.8 DESIGN TWO: A CROSS-REFERENCE GENERATOR 215
- 7.8.1 The Table Handler, 218
- 7.8.2 The Scanner, 221

7.9 STYLE GUIDE: THREADING TREES FOR EFFICIENCY 227

7.10 SUMMARY 229

7.11 EXERCISES 232

8 Hash Table Methods 234

8.1 GOAL OF THE CHAPTER 234

8.2 SEQUENTIAL AND BINARY SEARCH REVISITED 235

8.3 THE HASH TABLE 235

8.4 CHOOSING HASH FUNCTION 237
- 8.4.1 Truncation, 238
- 8.4.2 Division, 238
- 8.4.3 Mid-Square, 238
- 8.4.4 Folding or Partitioning, 240

8.5 RESOLVING COLLISIONS IN HASH TABLES 240
- 8.5.1 Linear Probing, 240
- 8.5.2 Nonlinear Probing, 243
- 8.5.3 Bucket Hashing, 243
- 8.5.4 Ordered Hashing, 244

8.6 DESIGN: HYBRID SEARCH STRATEGIES 245

8.7 SUMMARY 245

8.8 EXERCISES 246

9 Internal Sorting Methods — 248

- **9.1** GOAL OF THE CHAPTER 248
- **9.2** SORTING TERMINOLOGY 249
- **9.3** INTERNAL SORT ALGORITHMS WITH GROWTH RATE $O(N^2)$ 249
  - 9.3.1 Simple Section, 250
  - 9.3.2 Delayed Selection, 251
  - 9.3.3 Bubble Sort, 252
  - 9.3.4 Linear Insertion, 254
  - 9.3.5 Binary Insertion, 256
- **9.4** INTERNAL SORTS WITH GROWTH RATE $O(N \log(N))$ 257
  - 9.4.1 Merge Sort, 257
  - 9.4.2 Heap Sort, 260
  - 9.4.3 Quicksort, 271
- **9.5** OTHER INTERNAL SORT ALGORITHMS 275
  - 9.5.1 Shell Sort, 275
  - 9.5.2 Quadratic Selection, 275
  - 9.5.3 Radix Sort, 278
- **9.6** DESIGN: PRIORITY QUEUES 281
- **9.7** SUMMARY 282
- **9.8** EXERCISES 282

10 Sorting External Files — 283

- **10.1** GOAL OF THE CHAPTER 283
- **10.2** THE TWO-WAY MERGE 283
- **10.3** THE K-WAY MERGE 285
- **10.4** SIMPLE MERGE SORTING 285
- **10.5** THE "NATURAL" DISTRIBUTION 287
- **10.6** POLYPHASE SORTING 288
- **10.7** FIBONACCI DISTRIBUTION FOR POLYPHASE MERGING 290

10.8	**SUMMARY**	292
10.9	**EXERCISES**	292

Bibliography   **293**

Index   **297**

PREFACE

Textbook literature in data structures and algorithms falls generally into two categories: the mathematical or theoretical, and the programming-oriented or applied. Books in the former category place heavy emphasis on the axiomatic specification of data types and/or algorithm performance prediction ("big O" and the like). Those in the latter category emphasize the programming side, with relatively little mention of algorithm performance (usually concentrated in the chapter on sorting). Although there is helpful concentration on the implementation of data structures in a real-world programming language, Pascal in most recent cases, abstract data types are treated cursorily if at all.

This book attempts a compromise, a blend of the theoretical and the applied. It presents students with the means to understand and design data structures and their implementations, and emphasizes the information-hiding principle to encourage both stepwise decomposition and encapsulating or "burying" implementation details. Since this abstraction can seriously affect the growth rate or "big O" of an algorithm if not handled carefully, attention is paid to performance prediction, at least to the degree that students will learn the earmarks of linear, quadratic, and logarithmic growth rates.

It is important that students develop an understanding of the tradeoffs between space and time and between abstraction and program performance. They must become aware that while micro-efficiency is less crucial than it was in the days of slow processors and limited, expensive memory, macro-performance needs to be considered right along with other factors in designing and implementing algorithms. Indeed, using an algorithm with a relatively slow growth rate often makes feasible the very solution of a problem, while no amount of "bit-fiddling" or micro-optimization will save a poor algorithm from ultimate disaster.

Students who continue in the formal study of data structures and algorithms will certainly have ample opportunity to cover these performance issues more rigorously. Those who do not continue with a more theoretical course still need to experience at least some "consciousness-raising"; they should develop a sensitivity to estimating their program's likely performance and to the fact that there is rarely a "free lunch."

CURRICULUM DESIGN

George Washington University (this author's home institution) offers a one-semester course entitled *Programming and Data Structures*. It is a second or third course in programming, and is populated by sophomores, juniors, and beginning graduate students. The course is similar to that offered by many institutions with a practical bent, and corresponds roughly to a "heavy CS2" or a "light CS7." This book has been designed to be used straight through from beginning to end with this type of course. Experience shows that students and instructors alike are more comfortable with such a book than with one where entire chapters must be taken out of sequence or skipped altogether to fit a one-semester course. Since "structured programming" in the sense of block structuring, use of procedures and functions, and strictly-controlled use of the "goto" statement, is taught almost universally in entry-level programming courses, it is a safe assumption that students already are comfortable with a modern language like Pascal, PL/1, Algol-W, C, or Fortran-77.

WHY MODULA-2?

Students should see algorithms presented in a real programming language, not a "toy" synthesized for pedagogical purposes. This book uses Modula-2, a derivative of Pascal which emerged in 1982 from Niklaus Wirth's group at the Zurich Institute of Technology. Modula-2 embodies all of the virtues of Pascal, and inherits most of Pascal's inner syntax. Some of Pascal's better-known syntactic difficulties have been remedied.

As its name suggests, Modula-2's most important new feature is the *module*. Supporting separate compilation and information hiding in a way not possible in standard Pascal, the module facility encourages good design of systems of nontrivial size. The module idea extends all the way down to low-level features of the target computer, enabling Modula-2 to be reliably used as a systems implementation language with little or no required use of assembler routines.

Modula-2's most significant contribution to the pedagogy of data structures is that the module facility permits specification and implementation of abstract data types in a natural and clean fashion. Operations on a defined data type can be encapsulated along with the definition of that type, and the designer may stipulate those operations which are accessible to a user and those which are purely internal. This notion is in accord with current thinking about how programs should be developed for reliability and maintainability. Thinking in terms of modules or "software components" should be inculcated very early in a student's experience. Programming in Modula-2 facilitates this kind of thought process.

Compilers for Modula-2 are available for many of the minicomputer systems in wide use in computer science departments, and for most microcomputers, at surprisingly modest cost. Many institutions, then, will be able directly to use the modules and programs developed in this book. For those not yet able to do so, these modules and programs may be used as designs to be implemented in whatever programming language is readily at hand. While many of the programs are given in full, others (although fully tested) have been left incomplete, allowing students to fill in the missing pieces as a learning exercise.

DESIGN SECTIONS

Most chapters in this book include one or two Design sections. Except for the Design section in Chapter 1, which is strongly recommended for all readers, these sections are optional; the main flow of the book is not disturbed by omitting them. Each Design section is an application of the structures introduced in the current chapter, and is usually presented as an explanation of the application, some diagrams, and part of the Modula-2 code for the solution. These sections are, in general, purposely left unfinished so that students may complete them, either as self-study or as an assigned problem. Most of the problems are interesting and provocative enough to warrant assignment as programming projects to be coded and tested.

While the Design sections are independent of the main flow of the book, they do tend to build upon each other. For example, the simple text-handling package introduced in Chapter 1 is elaborated in the Chapter 4 Design section, then used in the Design sections of Chapters 5, 6, and 7. Similarly, the sparse vector and sparse matrix concepts introduced in Chapter 3 are used in a Design section in Chapter 4, and the very simple state-graph lexical scanner presented in the Chapter 6 Design section is elaborated in a corresponding section of Chapter 7. Finally, the scanner written in the Chapter 5 Design section to generate the Reverse Polish Notation form of an arithmetic expression is modified in Chapter 7 to produce an expression tree. Thus, while the Design sections demonstrate interesting uses of data structures, they serve to strengthen and reinforce each other. This is both appropriate and exemplary of the re-use of application code as well as packages.

STYLE GUIDES

An unusual feature of this book is the Style Guide section found at the end of most chapters. Although data structures textbooks should present important structures and algorithms in the most modern and powerful language forms available in a "real-world" language such as Modula-2, many students will find themselves in situations where programs must be coded in languages like Fortran whose support for data structures, even in the 1977 Standard version, is sorely lacking. The Style Guide sections demonstrate how one can do powerful programming with less-than-powerful languages, indicating how the ideas expressed in a chapter may be coded in languages where no built-in support is provided.

One Style Guide (in Chapter 3) discusses how to code structures which are logically records, even though the coding language (such as Fortran or Basic) doesn't support records; another (in Chapter 4) takes up the problem of simulating linked structures using arrays and array indices ("cursors") where no language support is available for pointers and dynamic allocation. Yet another Style Guide (in Chapter 7) discusses threaded implementations of binary trees, to support fast, non-recursive, traversal routines.

The Style Guides make useful reading for students interested in language-to-language comparisons, or needing to code in older languages, but the Style Guides can safely be skipped in a one-semester course focusing exclusively on Modula-2.

ADVICE AND CAVEATS

This book uses Modula-2, but it is not a "Modula-2 book" in the sense of teaching the language details one at a time. Students with some Pascal programming experience can learn Modula-2 very rapidly, by reading examples rather than rules; students with no Pascal experience may need to invest in a textbook dealing specifically with the Modula-2 language. Most of those in print at the time this book went to press are listed in the Bibliography.

Programs and modules appearing in this book have been thoroughly tested for both language and algorithmic correctness on two compilers: the University of Hamburg compiler on a Digital Equipment Corporation VAX/11-780, and *MacMETH*, a recent compiler for the Apple Macintosh developed at Zurich and marketed in the United States by Modula Corporation. The language style used in the programs is straightforward and non-esoteric, leading both to easy understanding of the programs and minimal risk of compiler-dependent difficulties.

Modula-2 is an evolving language and there are slight differences among implementations. In this book's programs, the most likely area of difference will be found in the names and location in the system modules of dynamic memory allocation procedures. In some compilers these are found in module STORAGE; in others they are in System. Also, Modula-2 is notoriously case-sensitive (X and x are different identifiers!), but the dynamic allocation procedure may be named

Allocate in one compiler but ALLOCATE in another. The code in this book is entirely consistent with *MacMETH's* library structure. There may be minor incompatibilities with other implementations, but it is futile to tabulate these because all the compilers are evolving over time. Students and teachers simply need to read their implementation documents and exercise a bit of care.

IN CONCLUSION

Through its early and consistent use of abstract data types as implemented in real-world modules, this book, I hope, will add to the pedagogy of data structures. This book also contributes to the emerging Modula-2 literature by highlighting Modula-2 as a general-purpose programming language for writing modern, well-structured systems using nontrivial data structures. I hope that the pedagogical features of student-oriented Modula-2 compilers will soon evolve to the point that Modula-2 is seen as a replacement for Pascal in the first programming course: at that point students will have a growth path from program-writing to system-building within a single standard coding language.

ACKNOWLEDGMENTS

This book is adapted from my 1985 book *Data Structures with Ada*. Many people contributed to both books in many ways. Keith Allen, Roie Black, Myron Calhoun and Stephen Lowe served as formal reviewers of the Ada book, and their influence still shows in this adaptation. Countless students in Computer Science 159 deserve a vote of thanks for patiently and without complaint learning data structures from manuscripts in various stages of completion. Graduate students and colleagues have read parts of the manuscript and offered valuable guidance; especially helpful was Melinda Moran. The George Washington University Engineering Computing Facility provided VAX time for testing programs. Jim Fegen and Ellen Greenberg, respectively, the acquisition and production editors at Prentice Hall, have been very encouraging and helpful. Finally, Ruth, Ben, and Keith Feldman have been loving and patient through the ups and downs of getting two books in print: they are true friends.

Chapter 1

ABSTRACTION AND ABSTRACT DATA TYPES

1.1 GOAL OF THE CHAPTER

Abstraction and abstract data types (ADTs) play an important role in modern program design and coding. Understanding these concepts fully is particularly important in a text devoted to data structures. In this chapter we shall explore the notions of abstraction and implementation, define abstract data types, and show examples of how the latter can be written as Modula-2 *modules*.

Specifically, rational numbers or fractions and text strings will be treated, and modules for them will be sketched out.

At the end of the chapter you will encounter the first of a series of style guides, where you will be given hints about how to implement concepts explained in the chapter, but using languages such as Fortran, Pascal, C, and PL/1, which do not have built-in facilities for building modules.

1.2 ABSTRACTION AND IMPLEMENTATION

Examples of abstraction abound in everyday programming, although you may not have recognized them as such. In this section we consider several common abstractions you may have used, namely, *numbers and variables, two-dimensional arrays*, and *blocked files*.

1.2.1 Numbers and Variables

Your ability to use things like integers and real numbers in programming depends on abstraction, because, in reality, nothing exists in the computer but groups of bits, operated on by machine instructions which "understand" the connection between that which you perceive as an integer (written in your program as a base 10 number!) and some group of eight or sixteen or thirty-two bits in the computer's memory.

In high-level-language (HLL) programming, we frequently use *variables*. A variable is a form of abstraction: the programmer may declare an integer variable X, for example, and assign to it the value 735, as shown in Figure 1–1(a). In translating the HLL source program into an object- or machine-language program, the compiler determines which location or address in computer memory will correspond to this variable and produces machine instructions which, when the object program is executed, will carry out the assignment of 735 to X. Assuming that the compiler associates X with locations (bytes) 1246 and 1247, Figure 1–1(b) depicts the result after the assignment is made.

Arithmetic on integer quantities is usually carried out by hardware instructions. To understand the kind of software-level abstraction we shall be doing in this book, consider arithmetic on real quantities. In many computers, no real or floating-point instructions are available in the hardware instruction set. Or they are options the purchaser chose not to pay for. So when you write, say, an assignment statement X := X × Y + 3.0 (in some programming language), the compiler for that language not only figures out which hardware memory locations to use to store the variables X and Y, but may also have to generate calls to *subroutines* to do the addition and multiplication operations.

The point is that through the use of the *abstraction* "variable name" and the *abstraction* "real number," you do not have to worry about the details of the internal storage or actual instructions used to *implement* the calculation you specify when you write an assignment statement. Abstraction is the way we arrive at a situation called *information hiding*, in which unnecessary details of a given data represen-

```
VAR X: INTEGER;--reserves space
       :
    X: = 735;--stores value
       :
```

x | 00000010 | 11011111 |

(a) Abstraction of integer assignment as seen by the programmer

(b) Sixteen-bit location (two bytes) with binary equivalent of 735 stored

Figure 1–1 Abstraction and implementation of integer assignment

tation or procedure are hidden from those who have no need (or desire) to see them.

We shall frequently contrast *abstraction* and *implementation*. The abstraction is essentially that which is made visible to the user, in the foregoing example the high-level-language (HLL) programmer; the implementation is all the messy details that have been hidden away. In the example, we have used an abstraction we might call RealNumbers, including the operations of *addition, multiplication*, and *assignment* of reals. There is also a *creation* operation, which we used explicitly by declaring X and Y to be real. The implementation of a real number as an area of memory divided into parts for the mantissa and exponent, and of the operations as subroutines to be called by your machine-language program, have been taken care of by the compiler designer.

In some languages, for example Fortran, the creation operation is implicit. A variable need not be declared because the compiler recognizes it as such by its first use. Further, in Fortran, the distinction between real and integer variables can be made implicitly: variables beginning with the letters I, J, K, L, M, and N are treated as integer variables, while those beginning with other letters are treated as real variables.

1.2.2 Two-dimensional Arrays

You have probably used two-dimensional arrays somewhere in your programming experience. You are probably aware, however, that the computer's memory is not two-dimensional, but is addressed as just a one-dimensional or linear sequence of bytes or words. Clearly, then, something must be sitting between your high-level-language program and the computer's linear memory which can interpret a statement like A[3,4] := B[4,7] + 1.0 correctly. As in the previous example, this something is the compiler, by means of which you can use abstraction to give you expressive power not present in the machine itself.

The abstraction used in two-dimensional arrays might be called RectangularArrays, and would include the operations of *retrieval* (the subscripted reference B[4,7] to the right of a := sign), *assignment* (the reference A[3,4] to the left of a := sign), and *creation* (the declaration of the arrays and their sizes at the beginning of the program). As before, the compiler designer has seen to the implementation of the rectangular arrays as areas of linear memory, and the assignment and retrieval operations as formulas, both expressed in machine language.

Note that the abstraction RectangularArrays implicitly makes use of the abstraction RealNumbers, since the values stored in the rectangular array are just that. This illustrates the capability of abstraction to "nest" itself many levels deep.

We shall return to arrays in detail in Chapter 3; in the meantime, study Figure 1-2, which shows the abstraction and implementation of a 3 × 4 integer array.

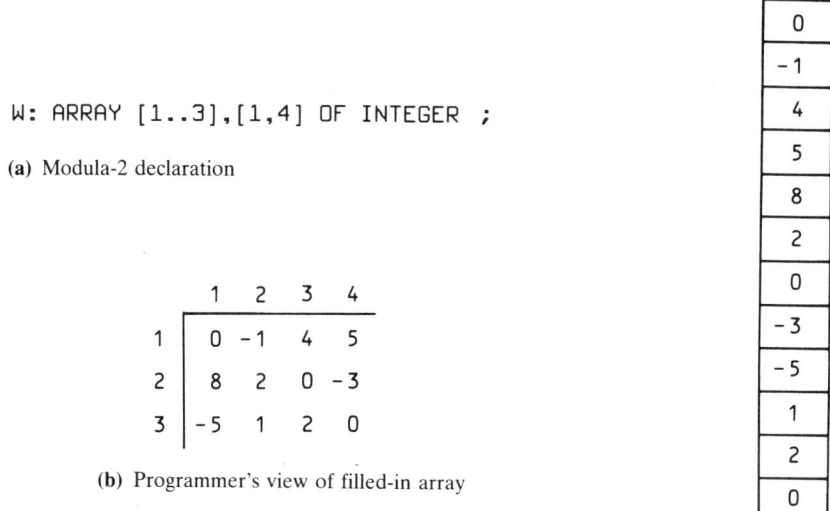

```
W: ARRAY [1..3],[1,4] OF INTEGER ;
```

(a) Modula-2 declaration

(b) Programmer's view of filled-in array

(c) Filled-in array as stored by compiler

Figure 1-2 Abstraction and implementation of two-dimensional array

1.2.3 Blocked Files

Yet another example can be drawn from the area of files on magnetic tape. Useful information on tape is typically recorded at a density of 1600 characters per running inch of tape (and often more), and a gap of about $\frac{3}{4}$ inch is left between groups of useful characters (to allow the tape motors space to accelerate and decelerate before and after reading). Thus, a file of, say, personnel records of 200 characters each would waste much more tape than it would use if each record were stored on its own "chunk" of tape, since a record would occupy only $\frac{1}{8}$ inch of tape followed by a $\frac{3}{4}$-inch gap.

For this reason, records are often *blocked* on tape or disk files. That is, a number of actual records are grouped together on one "chunk" of tape between gaps. If the "blocking factor" were, say, ten, then a block would occupy $1\frac{1}{4}$ inch of tape, with the same $\frac{3}{4}$-inch gap.

However, since tape is relatively cheap, this more economical use of storage is really a secondary reason for blocking. A more important reason is that there is a fair amount of overhead associated with each tape read or write operation: the time taken to set up the operation and to start and stop the tape-drive motor is significant compared with the time taken to transfer the information on the tape to and from main storage. Thus, time is saved—as is motor wear and tear—if a

Sec. 1.2 Abstraction and Implementation

lot of information can be read or written in a single operation, once the drive motor is up to speed.

Now suppose that you write a program in a high-level language to process the preceding file. Your program is written to process one record at a time, yet in actuality a number of records—a block—are being read from your tape file in one I/O operation. Should you be worried about this "mismatch?" Emphatically not! In fact, the operating system or the compiler designer has applied abstraction to hide these messy details from you, and your program ends up processing exactly the record it "wants." In effect, the abstraction LogicalRecord is implemented as a block of records written as one physical tape record: you use the operations of reading and writing records, which are implemented so that even though your program executes many "write" operations, an actual write to tape is executed only when a block of records has been assembled in an area of main storage, usually called a *buffer*.

In the terminology of operating systems, we refer to *logical* vs. *physical* records, files, devices, etc., where these terms corresponded, respectively, to abstraction and implementation. The tape example is illustrated in Figure 1–3.

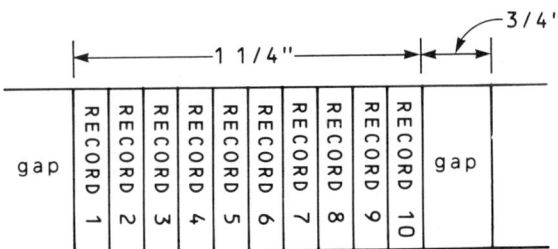

(a) Magnetic tape with 1600 character/inch storage density, showing storage of unblocked, 200-character records. Each input operation reads one physical record, thus one logical record.

(b) Magnetic tape with same density and record size, but with ten-record blocks. Each input operation reads one physical record but ten logical records.

Figure 1–3 Abstraction and implementation of magnetic tape file

1.3 THE ABSTRACT DATA TYPE (ADT)

Informally, an abstract data type (ADT) is a data type of some sort, together with all the operations that allow us to create and manipulate objects of that type. For example, the set of all the integers representable in some computer, together with all the integer assignment, arithmetic, and comparison operators we are used to, constitute an ADT. As it happens, most languages in wide use include this particular ADT as part of their standard facilities; an ADT could also be created by a programmer and "added on" to the language as a set of subroutines.

1.3.1 Data Types, Abstract Data Types, and Data Structures

Terminology in the computing field is often confusing, particularly where the same concept is known by several names or where different concepts have similar names. This lack of standardization comes about because computing is a very new field with many people participating, and in many cases there is no general agreement on terms. Since the area of ADTs is newer still, and different writers use different terms, it is important for us to define our terms very carefully and use them consistently. To get started, we point out the difference between the terms *data type*, *abstract data type*, and *data structure*.

The term *data type* (sometimes just *type*) is used by programming language designers to mean the set of values that a given variable is permitted to take on. A variable of type *integer*, for example, is allowed to be assigned integral values between two extremes, often related to the largest positive and negative integers representable in the underlying hardware. For a 16-bit computer, an integer variable typically will be allowed to take on values between $-32,768$ and $+32,767$.

Similarly, a variable of type *Boolean* may have only the values *true* and *false*.

Every programming language has a set of built-in or *primitive* types for its variables. In Fortran these primitives are INTEGER, REAL, LOGICAL, CHARACTER, and COMPLEX; in Basic they are INTEGER, REAL, and STRING; and in Pascal and Modula-2 they are INTEGER, REAL, BOOLEAN, and CHAR. Cobol permits character strings, decimal numbers, and several different kinds of binary (or computational) numbers as types.

Many recent programming languages, such as Pascal, Ada, and Modula-2, permit the definition of new types. Such a type may be just a restriction on the range of a primitive type, e.g., the integers between 1 and 10, or a combination or grouping of primitives, e.g., a *record* consisting of a student number of integer type, a student name of character-string type and a course average of real type. A type in which each object consists of only a single value is known as a *scalar type*; a type in which each object may consist of a set of values, such as the record just described, is called a *composite* or *structured* type.

An *abstract data type* or *ADT* includes not only a data-type definition, but also a specification of the operations that are valid for the new type. Nothing in

Sec. 1.3 The Abstract Data Type (ADT)

the type-definition facilities of Pascal or Modula-2 *requires* such a specification. Modula-2 *permits* it, however, as we shall see later. But even if one's programming language does not permit the direct definition of an ADT, the concept is extremely useful in designing programs and can thus be thought of as a kind of "pseudocode" or design language.

What, then, is a *data structure?* A data structure is a way of *representing* or implementing an ADT in a program. For example, in a hypothetical language that did not have reals as a primitive data type, we might define an ADT for reals, and then decide to represent a real number as a one-dimensional array with two elements, both integers, corresponding to the mantissa and exponent of the number, respectively. Alternatively, we could use a record structure if the language permitted records. Arrays and records are two commonly used data structures. We shall consider others in the course of this book as well.

TO SUMMARIZE:

- A *data type* is the set of values that are valid for a set of variables;
- An *abstract data type* is a data type together with a set of operations is valid for that type;
- A *data structure* is a concrete way of implementing an ADT.

1.3.2 ADTs and Complexity Reduction

Students of the human mind have discovered that it can deal with no more than a certain amount of complexity at a time, or else it gets very confused. Program design and coding is a complex activity, and students of programming as a human endeavor have discovered that many large programs have performed inadequately or been difficult to maintain because the humans dealing with them have had to cope with too much complexity at once. Much work has been done recently in developing techniques for building less complex programs by reducing the amount of detail that a given programmer has to deal with at a given time.

One of the main goals of this book is to illustrate how complexity in programming can be reduced by hiding away detailed information from program segments that have no need to know that information. This is often called *encapsulation* or *packaging*. An important way of encapsulating is through the use of ADTs, which enable mathematical concepts like real numbers, sets, graphs, trees, and so on to be treated as such. With encapsulation, we can concentrate on the mathematical properties of these entities, their visible structure, and the set of operations that can be performed on them, without concern for the details of how they should actually be represented in a program.

Of course, we cannot ignore these details entirely, since programs really do need to get written! Rather, we *postpone* consideration of unnecessary detail, defining our objects first as ADTs for use by higher level programs, and later as concrete data structures which implement those objects, and concrete functions and procedures which implement those operations. If we use our programming languages well, we are then able to find ways of encapsulating those implementations so that the details are hidden from individuals and programs that do not need them.

Encapsulation is relatively easy in recently created languages such as Ada and Modula-2, and it is also possible, though more difficult, in other languages. Often, searching out ways to do it in these languages pays off handsomely in reduced program complexity and enhanced program reliability.

1.4 THE LIBRARY MODULE: MODULA-2's MECHANISM FOR ADTs

The very name Modula-2 suggests the word "module." Indeed, the notion of modules is central to the language. For instance, a main program is called a module. But the important feature of Modula-2 for building ADTs is called the *library module*. A library module is designed to encapsulate *types, variables, constants*, and *procedures* in a unit which can be separately compiled and stored in a library.

In studying how library modules are used to build ADTs, we shall be making frequent reference to five important terms: *client program, export, import, definition module, and implementation module*. We consider these concepts in turn.

1.4.1 Libraries, Clients, and Import/Export

A program library is of no use unless some program wishes to use it. Let us call a program wishing to use the library a *client program*. Now what information does the client program need in order to make effective use of the library? Suppose the library consists entirely of a collection of subprograms. Then for each subprogram in the library, the client program needs to know just

- The *name* of the subprogram;
- The *number of parameters* expected by the subprogram;
- The *order* in which the parameters are expected to appear in a call or invocation of the subprogram; and
- The *type* of each parameter.

Thus, a client program needs to know the exact form of a subprogram call if it is to call the program correctly. But notice that the client does *not* need to know anything about *how* the subprogram performs its job; that is, it does not need to know the details of the algorithm the subprogram uses. Not only that, but the library may have some purely private subprograms, placed there for use only by other subprograms in the library, but not by clients.

Modern languages allow programmers to define their own types. It makes sense, then, to allow a type definition to be put in a library so that one or more client programs can use the definition to declare variables. But these client programs often need no further details about the type beyond its name. So just as we hide the detailed code of a subprogram, we ought to be able to hide the details of a type definition. Furthermore, the library may possess some types that are intended only for its own internal use.

Now we need some terminology to distinguish between those resources (subprograms, types, and perhaps also variables and constants) of a library that are intended for a client program's use from those that are the library's private property. Resources intended for client use are called *exported* resources; correlatively, the client program is said to *import* them from the library. As in the case of international trade, from which these terms are borrowed, not all resources are exported and imported; some are kept for internal or "domestic" use only.

1.4.2 Modula-2 Libraries: Definition and Implementation Modules

A library module consists of two separate (and separately compiled) files called the *definition module* and the *implementation module*. The definition module contains the definitions of all types, variables, constants, and procedures exported by the module, but no detailed code for the procedures (sometimes the details of type definitions are omitted as well); the implementation module contains the detailed code for all the procedures, perhaps some additional detail on exported types, and maybe some purely internal information as well.

In compiling a library, the definition module is compiled first. The compiler produces a file—usually called a symbol file—which consists of the definitions represented in a more efficient form. Then the implementation module is compiled, using the symbol file to check whether everything that was promised in the definition module was delivered to the implementation module. This ensures that client programs will be able to use the library with correct results.

Now, when a client program is compiled, the compiler uses the symbol files from the imported libraries to ensure that the client is making correct use of imported resources—for example, using the correct names for subprograms and supplying the right number and type of arguments for parameters in subprogram calls.

The Modula-2 compiler has the ability to make complete cross-checks of the interfaces between a library and the clients and other libraries that use it. This cross-checking allows large systems of modules and programs to be constructed far more systematically and reliably than would otherwise be possible.

1.5 CONSTRUCTING AN ADT IN MODULA-2: FRACTIONS

In this section we consider how to build an ADT in Modula-2 that support *fractions* or, more formally, *rational numbers*. In general, programming languages support integers and real numbers as primitive types, but not fractions. But there are certain applications where fractions are useful, e.g., when we want to represent the number $\frac{1}{3}$ exactly and not as 0.3333. . . . To make programs that use fractions easier to write, it makes sense to add support for fractions to the programming language in a systematic way.

1.5.1 Fractions as Mathematical Entities

Mathematically, a fraction is just an ordered pair of integers ⟨numer,denom⟩ which we understand to be the fraction numer/denom. For example, the fraction we usually write as $\frac{3}{7}$ is mathematically the ordered pair ⟨3,7⟩. Note that a fraction cannot have a denominator of 0, and that a fraction with a denominator of 1 is just an integer (a whole number).

We shall not require that a fraction be *proper* (numer < denom). Thus, $\frac{7}{4}$, for example, is perfectly acceptable. But we shall require a fraction to be *reduced*, because we do not want to have too many representations for the same number (e.g., $\frac{3}{6} = \frac{4}{8} = \frac{1}{2}$). A fraction can be reduced by first finding the largest number that divides both its numerator and its denominator (we call this its *greatest common divisor* or GCD), and then dividing numerator and denominator by it.

In mathematics, entities are defined by both the range of values they can assume and the set of operations defined on those values. What are the operations usually performed on fractions? Fractions are added to, subtracted from, multiplied and divided by each other, and converted to reduced form. They are compared to one another for equality or to see if one value is less than another.

Let us see how two of these operations work. To multiply two fractions, we multiply their numerators and their denominators and then reduce the result. For example,

$$\frac{3}{4} \times \frac{5}{9} = \frac{15}{36} = \frac{5}{12}$$

$$\frac{7}{5} \times \frac{10}{11} = \frac{70}{55} = \frac{14}{11}$$

$$\frac{3}{4} \times \frac{4}{3} = \frac{12}{12} = \frac{1}{1} \left(=1, \text{ but we keep it as } \frac{1}{1} \right)$$

Next, consider testing for equality. Recall that two fractions are equal not only if their numerators and denominators are respectively equal, but also if their reduced forms are equal. Thus, $\frac{1}{2} = \frac{2}{4} = \frac{24}{48}$. Now an easy way to discover whether two fractions are equal is to "cross-multiply" numerators and denominators, and then compare the products. For example, to see whether X = $\frac{2}{6}$ is equal to Y = $\frac{18}{54}$, we multiply X's numerator by Y's denominator, getting 108, and then multiply X's denominator by Y's numerator, getting 108 again. Since the two products are equal, X and Y are equal. Convince yourself that cross-multiplication can be used to test for "less than" or "greater than" as well.

Sec. 1.5 Constructing an ADT in Modula-2: Fractions

```
TYPE Fraction =
  RECORD
    Numerator:   INTEGER;
    Denominator: INTEGER;
  END;
```

Figure 1–4 Type definition for Fraction

1.5.2 Implementing Fractions in Modula-2

Let us now see how to represent fractions and operate upon them in Modula-2. A fraction is an ordered pair; this suggests the use of a record structure to represent it. We can use the type-definition facilities of Modula-2 to create a record type Fraction as shown in Figure 1–4. If we assume that a procedure FractReduce exists to reduce a fraction (this is left as an exercise), then it is easy to write a procedure FractMult to multiply two fractions and a procedure FractEqual that returns a Boolean function to test for equality. Modula-2 code for these procedures appears in Figure 1–5.

Although the preceding type definition and operations can be used in a program, they are not enough. What we really want to do is encapsulate all the important information about fractions in an ADT, so that *many* client programs could have easy access to it. We shall use the library module mechanism for this, but first we need a systematic classification of operations for our ADT.

1.5.3 Classifying Operations: Constructors, Selectors, and Predicates

First, how do fractions come about? In everyday mathematics, we write, for example, $\frac{1}{3}$ to create a fraction. In programming, however, merely writing $\frac{1}{3}$ isn't enough. Given the type definition for a fraction, one thing we could do is declare F to be a Fraction (note the use of uppercase to denote the type) and then explicitly store two values in it by writing F.Numerator:=1 and F.Denominator:=3. But

```
PROCEDURE FractEqual(X,Y: Fraction) : BOOLEAN;

  BEGIN
    RETURN X.Numerator * Y.Denominator =
           Y.Numerator * X.Denominator;
  END FractEqual;


PROCEDURE FractMult(VAR F: Fraction; X,Y: Fraction);

  BEGIN
    F.Numerator   := X.Numerator * Y.Numerator;
    F.Denominator := X.Denominator * Y.Denominator;
    FractReduce(F);
  END FractMult;
```

Figure 1–5 Modula-2 code for multiplication and equality testing of Fractions

```
PROCEDURE MakeFract(VAR F: Fraction; N,D: INTEGER);

   BEGIN
      F.Numerator   := N;
      F.Denominator := D;
   END MakeFract;
```

(a) MakeFract: Fraction constructor

```
PROCEDURE FractNumer(X: Fraction) : INTEGER;

   BEGIN
      RETURN X.Numerator;
   END FractNumer;


PROCEDURE FractDenom(X: Fraction) : INTEGER;

   BEGIN
      RETURN X.Denominator;
   END FractDenom;
```

(b) Numer and Denom: Numerator and denominator selector functions

Figure 1-6 Fraction constructor and selectors

this pair of statements will occur so frequently that it makes sense to build them into a procedure. Let us call this procedure MakeFract and specify it with two input parameters, for the numerator and denominator, and a VAR parameter to hold the fraction returned. Code for the procedure appears in Figure 1-6(a).

There is another advantage to our having a procedure for this: if, for some reason, we wish to change the implementing data structure for fractions, then only the code for MakeFract needs to be changed, instead of our having to search the client program for the explicit field-setting statements. MakeFract is an instance of a class of ADT operations called *constructors*. A constructor operation for a given type is one which creates an object of that type from its component parts. In the case of Fractions, the constructor creates a fraction from its numerator and denominator.

Often it is necessary to select a component from an object: a client program may need to work with just the numerator of a fraction F. In that case, the author of the client program could just write, say, X:=F.Numerator, thereby selecting the numerator and storing it in a variable X. But this means that the client would have to know that a Fraction is a record, with a field called Numerator. Perhaps we do not wish the client to have this knowledge; more important, even if we don't mind whether the client *knows* the structure of a fraction, we want to keep open the option of changing that structure without forcing an undue amount of change in every client program. Therefore, we write two *selector* operations, FractNumer and FractDenom, function procedures which select and return the numerator and

```
TYPE FractRecord =

RECORD
    Numerator:   INTEGER;
    Denominator: INTEGER;
END;


TYPE Fraction =

    POINTER TO FractRecord;
```

Figure 1-11 Revised type definition for Fraction, to allow a Fraction to be used as a function result

1.6.1 The Pointer in Modula-2

A *pointer* is an abstraction of the idea "address of. . ." Pointers exist in many programming languages; C, Pascal, PL/1, Ada, and Modula-2 all have them.

Pointers must be used with care. In fact, it has often been said that pointers are to data what go-to's are to code. But in appropriate situations pointers in Modula-2 are extremely useful. In the current situation, pointers are useful because they are scalars, and thus are allowed to be returned as the outputs of functions.

We can use pointers to improve the usefulness of our Fractions module. First, let us change the name of our fraction record to FractRecord and define a pointer type Fraction as a pointer to a record of type FractRecord. This is shown in Figure 1-11. Now a declaration such as VAR F1,F2,F3,F4,F5: Fraction no longer creates records, but only variables capable of holding the *addresses* of records. The records will come into being later.

We now write an improved MakeFract as a function procedure that returns a Fraction. We use the Modula-2 procedure Allocate, which, exported by a system module called System, is used to "create" instances or objects of a type at execution time. In Chapter 4 detailed consideration will be given to just how this happens; for now, all we need to know is that the space to be allocated is found in a special storage area called the *heap*, which is automatically provided by the compiler.

A call such as Allocate(P,SIZE(T)), where P is a pointer to objects of some type T, will allocate space for just one object and return the location of the object in P. The number of bytes to be allocated is given by SIZE(T), where SIZE is a standard Modula-2 function. (Note that, because Modula-2 is a new, evolving, and not entirely standard language, dynamic allocation is handled by different compilers in slightly different ways. For example, some compilers export the storage allocator from a module called STORAGE instead of from System; and some compilers use the function TSIZE in place of SIZE. Check your compiler documentation for details.)

Figure 1-12 shows the new MakeFract, which takes two integer numbers as arguments and returns something of type Fraction. The construct P^ is read "the thing pointed to by P" and is called a *dereferencing operation*. So the statement P^.Numerator := N means "Store N in the Numerator field of the thing pointed to by P."

grams *import* them. There are two ways to accomplish this export/import, both of which are shown in the figure. (1) We simply write, for example, IMPORT FractionIO, in which case we need to *qualify* all references to resources in the imported module—for instance, FractionIO.ReadFract(F). (2) We can write, for example, FROM Fractions IMPORT followed by an explicit list of all resources to be used by the client. This form of import "factors out" the qualification to the top of the program. The language allows either form; in this book we shall usually use the second, but in writing your own programs the choice is up to you.

1.6 FUNCTION/EXPRESSION NOTATION AND POINTERS

The Fractions module contains a number of constructor operations which are written as procedures, with VAR parameters to return the results. This notation does not correspond well with that of everyday mathematics, in which constructors are written as functions. To take a specific example, assume that F1, F2, F3, F4, and F5 have all been declared Fractions and assigned values. Suppose we wish to compute F1 := (F2 + F3) × (F4 − F5). Then, using the procedures in our module, we would have to declare two temporary variables, Temp1 and Temp2, and write the following statements:

    FractAdd(Temp1,F2,F3);

    FractSub(Temp2,F4,F5);

    FractMult(F1,Temp1,Temp2);

If this were a real-world problem with some lengthy calculations, we would be declaring a lot of temporary variables and writing a lot of procedure calls. Moreover, we would have to do so *very* carefully because the list of statements doesn't bear a close resemblance to the original equation, and thus is error prone. Our program would correspond better to the original mathematics if we could use functional notation and expressions:

    F1 := FractMult(FractAdd(F2,F3),FractSub(F4,F5))

It would be even better if we could use the mathematical symbols +, −, ×, and / for the operations we need, but Modula-2 doesn't allow this sort of thing. (Incidentally, this use of mathematical symbols *is* permitted by some languages—for example, Ada.) In any event, the functional notation is certainly better than the long series of procedure calls.

Why, then, didn't we just write the arithmetic operators (or MakeFract, for that matter) as function procedures returning Fractions? The reason is that there is a limitation in Modula-2: a function procedure is permitted to return only a scalar value, not a structured one. Since a Fraction is a record, we can't return one from a function procedure: the compiler will complain. What, then, do we do? Enter the pointer.

denominator, respectively. Then, if the implementation changes, we need only change the code for these functions; the client program can remain as is. This constancy in the midst of change is one of the benefits of encapsulation. The two selectors are FractNumer and FractDenom shown in Figure 1–6(b).

Writing constructor and selector operations may seem to be overkill for a structure as simple as a fraction. But since we need to establish good design principles that apply to more interesting structures, we use them even in this simple case.

Now, how shall we classify these arithmetic and comparison operations on Fractions? The arithmetic operations all produce new fractions, so they are constructors. On the other hand, the comparison operations do not produce new fractions, but rather, merely examine the values of their arguments. Strictly speaking, they are selectors. However, since operations which ask true–false questions about the state of objects occur so frequently, we will use the term *predicates* to refer to them. In Modula-2 we shall always write them as function procedures that return Boolean values.

1.5.4 Putting It All Together

To put everything we've done regarding fractions into a library module, we first write the definition module. Figure 1–7 shows the definition module for Fractions. The statement beginning with EXPORT QUALIFIED is required by many Modula-2 compilers. The names of *all* exported resources are written here. The next statement gives the type definition for Fraction, and subsequent lines give procedure headings for all the operations in the module. Since Modula-2 does not itself require a classification of the ADT operations into constructors, selectors, and predicates, we shall include this information in the form of comments.

Notice the BOOLEAN variable ZeroDenom in the figure. As the associated comment points out, this variable is set to TRUE or FALSE by MakeFract whenever it is called. The purpose of this is to signal to the calling program the occurrence of an erroneous situation or *exception* which makes it impossible to proceed, since a fraction with a zero denominator does not make mathematical sense. Exporting a variable declaration from a definition module makes the declaration *global*, i.e., able to be used by a client program as well as the procedures in the implementation module. The writer of the client program can incorporate code to test this variable to ensure that a call to MakeFract completed successfully.

Revising MakeFract so that ZeroDenom is always set to an appropriate value is left as an exercise. Also left as an exercise is the modification of the arithmetic constructors so that they handle a zero denominator correctly. A partially filled-in implementation module is shown in Figure 1–8; completing the module is left as an exercise.

```
DEFINITION MODULE Fractions;
(* Create and perform arithmetic on rational numbers *)

   EXPORT QUALIFIED
       Fraction,    MakeFract,  FractNumer,   FractDenom,
       FractAdd,    FractSub,   FractMult,    FractDiv,
       FractEqual,  FractLess,  FractGrtrEq,
       FractReduce, FractTrunc, ZeroDenom;

   (* whether the above statement is required, optional, or
      forbidden depends on the compiler. Check documentation *)

   TYPE Fraction =
      RECORD
         Numerator:   INTEGER;
         Denominator: INTEGER;
      END;

   VAR ZeroDenom: BOOLEAN;   (* Exception Variable *)

   PROCEDURE MakeFract(VAR Result: Fraction; N,D: INTEGER);
   (* Constructor: creates Fraction from two INTEGERs;
      N and D become numerator and denominator.
      If D=0, MakeFract will set the BOOLEAN variable
      ZeroDenom to TRUE.
      MakeFract does NOT put its result in reduced form.*)

   PROCEDURE FractNumer(X: Fraction) : INTEGER;
   PROCEDURE FractDenom(X: Fraction) : INTEGER;
   (* selectors for Numerator and Denominator respectively. *)

   PROCEDURE FractReduce(VAR X: Fraction);
   (* returns its argument in reduced form by
       dividing numerator and denominator by their
       greatest common divisor   *)

   PROCEDURE FractAdd (VAR Result: Fraction; X,Y: Fraction);
   PROCEDURE FractSub (VAR Result: Fraction; X,Y: Fraction);
   PROCEDURE FractMult(VAR Result: Fraction; X,Y: Fraction);
   PROCEDURE FractDiv (VAR Result: Fraction; X,Y: Fraction);
   (* perform arithmetic on Fractions,
       returning results in reduced form  *)

   PROCEDURE FractEqual  (X,Y: Fraction) : BOOLEAN;
   PROCEDURE FractLess   (X,Y: Fraction) : BOOLEAN;
   PROCEDURE FractGrtrEq (X,Y: Fraction) : BOOLEAN;
   (* predicate operators on fractions  *)

   PROCEDURE FractTrunc(VAR X: Fraction);
   (* find "whole number" part of a Fraction *)

END Fractions.
```

Figure 1-7 Definition module for Fractions module

Sec. 1.5 Constructing an ADT in Modula-2: Fractions

```
IMPLEMENTATION MODULE Fractions;

    TYPE Fraction =

      RECORD
          Numerator:   INTEGER;
          Denominator: INTEGER;
      END;

    PROCEDURE MakeFract(VAR Result: Fraction; N,D: INTEGER);

      BEGIN
         Result.Numerator   := N;
         Result.Denominator := D;
      END MakeFract;

    PROCEDURE FractNumer(X: Fraction) : INTEGER;

      BEGIN
          RETURN X.Numerator;
      END FractNumer;

    (* fill in FractDenom, FractReduce here *)

    PROCEDURE FractEqual(X,Y: Fraction) : BOOLEAN;

      BEGIN
          RETURN X.Numerator * Y.Denominator =
                 Y.Numerator * X.Denominator;
      END FractEqual;

    (* fill in FractLess, FractGrtrEq here *)

    PROCEDURE FractMult(VAR Result: Fraction; X,Y: Fraction);

      BEGIN
         Result.Numerator   := X.Numerator * Y.Numerator;
         Result.Denominator := X.Denominator * Y.Denominator;
         FractReduce(Result);
      END FractMult;

    (* fill in remaining operations here *)
END Fractions.
```

Figure 1–8 Partial implementation module for Fractions

1.5.5 Using the Library Module in a Client Program

Suppose we have compiled two program modules—Fractions from the previous section, and a module we shall call FractionIO, the writing of whose details is an exercise. FractionIO handles the input and output from Fraction in a way consistent with Modula-2 input–output (I/O) style: ReadFract reads a fraction from

```
DEFINITION MODULE FractionIO;
(* Fraction Input/Output *)

   FROM Fractions IMPORT Fraction;

   EXPORT QUALIFIED ReadFract ,WriteFract;

   PROCEDURE ReadFract (VAR X: Fraction);
   PROCEDURE WriteFract(X: Fraction);

END FractionIO.
```

Figure 1–9 Definition module for Fraction IO

the terminal keyboard, and WriteFract writes one to the terminal screen. The VAR parameter in ReadFract is used to return the result to the calling program. A definition module for FractionIO appears in Figure 1–9.

Figure 1–10 gives a program that uses the preceding two modules. Notice again that a main program is also called a module in Modula-2.

We have mentioned that just as modules *export* resources, so do client pro-

```
   MODULE TestFractions;

      FROM InOut IMPORT WriteString, WriteLn;

      FROM Fractions IMPORT Fraction, MakeFract, FractMult,
                   FractReduce, FractEqual;

      FROM FractionIO IMPORT ReadFract, WriteFract;


   VAR
      F1, F2, F3, F4: Fraction;


   BEGIN

      MakeFract(F1,1,3);
      MakeFract(F2,-24,48);

      IF FractEqual(F1,F2) THEN
         WriteString("Equal");
         WriteLn;
      ELSE
         WriteString("Unequal");
         WriteLn;
      END;

      WriteString("Please enter a fraction: "); WriteLn;
      ReadFract(F3);

      FractMult(F4,F1,F2);
      WriteString("Result is"); WriteFract(F4); WriteLn;

      FractReduce(F2);
      WriteString("Result is"); WriteFract(F2); WriteLn;

   END TestFractions.
```

Figure 1–10 A program that uses Fractions

```
PROCEDURE MakeFract(N,D: INTEGER) : Fraction;

   VAR
      Z: Fraction;

   BEGIN
      ZeroDenom := FALSE;

      IF D = 0 THEN   (* Set Exception Variable,
                         return undefined value *)
         ZeroDenom := TRUE;

      ELSE

         Allocate(Z,SIZE(FractRecord));
         Z^.Numerator   := N;
         Z^.Denominator := D;
      END;

      RETURN Z;

   END MakeFract;
```

Figure 1-12 Improved MakeFract, now a function procedure

Using the new MakeFract, we can write, for example,

$$F := \text{MakeFract}(1,3);$$

and, assuming that F has been declared to be of type Fraction, never realize or care what is going on at the lower level.

Figure 1-13 shows the Numer and Denom selectors, now modified to use the new structure for a Fraction; Figure 1-14 gives similarly modified versions of FractMult and FractEqual.

1.6.2 Encapsulating All the Details: Opaque Export

We gain yet another advantage by making Fraction a pointer type: Modula-2 allows us to export it in the definition module merely by writing TYPE Fraction; the rest of the type definition is written in the implementation module. Now *all* information about the details of Fraction are hidden from client programs. Such complete hiding is known as *opaque export*. A client program, having no knowledge of the

```
PROCEDURE FractNumer(X: Fraction) : INTEGER;

   BEGIN
      RETURN X^.Numerator;
   END FractNumer;


PROCEDURE FractDenom(X: Fraction) : INTEGER;

   BEGIN
      RETURN X^.Denominator;
   END FractDenom;
```

Figure 1-13 Modified selector functions

```
PROCEDURE FractMult(X,Y: Fraction) : Fraction;

   VAR
     N,D: INTEGER;
     Z: Fraction;

   BEGIN
     N := FractNumer(X)*FractNumer(Y);
     D := FractDenom(X)*FractDenom(Y);
     Z := FractReduce(MakeFract(N,D));
     RETURN Z;
   END FractMult;


PROCEDURE FractEqual(X,Y: Fraction) : BOOLEAN;

   BEGIN
      RETURN FractNumer(X)*FractDenom(Y)
           = FractNumer(Y)*FractDenom(X);
   END FractEqual;
```

Figure 1-14 Modified arithmetic constructor and predicate functions

implementation details of Fraction, is now prevented by the compiler from making any explicit reference to the numerator and denominator fields; instead, it *must* use selectors to select values. Now we have really encapsulated everything!

Figure 1-15 shows a partially completed revised definition module for Fractions, and Figure 1-16 shows a partial implementation module. We leave it as an exercise to fill in the missing code and to write a new client program which uses the new notation for Fraction arithmetic.

1.6.3 What Happens to the Allocated Memory?

Everything would work splendidly, except for one problem. Since the arithmetic constructors all call MakeFract, the statement

   F1 := FractMult(FractAdd(F2,F3),FractSub(F4,F5))

would result in Allocate being called twice to get space for the intermediate results FractAdd(F2,F3) and FractSub(F4,F5). These two records are used only very briefly; what happens to them after we are finished with them?

The answer is that they remain allocated and will not be reused. Accordingly, continual use of expressions like this will continually allocate temporary space which will not be recovered. Eventually, our program will run out of dynamic memory space (usually called heap space) and "bomb," even though we're not actually using all that space.

This is not a happy state of affairs, but suitable solutions exist. However, to discuss them now would force us to leap ahead to some material more appropriately considered later in the book. We shall return to this subject in Chapter 4; in the meantime, you can assume that your compiler gives you a large enough

Sec. 1.6 Function/Expression Notation and Pointers

```
DEFINITION MODULE Fractions;
(* Create and perform arithmetic on rational numbers.*)

    (* Check whether your compiler requires EXPORT QUALIFIED *)

    TYPE Fraction;    (* opaque type; details hidden *)

    VAR ZeroDenom: BOOLEAN;    (* Exception Variable   *)

    PROCEDURE MakeFract(N,D: INTEGER) : Fraction;
    (* Constructor: creates Fraction from two INTEGERs *)

    (* Fill in headings for field selectors here *)


    PROCEDURE FractReduce(X: Fraction) : Fraction;
    (* returns its argument in reduced form by
        dividing numerator and denominator by their
        greatest common divisor   *)

    PROCEDURE FractEqual  (X,Y: Fraction) : BOOLEAN;
    (* fill in remaining predicates *)

    PROCEDURE FractTrunc(X: Fraction) : INTEGER;

    PROCEDURE FractMult   (X,Y: Fraction) : Fraction;
    (* fill in remaining procedure headings for arithmetic *)

END Fractions.
```

Figure 1–15 Revised Fractions definition module, partially completed

```
IMPLEMENTATION MODULE Fractions;

    FROM System IMPORT Allocate;

    TYPE FractRecord =

        RECORD
           Numerator:   INTEGER;
           Denominator: INTEGER;
        END;

    TYPE Fraction =

        POINTER TO FractRecord;

    (* Fill in code for all modified operations here *)

END Fractions.
```

Figure 1–16 Partially completed implementation module for Fractions

heap space to absorb the wasted memory. Since functional notation is so convenient, we shall continue to use it, realizing even as we do so that it involves the risk of running out of memory.

1.7 DESIGN: AN ABSTRACT DATA TYPE FOR TEXT STRINGS

Modula-2 does not have much built-in support for variable-length character strings. String *literals* (e.g., "AbcDe") may be written, but a string is just an array of characters, and such an array must always be declared with a fixed length. On the other hand, it is common in applications to use string objects with a fixed *maximum* length but a variable *actual* length. There is nothing built into Modula-2 to keep track of how many useful characters there actually are in a string at any given moment. To get this capability, one must build on top of the language.

Accordingly, let us design an ADT for text strings which will provide the desired capability. We shall create a type Text, all of whose objects will have a fixed maximum length (say, 32 characters) but a variable actual length. This will give us approximately the same capability that is available in PL/1 or UCSD Pascal. Figure 1–17 shows a type definition for Text; note that Text is a pointer to TextRecord, so that we can use functional notation and opaque export. TextRecord has a value field Val which is a character array capable of holding TextMaximum (=32) characters, and a length field Len which is a nonnegative integer no larger than TextMaximum.

What operations should apply to Text objects? A definition module appears in Figure 1–18. Given a single character C and a normal Modula-2 string (character array) S, two constructors, CharToText(C) and StringToText(S), create a Text object. Also, selectors TextLength(T) and TextToString(T) return the current length and current value, respectively, of T. Note that TextToString must be a procedure with a VAR parameter, because a function procedure is not allowed to return an array.

Another operation on Text objects is *concatenation*, represented by the constructor TextConcat(T1,T2). The concatenation of two text objects T1 and T2 returns a Text object with the characters of T1 followed by those of T2. Thus, TextConcat(StringToText("ABC"),StringToText("DEF")) will return a Text object of length 6 and value "ABCDEF". Along with TextConcat, we have included TextAppendChar(T,C), which appends the character C to the Text object T.

```
TYPE   STRING = ARRAY[0..TextMaximum] OF CHAR;

TYPE   TextRecord =

       RECORD
         Len: INTEGER;
         Val : STRING;
       END;

TYPE   Text = POINTER TO TextRecord;
```

Figure 1–17 Type definition for variable-length Text objects

Sec. 1.7 Design: An Abstract Data Type for Text Strings

```
DEFINITION MODULE Texts;

(* Check to see whether your compiler requires EXPORT QUALIFIED *)

    CONST TextMaximum = 32;
    TYPE STRING = ARRAY[0..TextMaximum] OF CHAR;

    TYPE  Text;       (* opaque export again *)

    TYPE Relation = (Less, Equal, Greater);
    PROCEDURE StringToText(S: ARRAY OF CHAR) : Text;
    PROCEDURE CharToText(C: CHAR) : Text;
    PROCEDURE NullText(): Text;
    (* basic constructors *)

    PROCEDURE TextLength  (T: Text) : INTEGER;
    PROCEDURE TextToString  (VAR S: ARRAY OF CHAR; T: Text);
    (* basic selectors *)

    PROCEDURE TextEmpty (T: Text) : BOOLEAN;
    (*       predicate          *)

    PROCEDURE TextHead(T: Text) : CHAR;
    PROCEDURE TextTail(T: Text) : Text;

    PROCEDURE TextInit(VAR T: Text); (* copy NullText() to T *)
    PROCEDURE TextCopy (VAR DESTINATION: Text; SOURCE: Text);

    PROCEDURE TextConcat (T1, T2: Text) : Text;
    PROCEDURE TextAppendChar(T: Text; C: CHAR) : Text;
    (*       concatenation        *)

    PROCEDURE TextCompare(T1, T2: Text) : Relation;
    (*       lexical comparison       *)

    PROCEDURE TextLocate(Sub: Text; Within: Text) : INTEGER;
    (* search Within for Sub; return location or 0 if not there *)

    PROCEDURE TextSubstr(T: Text; Start, Size: INTEGER) :  Text;
    (*       return substring; truncate if overflow     *)

END Texts.
```

Figure 1-18 Definition module for Texts

The definition module for Texts includes a comparison operation that assumes dictionary or *lexical* order, so that "BCD" < "BCDE" (obviously), but also "BCD" < "CD" (perhaps not so obviously). Here, instead of exporting several comparison operations as in Fractions, we export an enumeration type Relation = (Less, Greater, Equal) and one operation TextCompare, which returns a value of type Relation. This is done because in text processing we often compare two text objects and then take one of three actions depending on the result. If our operation were just a Boolean predicate, two separate scans of the strings might be required to determine which of the three actions to take, and this might use a good bit of computation time if the text objects were really long strings. So we have chosen the form of the operation as a way of reducing computation time.

```
PROCEDURE StringToText(S: ARRAY OF CHAR) : Text;
   VAR
      T:     Text;
      Count: INTEGER;
   BEGIN
      Allocate(T,SIZE(TextRecord));
      Count := 0;
      LOOP
         T^.Val[Count] := S[Count];
         IF (S[Count] = 0C) OR (Count = TextMaximum) THEN
            EXIT;
         ELSE
            INC(Count);
         END;
         IF Count = HIGH(S) + 1 THEN
            T^.Val[Count] := 0C;
            EXIT;
         END;
      END;
      T^.Len := Count;
      RETURN T;
   END StringToText;


PROCEDURE TextToString  (VAR S: ARRAY OF CHAR; T: Text);
   VAR
      Count: INTEGER;
   BEGIN
      IF T^.Len = 0 THEN
         S[0] := 0C;
      ELSE
         Count := 0;
         LOOP
            S[Count] := T^.Val[Count];
            IF Count = HIGH(S) THEN (* T longer *)
               EXIT;
            END;
            IF S[Count] = 0C THEN    (* S longer *)
               EXIT;
            END;
            IF Count+1 = TextLength(T) THEN (* S longer *)
               S[Count+1] := 0C;
               EXIT;
            END;
            Count := Count+1;
         END;
      END;
   END TextToString;
```

Figure 1–19 Two Operations on Text objects

Also included are a selector TextHead(T), which returns the first character of its Text argument, and a constructor TextTail(T), which returns T with its first character removed. Other useful operations are TextLocate(T,S), which searches a target Text object T for the presence of a substring Text object S and returns the position in the target where the substring begins; and TextSubstr(T,Start,Size), which returns the Text substring of length Size, counted from position Start of T. Accordingly, TextLocate(StringToText("BC"),StringToText("ABCDEF")) returns 1, and TextLocate(CharToText('G'),StringToText("AB")) returns 0 because

'G' isn't in "AB". Also, TextSubstr(StringToText("ABCDEF",2,4)) returns a Text object with value "CDEF" and length 4.

Some additional operations are included in the definition module; the associated comments should make their function clear. In Figure 1–19 we show Modula-2 implementations of several of these operations; it is left as an exercise to finish the implementation module. Notice, in reading the code, that we have assumed that Modula-2 strings are always terminated by the ASCII null character, represented in Modula-2 by "0C". This is consistent with the way most compilers treat strings. Figure 1–20 illustrates some Text objects and operations. The nulls are depicted by the Greek character ϕ (phi). Left as an exercise is the design of appropriate exception variables and the modification of the appropriate operations in the module.

1.8 STYLE GUIDE: USING ADTs IN OLDER LANGUAGES

Support for building ADTs as library modules is built into the most recently developed programming languages, in particular Ada and Modula-2. Some extended versions of Pascal, for instance UCSD Pascal and IBM's Pascal/VS, and of C, for instance C++, also incorporate such constructs. But a good deal of programming is still done in older languages, like the original C and Pascal, Fortran, and PL/1.

The designers of our most capable recent programming languages did not get their ideas for support for ADTs "out of the air." These ideas came from the recognition that ADTs help to reduce the complexity of programs, and only after the ADT idea was used as a design technique was it incorporated into actual languages.

It is thus apparent that even though one may be *coding* in the older languages, one can still use the ADT idea as a *design* tool. One of the most important aspects of "ADT thinking" is import/export, which is really just a formalization of the old idea of clear and clean specification of interfaces. Even if the compiler gives you no aid in enforcing these interfaces, you can design programs as though it did. Here are some guidelines:

1. Write down the specification of your ADT in such manner that it looks like a definition module.
2. Understand clearly, and document carefully, just what type it is that your ADT is exporting.
3. Classify the operations on the type you are exporting into constructors, selectors, and predicates, and try to make sure that *all* operations exported from the quasi-ADT fall into one of these categories. If an operation is not obviously a constructor, selector, or predicate, you're probably trying to make it do too many things at once. This is poor decomposition, and will probably lead to unnecessarily difficult and expensive maintenance later on.

VAR T1,T2,T3: Text;

T1: = StringToText("ABC");

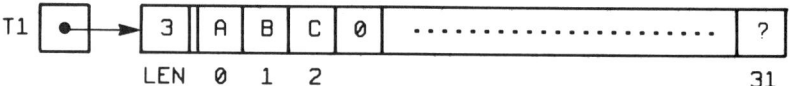

T2: = StringToText("XYZW");

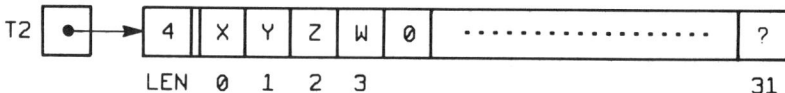

T3: = TextConcat(T1,T2);

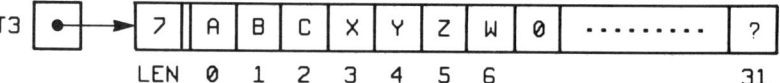

T2: = TextConcat(T2,StringToText("HIJKL"));

T1: = TextSubstr(T2,3,5);

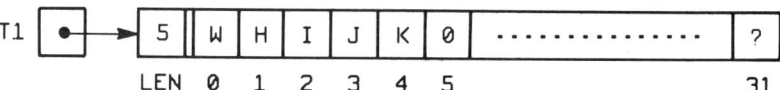

T3: = NullText();

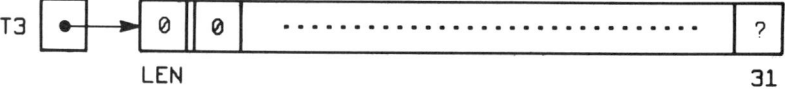

Figure 1-20 Some typical operations on Text objects

1.9 SUMMARY

This chapter has presented the important concepts of *abstraction, implementation, data type, abstract data type* or *ADT*, and *data structure*, as well as the notion of *import/export* and a classification of operations into *constructors, selectors,* and *predicates.*

Abstraction is the practice of viewing computer solutions in terms that are natural or appropriate to the problem being solved; implementation is the conversion of the abstract solution into a concrete one suitable for coding in a real programming language.

A data type is a set of appropriate values (like fractions, reals, integers, and text strings). An ADT is a data type plus a set of operations appropriate for that type. A data structure is a programming mechanism suitable for the representation of the type in a real program.

ADT operations can be classified into constructors, which create new objects of the type; selectors, which are used to select components of an object without altering the object or creating a new one; and predicates, which ask true–false questions about objects.

We have introduced the library module feature of Modula-2 which can be used as a mechanism for implementing ADTs. Library modules permit the designer to indicate explicitly which parts of the implementation are to be exported, or visible to the "outside world," and which parts contain details which are more appropriately hidden away from the user's view. A library module consists of a definition module and an implementation module; the former containing a list of all the exported resources, the latter the detailed code.

The pointer, an abstraction of a machine address, was introduced as a mechanism for facilitating opaque export and the use of functional notation. It will be seen in later chapters to have other interesting uses.

Designs and (partial) Modula-2 implementations for two ADTs—fractions and text strings—have shown you the way to get started with ADTs and library modules, and the Style Guide section gave some ideas on how you might write them in older languages. The rest of the book will continue in this vein, presenting a number of interesting ADTs and alternative strategies for their implementation.

CHAPTER 1 EXERCISES

1. Write a procedure FractReduce which will put a Fraction in lowest terms. HINT: first find the greatest common divisor.
2. Revise the MakeFract constructor, as discussed in section 1.5, so that the exception variable ZeroDenom is always set to an appropriate value.
3. Revise the Fraction arithmetic constructors so that they handle a zero denominator appropriately.

4. Write the implementation module for FractionIO.
5. Complete and test the implementation module for Fractions.
6. Using the pointer technique discussed in section 1.6, revise the Fractions and FractionIO modules to use function procedures wherever appropriate.
7. Complete and test the module Texts as described in section 1.7.
8. Design, code, and test a module for handling complex numbers. Recall that a complex number is an ordered pair of real numbers ⟨R,I⟩, where R is called the *real part* and I is called the *imaginary part*. Look up in a mathematics book the mathematical definitions of the arithmetic operations on complex numbers.
9. Design, code and test a module to handle numbers represented in scientific notation, e.g. 0.3×10^5. Store the fraction and exponent as fields of a record and implement arithmetic constructors and comparison predicates for numbers in this form.
10. Make a list of other abstractions and implementations you have encountered in your programming experience.

Chapter 2
ALGORITHMS, RECURSION, AND PERFORMANCE PREDICTION

2.1 GOAL OF THE CHAPTER

Any pragmatic study of data structures must also include a study of *algorithms*, the methods by which problems are solved on a computer. Data structures without algorithms are useless and empty curiosities.

In this chapter we shall study two important aspects of algorithms. The first is the use of *recursion* to solve certain computing problems. A *recursive algorithm* moves ahead by applying *itself* to a smaller part of the problem; in terms of programming, we say that the algorithm, written as a function or procedure, "calls itself." Several useful but easy-to-understand recursive algorithms will be presented, together with Modula-2 programs implementing them.

The second important aspect of algorithms involves *performance prediction*. Here, we consider techniques and rules of thumb to estimate the computation time of an algorithm—or more specifically, the *variation* of the computation time as a function of the *size* of the problem being solved. An important bit of terminology in performance prediction is "big O" notation. This notation is a way of representing the 'order of magnitude" or "growth rate" of an algorithm that is given in its variation of computation time with problem size.

The most common variations or "big O's"—the *constant, logarithmic, linear, quadratic,* and *Nlog(N)* growth rates—will be encountered later in the book.

2.2 ALGORITHMS AND ALGORITHM DESIGN

Given that you are reading this book, you probably have some experience in programming. Also, you probably know intuitively that an algorithm is a method used to solve a problem on a computer. Formally, we say that

> *An algorithm is*
> *a finite sequence of instructions*, each of which
> has a *clear meaning* and can be
> performed with a *finite amount of effort*
> in a *finite length of time*.

The finiteness conditions imply that an algorithm never goes into an infinite loop, no matter what input we give it.

One might ask, why, in a text devoted to the study of data structures, is there a chapter on algorithms? The answer is that algorithms are the methods used in systematic problem solving to move the solution from its starting point to its end. Thus, while we indeed concentrate on data structures in this book, we cannot ignore the study of algorithms, since data structures by themselves are useless. Without methods for storing data in them, retrieving data from them, and performing computational operations on the data stored in them, data structures are nothing but empty curiosities.

Furthermore, the computation time and memory space required by data structures and the algorithms that operate on them are important and sometimes scarce resources. Indeed, we shall speak frequently in this book of "trade-offs." A *trade-off* is a situation in which alternative solutions to a problem are considered in terms of their resource requirements: one solution may require more time but less space than another; a third might require more time *and* more space than either of the others, but the programs might be simpler and easier to maintain, hence more economical in *human* terms. We thus speak in terms of *trading off* space for time, or performance for clarity, or computer resources for human ones.

No book can give you a single right answer that will serve in every case. When you are faced with a trade-off situation in deciding on a computer solution to some problem, you must base your decision on the specific circumstances that exist at the time. What a book can supply is a set of tools for you to use in analyzing all the factors and trade-offs; the analysis itself, and the final decision, is up to you and your colleagues.

It is plain that data structures and algorithms are interrelated and cannot be studied completely apart from one another. Since, presumably, you already have some experience in writing algorithms, we shall not go back to "first principles" in this chapter. Rather, we shall focus attention on two important concepts relating to algorithms: One is the important and useful mathematical notion of *recursion*, a tool we shall use frequently in this book and the other is *performance prediction*, a tool to "give us a handle" on the time requirements of a problem solution.

2.3 RECURSIVE ALGORITHMS

In this section, we introduce a concept in algorithm design called *recursion*. A *recursive algorithm*—an algorithm that uses *recursion*—is an algorithm that is defined in terms of itself; the solutions of many interesting programming problems are stated clearly and elegantly in recursive form, and you will see many recursive algorithms in this book.

A classical simple example of recursion is the definition of the *factorial* of a positive integer N. Written $N!$ and read "N-factorial," the concept is easily understood as the product $1 \times 2 \times \ldots \times N$. Thus, $3! = 6$, $4! = 24$, $5! = 120$, and so on. We can write a formal definition as follows:

TO FIND $N!$:

1. If $N = 1$, then $N! = 1$;
2. Otherwise $N! = N \times (N - 1)!$

Notice that, although we have defined the "!" operation in terms of itself, the definition is not circular, because the "!" is applied to a smaller and smaller number each time, until it is applied to 1. Figure 2-1 shows how to apply the definition to calculate 5!; try it on some other number to be sure you understand how the recursion works. You will soon discover that $N!$ gets very large very quickly: even an innocent-looking calculation like 10! produces a rather big number. In fact, if you were to write a program to calculate $N!$ and run it on a computer using 16 bits to represent an integer, your program would fail because the largest number a 16-bit computer can hold is $32,767 < 8!$

Observe that a workable recursive algorithm must always reduce the size of the "data set," or the number that it is working with, each time it is recursively called, and must always provide a terminating condition, such as the first line in our factorial algorithm. Otherwise the algorithm may never terminate, getting itself stuck in an infinite recursion.

Modern programming languages like Modula-2 permit direct use of recursion in defining functions and procedures. Recursive algorithms seem mysterious to programmers trained only in Fortran, Cobol, or Basic, because those languages

```
5! = 5 x 4!
   = 5 x (4 x 3!)
   = 5 x (4 x (3 x 2!))
   = 5 x (4 x (3 x (2 x 1!)))
   = 5 x (4 x (3 x (2 x 1)))
   = 5 x (4 x (3 x 2))
   = 5 x (4 x 6)
   = 5 x 24
   = 120
```

Figure 2-1 Recursive calculation of 5!

```
PROCEDURE Factorial(N: INTEGER): INTEGER;

  BEGIN

    IF N <= 1 THEN
       RETURN 1;
    ELSE
       RETURN N * Factorial(N-1);
       (* NOTE RECURSIVE CALL OF FACTORIAL! *)
    END;

  END Factorial;
```

Figure 2–2 Recursive Modula-2 function to find N!, N >= 1

do not allow recursion, but recursion is really a very simple and elegant concept directly applicable in Modula-2, Pascal, Ada, PL/1, C, and others. Figure 2–2 shows a recursive factorial function in Modula-2.

Finding the factorial of a positive integer is a simple example of a recursive algorithm. We next consider Modula-2 programs that use recursive algorithms to (1) find the reverse of a string, (2) find the permutations of the elements in a set, (3) perform a recursive binary search, and (4) perform a recursive merge/sort.

2.3.1 Reverse of a String

A *palindrome* is a phrase that reads the same forward and backward. Two examples of English palindromes (neglecting differences in upper and lowercase letters) are "Radar" and "Able was I ere I saw Elba." Similarly, the statement, "Madam, I'm Adam," which might have been spoken by the Biblical first man when he met his wife Eve, is a palindrome if we neglect spaces and punctuation as well.

One way to discover whether a string of characters is a palindrome is to find the reverse of the string. The string is a palindrome if its reverse is identical to it.

We can find the reverse of a string very easily by the following algorithm:

TO FIND THE REVERSE OF A STRING:

1. If the string contains only one character, its reverse is identical to it, and we're finished.
2. Otherwise, remove and save the first character.
3. *Find the reverse* of the remaining string, and then concatenate the saved character onto the right-hand end.

Notice that we've *found the reverse* of a string by removing the first character and *finding the reverse* of what's left. This is a typically *recursive* algorithm: to carry it out on the whole set of data, we need to carry it out on a smaller set of data. Figure 2–3 demonstrates this algorithm.

Suppose now that we have available a text-string package such as that de-

Sec. 2.3   Recursive Algorithms

```
To find the reverse of                              "ABCD"
append 'A' to the reverse of                        "BCD"
    Now, to find the reverse of                     "BCD"
    append 'B' to the reverse of                    "CD"
        Now, to find the reverse of                 "CD"
        append 'C' to the reverse of                "D"
            The reverse of ''D'' is                 "D"
        Appending 'C' gives                         "DC"
    Appending 'B' gives                             "DCB"
Appending 'A' gives                                 "DCBA"
```

Figure 2-3   Finding the reverse of a string, recursively

scribed in Chapter 1 which gives us the predefined functions (1) TextHead(S), which returns the first character of string S; (2) TextTail(S), which returns the string S with the first character removed; and (3) AppendChar, which adds a character to the end of string S. Figure 2-4 shows a function StringReverse(S) which returns the reverse of string S. Note the line in which StringReverse is called by StringReverse; this is known as a *recursive call* of a function.

Figure 2-4 also shows a function Palindrome(S), which uses StringReverse to determine whether its text-string argument S is a palindrome.

2.3.2 Permutations of a Set

Consider a wealthy family of two parents and two college-age children. Suppose that each family member has an automobile and that their house has a four-car garage. Letting the cars be called *A*, *B*, *C*, and *D*, and the garage stalls 1, 2, 3, and 4, what are the different ways in which the cars can be parked in the garage?

Suppose *A* parks in stall 1. Then we can list all the ways that are left to *B*, *C*, and *D*. Suppose, then, that *B* parks in stall 2. Clearly, *C* and *D* can now park only in two different ways: *C* in stall 3 and *D* in stall 4, or the other way around.

But suppose *C* parks in stall 2. Then it is *B* and *D* that use stalls 3 and 4 in

```
PROCEDURE IsPalindrome(S: Text): BOOLEAN;

BEGIN
  IF TextCompare(S,StringReverse(S)) = Equal THEN
    RETURN TRUE;
  ELSE
    RETURN FALSE;
  END;
END IsPalindrome;
```

Figure 2-4   String Reverse and Palindrome functions

```
    1 2 3 4
    A B C D
    A B D C
    A C B D
    A C D B
    A D B C
    A D C B
    B A C D
    B A D C
    B C A D
    B C D A
    B D A C
    B D C A
    C A B D
    C A D B
    C B A D
    C B D A
    C D A B
    C D B A
    D A B C
    D A C B
    D B A C
    D B C A
    D C A B
    D C B A
```

Figure 2-5 Permutations of the set {A,B,C,D}

one of two ways. And if D parks in stall 2, then it is B and C sharing stalls 3 and 4.

Clearly, then, there are six possibilities once A has parked in stall 1. It is easy to see that another six possibilities come from B's parking in stall 1, and twelve more from C and D's parking there. Thus, there is a total of 24 possibilities, all shown in Figure 2-5.

The exercise we have just performed is an example of finding the *permutations* of the elements of a set. In this case, the set is the family's automobiles, and a permutation is an assignment of them to the stalls in the garage. If the set has N members, the number of permutations is $N! = 1 \times 2 \ldots \times N$.

Let us try to write an algorithm to print out the permutations of the members of a set. Letting the set be $\{A,B,C,D\}$, we can say:

TO PRINT ALL PERMUTATIONS OF {A,B,C,D}:

1. Print A, followed by *all permutations of* $\{B,C,D\}$.
2. Print B, followed by *all permutations of* $\{A,C,D\}$.

Sec. 2.3 Recursive Algorithms

   3. Print C, followed by *all permutations of* {B,A,D}.
   4. Print D, followed by *all permutations of* {B,C,A}.

We have thus interchanged A with B, C, and D in turn (as though B, C, and D had parked, in turn, in stall 1).

Now, to print out all permutations of {B,C,D}, we have a problem just like the large one, but smaller. And similarly, printing out the permutations of {C,D} is just a smaller version of *that* problem. This methodology—where the same algorithm can be applied repeatedly to smaller and smaller sets—lends itself to a recursive solution.

Accordingly, let us construct a recursive procedure to print the permutations of an ordered set S with members numbered 1 through N. Represent this set, for simplicity, as an array, and assume we have available a procedure PrintSet(S) which prints the entire set, and a procedure Interchange(S,k,i) which interchanges the *i*th and *k*th members of S. Then our recursive procedure, called PrintPermutations(S,k), prints the permutations of the *k*th through *N*th members of S. The detailed Modula-2 text is shown in Figure 2–6; examine it carefully to understand how it works.

Figure 2–7 shows a program TestPerms which makes use of the preceding definitions and operations and incorporates the PrintPermutations procedure.

```
PROCEDURE PrintPermutations(S: CharSet; k: INTEGER);

   VAR
      N, i: INTEGER;
      S1:   CharSet;
      (* THE LOCAL VARIABLE IS USED HERE SO THE INPUT SET
         S DOESN'T GET CHANGED. *)

   BEGIN

      S1 := S;   (* ARRAY COPYING IS ALLOWED *)

      N := HIGH(S1);
      IF k = N
         THEN PrintSet(S1)
      ELSE
         FOR i := k TO N DO

            Interchange(S1,i,k);
            PrintPermutations(S1,k+1);
            (* THIS RECURSIVE CALL PRINTS ALL PERMUTATIONS OF
               THE CharSet WITH THE 1st THROUGH k-th MEMBERS
               CONSTANT AND THE k+1st THROUGH N-th VARYING. *)

         END;
      END;

   END PrintPermutations;
```

Figure 2–6 Recursive procedure to print all permutations of a set S. The set is stored as an array.

```
MODULE TestPerms;
  FROM InOut IMPORT Write, WriteLn;

  CONST SetMaximum = 5;
  TYPE CharSet = ARRAY [1..SetMaximum] OF CHAR;

  VAR P: CharSet;

  PROCEDURE PrintSet(S: CharSet);

    VAR
      i: INTEGER;
    BEGIN
      FOR i := 1 TO SetMaximum DO
        Write(S[i]);  Write(' ');
      END;
      WriteLn;
    END PrintSet;

  PROCEDURE Interchange(VAR S: CharSet; i,k: INTEGER);

    VAR
      temp: CHAR;
    BEGIN
      temp := S[i];
      S[i] := S[k];
      S[k] := temp
    END Interchange;

            (* CODE FOR PrintPermutations GOES HERE *)
BEGIN
   P[1] := 'A';
   P[2] := 'B';
   P[3] := 'C';
   P[4] := 'D';
   P[5] := 'E';
   PrintPermutations(P,1);
END TestPerms.
```

Figure 2–7 Program illustrating PrintPermutations

2.3.3 Recursive Binary Search

Imagine that you've written up a list of your friends, giving their names in alphabetical order together with their telephone numbers. Because you're very popular, you have many friends and the list is quite long, running over a number of pages.

Let us consider a clever way to look up a friend's phone number in this long list. (Actually, it's a way better suited to a computer than a person, but that's because people look things up intuitively instead of using algorithms!)

TO LOOK UP A NAME:

1. Divide your list in half, and look at the name right in the middle. If it's the one whose number you're searching for, you're done.

Sec. 2.3   Recursive Algorithms

2. If your friend's name is *earlier* in the alphabet than the middle name, ignore all the names from the middle name to the end, and *look up the name* only in the first half of the list. That is, divide *it* in half, and look at the middle name, etc.
3. If your friend's name is *later* in the alphabet than the middle name, ignore the first half of the list, and *look up the name* in the second half.

If you use this procedure, eventually one of two things will happen: either you'll find your friend's name in the list, or you'll divide the list in half so many times that only one name will remain and it's not the one you wanted! This means that the friend you were looking for isn't in your list.

Like the reversal and permutation algorithms, the binary search algorithm is recursive: the method applied to the full list is applied to half the list, then to half of half, etc. Let us construct a program for this algorithm. We'll let the list be implemented as an array with subscripts 1 . . N. The function, which will be called LookUpName(L,Lower,Upper,Name), looks up Name in the section of the array bounded by L[Lower] and L[Upper] and returns the location of Name if it can find it, and zero if it can't.

The text for LookUpName appears in Figure 2–8. Try finding the locations of some names in the table given in Figure 2–9.

```
PROCEDURE LookUpName(L:              list;
                     Lower, Upper:   INTEGER;
                     Name:           NameType) : INTEGER;

   VAR
      Middle: INTEGER;

   BEGIN
      Middle := (Lower + Upper) DIV 2;

      IF    Name = L[Middle] THEN
         RETURN Middle
            (* WE FOUND IT! *)

      ELSIF Lower = Upper    THEN
         RETURN 0
            (* SUBARRAY ONLY HAS ONE NAME AND IT'S
               NOT THE ONE WE WERE LOOKING FOR! *)

      ELSIF Name < L[Middle] THEN
         RETURN LookUpName(L,Lower,Middle-1,Name)
            (* LOOK ONLY IN THE FIRST HALF -
               RECURSIVE CALL! *)

      ELSE
         RETURN LookUpName(L,Middle+1,Upper,Name)
            (* LOOK ONLY IN THE SECOND HALF -
               RECURSIVE CALL! *)

      END;

   END LookUpName;
```

Figure 2–8   Recursive function to find a name in an alphabetized list

```
 1  ALAN
 2  ALEX
 3  BEN
 4  BILL
 5  BRAD
 6  EUGENE
 7  JENNY
 8  JESSICA
 9  JONATHAN
10  JUSTIN
11  KEITH
12  KEVIN
13  KRISTIN
14  MARTHA
15  SHARON
16  SHERRY
```

Figure 2-9 Alphabetized table of names

We call the algorithm that LookUpName implements a *recursive binary search*. It is an example of a whole class of algorithms known as "divide and conquer" which work by dividing and subdividing the set of data into two parts.

2.3.4 Recursive Merge/Sort

The last example of recursion involves sorting the elements of a list into an ascending sequence. We merely sketch out a *recursive merge/sort* algorithm, leaving the details until the chapter on sorting.

The algorithm depends upon our knowing how to merge two sorted lists into a single sorted list, informally, the two sorted lists {B,G,H,P} and {A,F,K,L,R,Z} can be merged into a single list {A,B,F,G,H,K,L,P,R,Z}, much as you might merge two sorted decks of 3 × 5 cards into a single deck. Without being concerned with the details of the merge operation, consider how the two original sorted lists came to be sorted: why not by the very same process? In other words, if we start with a single unsorted list, we can write an informal algorithm as follows:

TO SORT A LIST

1. If the list contains only one item, it is sorted already.
2. Otherwise, divide the list in half, *sort* the two halves, and then merge them.

Aha! Another recursive algorithm! The method moves forward by dividing its problem in half and then applying itself to the two halves of the list. Thus, the recursive merge/sort is another example of a divide-and-conquer algorithm. An illustration of its use is shown in Figure 2-10.

You have seen five recursive algorithms in this section, and will see many more throughout this book. Recursion is not a mysterious or magical concept,

```
Original      (Z A C F Q B G K P N D E M H R T)
Divide        (Z A C F Q B G K) (P N D E M H R T)
Divide        (Z A C F)(Q B G K) (P N D E)(M H R T)
Divide        (ZA)(CF) (QB)(GK) (PN)(DE) (MH)(RT)
Sort pairs    (AZ)(CF) (BQ)(GK) (NP)(DE) (HM)(RT)
merge         (A C F Z) (B G K Q) (D E N P) (H M R T)
merge         (A B C F G K Q Z) (D E H M N P R T)
merge         (A B C D E F G H K M N Q P R T Z)
                                                sorted!
```

Figure 2-10 Recursive ("top-down") merge/sort

rather just another tool in the algorithm designer's tool kit. It is time now to move to another important topic in algorithms, namely *performance prediction*.

2.4 PERFORMANCE PREDICTION AND "BIG O" NOTATION

In considering the trade-offs among alternative solutions to problems, an important factor is the expected computation time of each of the alternatives. It is difficult to predict the *actual* computation time of an algorithm without knowing the intimate details of the underlying computer, the object code generated by the compiler, and other related factors. But we can *measure* the time for a given algorithm, language, compiler, and computer system by means of some carefully designed performance tests, usually called *benchmarks*.

It is also helpful to know the way the running time will *vary* or *grow* as a function of the problem size—the number of elements in an array, the number of records in a file, and so forth. Programmers sometimes discover that programs that have run in perfectly reasonable time for the small test sets they have used take extraordinarily long when run with real-world-sized data sets or files. These programmers were deceived by the *growth rate* of the computation.

To take an example, it is common to write programs whose running time varies with the square of the problem size. Thus, a program taking, say, one second to complete a file-handling problem with 10 records in the file will require not two, but four, seconds for 20 records. Increasing the file size by a factor of 10, to 100 records, will increase the run time to $10^2 = 100$ seconds; and 1,000 records will require 10,000 seconds, or about three hours, to complete! Finally, 10,000 records (the number of accounts in a fair-sized bank, or the number of students in a fair-sized university) will need almost two weeks to finish! This is a long time compared to the one second taken by the 10-record test.

This example shows that it makes sense to know something about growth rates, lest program running time grow in disturbing ways when problems grow to meaningful size. Sometimes there is no choice: there may be no alternative solution to that program running in "squared" time. But a programmer with some experience in performance estimation is at least in a position not to be surprised.

2.4.1 Algorithm Growth Rates

Getting a precise estimate of the computation time of an algorithm is often difficult, but it helps to "get a handle on it." We do this by trying to write a formula for the computation time in terms of the problem size N. This will have a factor in it that depends upon the programming language, compiler, and underlying computer. It is often a good assumption that this system-dependent factor is reasonably constant—not varying with the problem size—so that we can "factor it out." (Obviously it's nice to have a small system-dependent constant as well as a small growth rate, but reducing the size of the constant is hard to do in a general way precisely because it's system dependent.)

We give the name *growth rate* to that part of the formula that *does* vary with problem size. In discussing the growth rates of algorithms, it is fashionable to use O-notation (read "growth rate," "big O," or "order of magnitude"). The most common growth rates encountered in data-structure work are the following:

1. $O(1)$, or *constant*;
2. $O(\log(N))$, or *logarithmic* (the logarithm is usually taken to the base 2);
3. $O(N)$, or *linear* (directly proportional to N);
4. $O(N \log(N))$ (usually just called $N \log N$);
5. $O(N^2)$, or *quadratic* (proportional to the square of N).

To give you an idea of the computation time of typical file sizes, Figure 2–11 shows the values of each of these functions for a number of different values of N. The values happen to be powers of 2, but this is just to make the computation of logarithms convenient.

From the figure, you can see that as N grows, $\log(N)$ remains quite small with respect to N and $N \log(N)$ grows fairly large, but not nearly as large as N^2. In studying sorting in Chapter 8, you'll discover that most sorting algorithms have growth rates of $N \log(N)$ or N^2. In the next section we'll look at some common algorithmic structures and discuss how to estimate their growth rates.

2.4.2 Estimating the Growth Rate of an Algorithm

While there are no absolute "cookbook" rules that will always work to estimate performance, we can "get a handle on it" by taking advantage of the fact that algorithms are developed in a structured way—that is, they combine simple statements into complex blocks in four useful ways:

1. *Sequence*, or writing one statement below another;
2. *Decision*, or the well-known if-then or if-then-else;

Sec. 2.4   Performance Prediction and "Big O" Notation

N	1	Log(N)	N Log(N)	N²
1	1	0	0	1
2	1	1	2	4
4	1	2	8	16
8	1	3	24	64
16	1	4	64	256
32	1	5	160	1024
64	1	6	384	4096
126	1	7	896	16384
256	1	8	2048	65536
512	1	9	4608	262144
1024	1	10	10240	1048576
2048	1	11	22528	4194304
4096	1	12	49152	16777216
8192	1	13	106496	67108864
16384	1	14	229376	268435456
32768	1	15	491520	1073741824

Figure 2–11   Table of some common algorithm growth rates

3. *Loop*, including counting loops, *while* loops, *until* loops, and the general loop structure;
4. *Subprogram call*.

Figure 2–12 shows the Modula-2 notation for a number of different variations on these structures.

Let us now take a look at some typical algorithm structures and estimate their "big Os." We denote the problem size by N throughout.

Simple statement.   An assignment statement is an example of a simple statement. If we assume that the statement contains no function calls (whose execution time may, of course, vary with problem size), the statement takes a fixed amount of time to execute. We denote this type of performance by $O(1)$, because if we factor out the constant execution time, we're left with 1.

Sequence of simple statements.   A sequence of simple statements obviously takes an amount of time equal to the sum of the times it takes each individual statement to execute. If the performances of the individual statements are $O(1)$, then so is that of the sum.

Decision.   For purposes of estimating performance, we consider both the *then* part and the *else* part of a conditional structure to be independent, arbitrary structures in their own right. Then, to estimate conservatively, we take the larger of the two individual "big Os" as the "big O" of the decision.

```
Temp := A;                          IF x > Max
A    := B;                             THEN Max := x;
B    := Temp;                       END;
```

(a) Sequence   (b) Decision

```
IF x > y                            IF x >= y AND x >= z
   THEN Max := x;                      THEN Max := x;
   ELSE Max := y;                   ELSIF y >= x AND y >= z
END;                                   THEN Max := y;
                                    ELSE
                                       Max := z;
                                    END;
```

(c) IF-THEN-ELSE   (d) IF-THEN-ELSEIF-ELSE

```
FOR i := p TO q DO                  WHILE x > 0 DO
   x := x + i;                         y := y + 3;
END;                                   x := x/2;
                                    END;
```

(e) counting loop   (f) WHILE loop

```
REPEAT                              LOOP
   x := x + k;                         x := x + k;
   y := y - z;                         IF x >= 100 THEN EXIT;
UNTIL x >= 100;                        y := y - z;
                                    END;
```

(g) REPEAT-UNTIL loop   (h) LOOP-EXIT-END loop

Figure 2-12 Some Modula-2 control structures

A variation of the decision structure is the *case* structure, really just a multiway *if-then-else*. Thus, in estimating the performance of a case, we just take the largest "big O" of all of the case alternatives. We estimate the performance of the *if-then-elseif-else* structure, provided by Modula-2 (see Figure 2-12) and several other languages, in a similar manner.

Sec. 2.4 Performance Prediction and "Big O" Notation

```
FOR counter := 1 TO 5 DO
    ...
        something with O(1)
    ...
END;
```

(a) Trip count is constant

```
FOR counter := 1 TO N DO
    ...
        something with O(1)
    ...
END;
```

(b) Trip count depends on N

Figure 2–13 Two simple Modula-2 counting loops

That performance estimation can get very complicated and "tricky" can be seen by realizing that the condition tested in a decision may involve a function call, and that the timing of the function call may itself vary with problem size.

Counting loop. A *counting loop* is a loop in which the loop counter is incremented (decremented) each time the loop is executed. This is different from some loops we shall consider later, where the counter is multiplied or divided by a given value.

How do we measure the performance of a simple counting loop? Suppose the body of the loop contains only a sequence of simple statements. Then the performance of the loop is just the number of times the loop executes. Let us use the term *trip count* to mean "the number of times a loop executes." If the trip count is constant—i.e., independent of problem size—then the performance of the whole loop is $O(1)$. On the other hand, if the loop is something like

$$\text{FOR counter} := 1 \text{ TO N DO}$$

the trip count depends on N, so the performance is $O(N)$. These two loop structures, where the body contains only simple statements, are shown in Figure 2–13.

Figure 2–14, on the other hand, shows a double counting loop. The outer loop's trip count is clearly N. But the inner loop executes N times for each time the outer loop executes, so the body of the inner loop will be executed $N \times N$ times, and the performance of the entire structure is $O(N^2)$.

In Figure 2–15, a structure is shown that looks deceptively similar to that of

```
FOR OuterCounter := 1 TO N DO
    FOR InnerCounter := 1 TO N DO
        ...
            something with O(1)
        ...
    END;
END;
```

Figure 2–14 A double counting loop in Modula-2

```
FOR OuterCounter := 1 TO N DO
    FOR InnerCounter := 1 TO OuterCounter DO
    
        ...
        something with O(1)
        ...
    
    END;
END;
```

Figure 2-15 Another double counting loop

Figure 2-14. Again, the outer loop has a trip count of N. But this time the trip count of the inner loop depends not only on N but also on the value of Outer-Counter!: if OuterCounter = 1, the inner loop has a trip count of 1; if OuterCounter = 2, the inner loop trip count is 2; and so on, so that, in general, if OuterCounter = N, the inner loop trip count is N.

How many times will the body of the inner loop be executed? The number of times is given by the sum

$$1 + 2 + 3 + \ldots + (N - 1) + N$$

This summation, as you probably have learned somewhere in an algebra course, is $N(N + 1)/2 = ((N^2) + N)/2$. We shall say that the performance of a structure with this characteristic is $O(N^2)$, since for large N the contribution of the $N/2$ term is negligible.

It is interesting that making the inner loop trip count depend on OuterCounter does not alter the "big O" of the previous loop, since we neglect the term in N.

The structure in Figure 2-16 is similar to that of Figure 2-15, except that the trip count of the inner loop decreases rather than increases. That is to say, if OuterCounter = 1, the inner loop has a trip count of N; if OuterCounter = 2, the inner loop trip count is $N - 1$; and so on until, if OuterCounter = N, the inner loop trip count is 1.

The number of times the body of the inner loop is executed is the sum

$$N + (N - 1) + (N - 2) + \ldots + 1$$

which, as is plain, is the same sum as before, i.e., $N(N + 1)/2 = ((N^2) + N)/2$. So this structure also has performance $O(N^2)$.

Examine the loop structures in Figure 2-17 and convince yourself that in all cases the performance is $O(N^3)$.

From these examples, we can generalize as follows: a structure with k nested counting loops—loops where the counter is just incremented or decremented by

```
FOR OuterCounter := 1 TO N DO
    FOR InnerCounter := OuterCounter TO N DO
    
        ...
        something with O(1)
        ...
    
    END;
END;
```

Figure 2-16 Yet another double counting loop

Sec. 2.4 Performance Prediction and "Big O" Notation

```
FOR OuterCounter := 1 TO N DO
    FOR Middle Counter := 1 TO N DO
        FOR InnerCounter := 1 TO N DO

            ...
            something with O(1)
            ...

        END;
    END;
END;
                    (a)


FOR OuterCounter := 1 TO N DO
    FOR MiddleCounter := 1 TO OuterCounter DO
        FOR InnerCounter := 1 TO MiddleCounter DO

            ...
            something with O(1)
            ...

        END;
    END;
END;
                    (b)


FOR OuterCounter := 1 TO N DO
    FOR Middle Counter := 1 TO OuterCounter DO
        FOR InnerCounter := MiddleCounter TO N DO

            ...
            something with O(1)
            ...

        END;
    END;
END;
                    (c)
```

Figure 2-17 Some triple counting loops

1—has performance $O(N^k)$ if the trip count of each loop depends only on the problem size. A growth rate $O(N^k)$ is called *polynomial*.

Multiplicatively controlled loop. A *multiplicatively controlled loop* is a loop whose controlling variable is multiplied or divided by a constant each time the loop is executed. Most programming languages do not have a structure designed specifically to accommodate multiplicative control; generally, they just use a *while* or *until* loop.

Whatever the specific structure used, every loop needs each of the following mechanisms:

- An *initialization step* which gives the starting value(s) of the control variable(s);

```
Control := 1;                        Control := N;
WHILE Control <= N DO                WHILE Control >= 0 DO
   ...                                  ...
   something with O(1)                  something with O(1)
   ...                                  ...
   Control := 2 * Control;              Control := Control DIV 2;
END;                                 END;
```

Figure 2-18 A multiplicatively-controlled loop

Figure 2-19 Another multiplicatively-controlled loop

- A *termination condition* which is tested during each iteration and which indicates the circumstances under which the loop stops executing; and
- A *modification step* indicating how the control variable(s) should be changed to move the loop along from its starting point to its ending point.

The difference between a *while* structure and an *until* structure is that in the former the termination condition is tested before each iteration, whereas in the latter the condition is tested after each iteration.

Consider the structure in Figure 2-18. In this loop, whose performance clearly depends on the problem size N, the variable Control is multiplied by the constant 2 until Control gets to be larger than N. Since Control's starting value is 1, after k iterations we have

$$\text{Control} = 2^k$$

The number of iterations k can be found by taking logarithms of both sides of the equation, yielding

$$\log_2(\text{Control}) = k$$

Since the loop stops when Control $> N$, the performance of the algorithm it implements is $O(\log_2(N))$.

Looking at the structure a bit more generally, suppose we multiply Control by some other constant, say, *Factor*. Then, after k iterations,

$$\text{Control} = \text{Factor}^k$$

and so, by an argument analogous to the preceding, the performance is $O(\log(N))$, where the logarithm is taken to the base Factor. It turns out in general that in considering the "big O" of an algorithm, it doesn't matter what base we use for logarithms. This is because the logarithm of a number to one base is just a constant times the logarithm of the same number to a different base. Consequently, since constant factors are "factored out" of a "big O," the base doesn't matter. An exercise is left for you to fill in the details.

Consider next, Figure 2-19, in which the control variable of the loop given is divided by a factor (2 in this case) instead of multiplied. This is similar to the previous example. There, Control was started at a small value and multiplied repetitively until it reached a maximum. Here, Control is started at a large value and divided repetitively until it reaches a minimum. What is the "big O" of this structure? Instead of repeating the preceding analysis, we shall merely state that it is $O(\log(N))$ and leave the details to be fleshed out in an exercise.

Now look at the two structures in Figure 2-20, which are analogous to the nested counting loops considered earlier. Again, we shall state that the performance of these structures is $O(N \log(N))$; and leave the proof to an exercise.

Subprogram call. We can handle the performance of a subprogram call by observing that the subprogram is an algorithm with its own "big O," and then imagining that this algorithm appears "in line" with the calling program. Then, if the subprogram call appears inside a decision statement, its "big O" is used in determining the maximum of the "big Os" of the different branches of the decision. If, on the other hand, the subprogram call appears inside a loop, its "big O" is essentially multiplied by the trip count of the loop.

If the subprogram (call it A) in turn calls another subprogram (B), then we use, first, B's "big O" in calculating A's, then A's in calculating the calling program's "big O," and so on for deeper nesting of subprograms.

Things get tricky if A and B are the *same* subprogram—that is, if a recursive call is involved. Then, in calculating A's "big O," the depth of recursion—the

```
FOR Counter := 1 TO N DO

    Control := 1;
    WHILE Control <= N DO
        ...
        something with O(1)
        ...
        Control := 2 * Control;
    END;

END;
```

(a)

```
Control := N;
WHILE Control >= 0 DO

    FOR Counter := 1 TO Control DO
        ...
        something with O(1)
        ...
    END;

    Control := Control DIV 2
END;
```

(b)

Figure 2-20 Two N × log(N) loop structures

number of times that *A* is called recursively—is usually itself a function of the problem size, so we need to do the same sort of analysis we performed on the preceding structures in order to get a handle on the depth of recursion.

2.4.3 Some Examples of Performance Prediction

Let us return to the algorithms of section 2.3 and estimate their performance. The analysis we shall use applies to recursive programs in general.

Factorial. In calculating $N!$, 1 is subtracted from the argument to Factorial each time a recursive call is made. Since multiplying one number by another results in a performance of $O(1)$, the performance of the factorial program is just that of a one-level counting loop, viz., $O(N)$.

Reverse of a string. Each time StringReverse is called, its argument—the string—is shortened by one character. Thus, the number of recursive calls is directly proportional to the length of the string. So, at first glance, the performance of the program seems to be $O(N)$.

However, there is a subprogram call involved: concatenation of a character to the end of a string. If that operation doesn't depend on the string length, then the performance of StringReverse is indeed $O(N)$. On the other hand, if for some reason the concatenation operation had to "walk across" the whole string to add the new character onto the end—a linear operation—then we would have an overall performance of $O(N^2)$, because a linear operation would be done a linear number of times. Without knowing more about concatenation, we cannot go any further.

Permutations of a set. In section 2.3.2, we calculated that a set of size N has $N!$ permutations. Thus, the program PrintPermutations has growth rate $O(N!)$, much larger than any of the other growth rates we've seen. Since algorithms with factorial growth rate are almost impossibly slow for large values of N that are interesting, we try to avoid them.

On the other hand, sometimes we can't avoid factorial growth: whichever algorithm we choose for printing all permutations, we cannot escape the mathematical fact that a set of size N has $N!$ permutations. Thus, anyone claiming an algorithm that can print $N!$ values with performance better than $O(N!)$ is claiming something magic, not mathematical!

Recursive binary search. Recall that in a binary search we divide a sorted list in half, then in half again, and so on, until we either find what we are looking for or are left with only one element that isn't the one we want. But this procedure is just a recursive version of the loop structure in Figure 2–19, whose performance we found to be $O(\log(N))$.

Sec. 2.5   Design: Maintaining a Dynamic Table   49

Recursive merge/sort.   Like a recursive binary search, a recursive merge/sort is a divide-and-conquer algorithm, involving repeated halving of the list to be sorted. Since the number of times we divide the list in half is $\log(N)$, the performance would appear to be $\log(N)$. But here again, there is a lower level subprogram called merge whose implementation is unknown to us.

To look ahead, the merge operation is usually linear in performance: all items in both lists are copied once. Accordingly, the topmost level will merge two lists of length $N/2$, requiring the copying of N values; the second level will do two merges, each on a list of length $N/4$; and so on. So if we add up all the merging done at a given level of recursion, we always get exactly N operations. Since there are $\log(N)$ levels of recursion, we arrive at a growth rate for the recursive merge/sort of $O(N \log(N))$. The program is a recursive version of the loop of Figure 2–20b.

2.5 DESIGN: MAINTAINING A DYNAMIC TABLE

Let us again consider the problem of maintaining a table like the phone list discussed earlier. Suppose that each item to be stored in the table contains a *key part* and a *value part*. The key part is the field we shall use to look items up in the table; the value part is everything else. For the phone list, the keys are your friends' names, while the values are their phone numbers.

Now if we consider the table to be an ADT, then five operations apply:

- *Create*, or set up a new table with no items in it;
- *Update*, or add a new item to the table;
- *Search*, or look up an item in the table, returning the value part corresponding to a given key;
- *Delete*, or remove an item with a given key from the table;
- *Report*, or print out, or display all the entries currently in the table, in order of their keys.

Figure 2–21 shows a sketch of a definition module for this ADT. We leave out the details of the type definitions for the keys and values, as they are irrelevant to the discussion.

Let us consider two possible implementations of this table, both using an array, and then determine the performance of the various operations. We shall not give detailed programs, since all we are really interested in is "reasoning out" the performance issues. We assume that the array in question can hold up to MaxItems items, and that the actual number of items in the array at a given time is ActualItems. The Create operation in both implementations just involves declaring the array, which takes at most a constant amount of execution time, and

```
DEFINITION MODULE TableHandler;

    TYPE KeyType   = ... ;  (* to be determined *)
    TYPE ValueType = ... ;  (* same here        *)
    TYPE TableType = ... ;  (* this one too     *)

    PROCEDURE Create(VAR T: TableType);
    PROCEDURE Update(VAR T: TableType;
                         K: KeyType;
                         V: ValueType);
    PROCEDURE Search(T:  TableType;
                     K:  KeyType) : ValueType;
    PROCEDURE Delete(VAR T: TableType;
                         K: KeyType);
    PROCEDURE Report(T: TableType);

END TableHandler;
```

Figure 2–21 Sketch of TableHandler definition module

setting ActualItems to indicate that the table is empty, also a constant-time operation. Completing the module requires supplying the type definitions. Figure 2–22 shows some suggested types. KeyType and ValueType might be imported from another module or declared right in TableHandler, depending on the application.

2.5.1 First Implementation: Unordered Array

In the first implementation of the table, we leave the array unordered, updating it just by keeping track of the number of positions currently occupied and then inserting a newly arrived item in the next available position. An update operation thus has performance $O(1)$ (constant), since the number of operations required to store an item in the next available position in an array doesn't depend either on the size of the array or on how many items are already stored there.

How about a search operation? Since the items are not in any particular

```
TYPE KeyType   = CHAR;     (* for example *)
TYPE ValueType = INTEGER;  (* maybe       *)

CONST MaxItems = 100;      (* arbitrarily chosen *)

TYPE Item =
   RECORD
      Key: KeyType;
      Val: ValueType;
   END;

TYPE TableType =
   RECORD
      Store:       ARRAY[1..MaxItems] OF Item;
      ActualItems: INTEGER;
   END;
```

Figure 2–22 Types associated with TableHandler

Sec. 2.5 Design: Maintaining a Dynamic Table   51

order in the table, we need to start at one end of the (occupied portion of the) table and check the key of every item, until either we find the one we wanted or we reach the other end of the occupied portion. Sometimes we find our item on the first attempt; sometimes we need to search the entire table. On average, we check half the items. Since both the worst and average situations depend directly on the number of items in the table, we can say that a search operation is linear with performance $O(N)$ (actually $O(\text{ActualItems})$).

A delete operation also is linear in performance. We delete an item corresponding to a given key by searching for it as in a search operation and then removing it by moving all items below it up one position, as shown in Figure 2–23(a). Since the number of operations for both the search and the move depends directly on ActualItems, we have a linear operation.

We can speed up the delete operation by recognizing that since the items in the table are not in order, we lose nothing by just copying the latest item into the position occupied by the one to be deleted, as shown in Figure 2–23(b), and then recovering the vacated space by decrementing ActualItems. However, while the average time for the delete is surely reduced by this optimization, the growth rate is unchanged because the search part of the operation is still linear.

A report operation involves sorting the table in some way, since we want the table to be printed out in key order. The details of the sorting process will be left for chapter 8, where many sorting algorithms are presented and compared. For completeness here, we just mention that the growth rate of a sorting algorithm is, in most cases, either $O(N \log(N))$ or $O(N^2)$.

2.5.2 Second Implementation: Array Ordered by Key

This implementation corresponds to the kind of table discussed in the phone list example of section 2.3.3, where we considered the binary search algorithm, which carries out a search with performance $O(\log(N))$. In such an implementation, update and delete operations turn out to have linear performance. To see this, recall that, in an update, we need to insert the newly arrived item in its proper place in the array, to preserve the ordering. We find this place by a modified search operation: since the item is not yet in the table, the search will always fail, but instead of reporting just that fact, we make the search also report the last location it tested, which will tell us exactly where the new item needs to go. This will work correctly except where the new key is smaller than the previously smallest key or greater than the previously greater one, so we test those two possibilities as special cases before beginning the binary search.

Once the proper location has been found, we need to make room for the new item by moving the ones with greater keys *forward* one position. So the performance has a logarithmic component and a linear one. However, the logarithmic component can be ignored since, as ActualItems increases, it is so much greater than its logarithm. Thus, an update operation has linear performance.

In this implementation, a delete is really just like a Search except that the

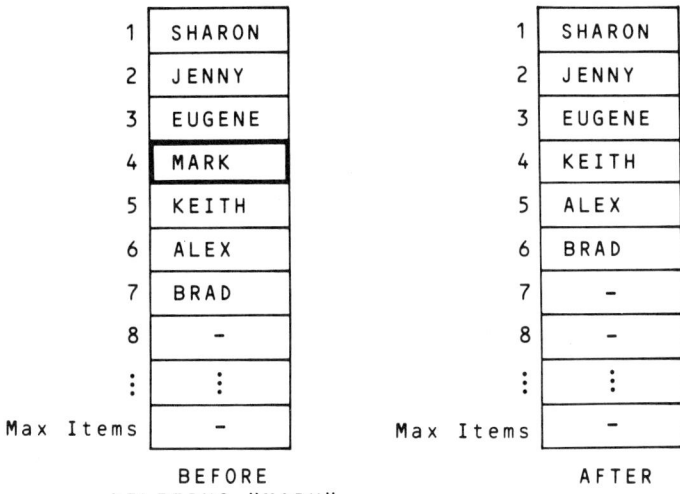

(a) Deletion by moving all items up

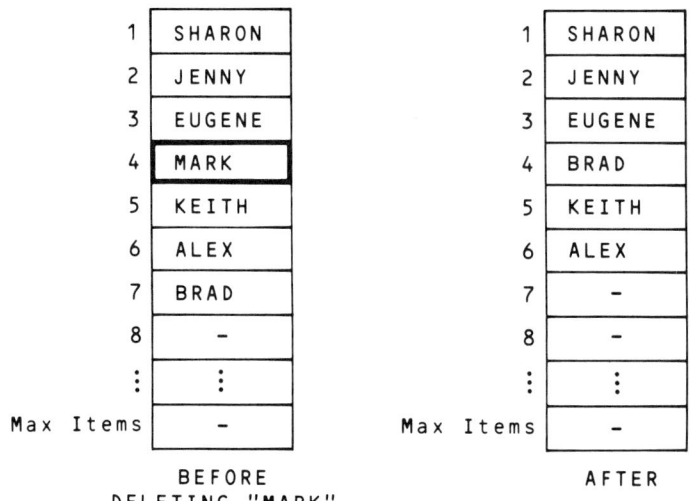

(b) Deletion by moving last item up

Figure 2–23 Deletion from unordered table

Sec. 2.6　Summary

	UNORDERED	ORDERED
Create	O(1)	O(1)
Update	O(1)	O(N)
Search	O(N)	O(Log(N))
Delete	O(N)	O(N)
Report	O(N x Log(N))	O(N)

Figure 2-24 Comparative performance of table operations for unordered and ordered array storage of a dynamic table

item to be deleted is removed by moving all items with greater keys *back* one position in the array. Since we can't use the speed-up process of the first implementation (why?), the move is linear, so the whole delete operation is again linear.

Finally, consider the report operation. Since the table is already ordered, a report can just start at the beginning and step through the array, printing each item as it encounters it. So the report operation is also linear.

Figure 2-24 gives a tabular summary of the growth rates of the various operations in both implementations.

2.6 SUMMARY

In this chapter we have discussed algorithms, recursion, and performance prediction.

An algorithm is a method used to solve a problem in a systematic way. It consists of a finite number of steps that will complete its work in a finite amount of time with a finite effort, regardless of the amount or nature of the input given to it.

A recursive algorithm is one which invokes itself—i.e., one whose own name appears in its definition. Infinite recursion is avoided by making certain that the algorithm has a specific step in its definition indicating the conditions for stopping the recursion, and that each recursive call operates on a "data set" smaller than the previous one.

Recursive algorithms were presented for (1) calculating the factorial of a number, (2) reversing a character string, (3) calculating the permutations of a set, (4) performing a binary search, and (5) performing a merge/sort. The Modula-2 versions of these algorithms made clear that recursive programs can be written straightforwardly in this programming language.

Performance prediction is the process of estimating how the computation or running time of an algorithm or program varies with the "problem size." Performance, also called growth rate, is often expressed in "big O" notation.

While there is no easy, guaranteed way to calculate algorithm or program performance, there are certain techniques and rules of thumb that are helpful in "getting a handle on it." In particular, performance prediction is facilitated when programs are written according to "structured coding" conventions, because such programs have well-defined loop and decision structures.

The most common growth rates in data structure operations are, in order of steepness, constant or $O(1)$, logarithmic or $O(\log(N))$, linear or $O(N)$, $O(N \log(N))$, and quadratic or $O(N^2)$. Several of these were exemplified in various operations having to do with different implementations of a table as an array.

CHAPTER 2 EXERCISES

1. Give a recursive definition of the integer addition operation. Write and test a recursive function to produce the sum of two integers (HINT: Use the built-in "+" operation *only* to add 1 to a number).
2. Give a recursive definition of the integer multiplication operation. Write and test a recursive function to carry out the definition (HINT: Multiplication is repeated addition).
3. Give a recursive definition of the integer exponentiation operation. Write and test a recursive function to carry out the definition (HINT: Exponentiation is repeated multiplication).
4. The Fibonacci numbers of order 1 are a sequence of positive integers starting with 1,2,3,5,8, . . . , in other words, each number except the first two is the sum of the two previous numbers. Give a recursive definition of this sequence; write a recursive procedure to print out the first twenty-five numbers.
5. What is the "big O" of the usual algorithm to set to zero all the elements of a 2-dimensional square array with N rows and N columns?
6. Show that in computing a "big O" which turns out to have a logarithmic component, i.e. something of the form log(. . .), the base we use to represent the logarithm does not matter.
7. Show that the growth rate of the algorithmic structure given in Figure 2–19 is $O(\log(N))$.
8. Show that the growth rates of the structures in Figure 2–20 is $O(N \times \log(N))$.
9. Estimate the performance of the concatenation and substring search algorithms from section 1.7.
10. Consider the problem of searching for a key k in an unordered array where duplicate keys are permitted. Discuss the performance of each of the following cases:
    (a) k does not appear in the array;
    (b) k appears once in the array;
    (c) k appears several times in the array (not necessarily in adjacent locations!) but only the location of the first appearance is desired;
    (d) k appears several times in the array and it is desired to report the locations of all appearances.
11. Repeat the previous problem for an ordered array.

Chapter 3
ARRAYS, VECTORS, MATRICES, AND LISTS

3.1 GOAL OF THE CHAPTER

In this chapter we consider the ways in which the familiar mathematical abstractions of vectors and matrices can be represented in programs. We first look at how these abstractions are handled in programming languages like Fortran, Pascal, and Modula-2, then examine some interesting operations commonly performed on vectors and matrices, and finally, consider how vectors or matrices with many zero or empty values can be efficiently stored. Such vectors or matrices are usually called *sparse*; their study provides the context for the introduction of linked data structures.

The design section of the chapter develops a general Modula-2-based scheme for using functional notation for array operations, and for handling multidimensional arrays in a fashion more general than that provided by the language. In the style guide section some hints are presented on how to simulate arrays of records in languages that lack record types.

3.2 CLASSICAL REPRESENTATION OF VECTORS AND MATRICES

The first generation of computers, developed in the 1950s, were intended chiefly for the solution of scientific and engineering—that is, *mathematical*—problems. Indeed, the first devices resembling what we would call digital computers, built in the mid-1940s, were designed mainly to do calculations resulting in tables used for artillery control. Later, it was realized that computers could be very powerful in data processing and other less mathematical applications like language translation and the maintenance of large-scale information systems.

In the mid-1950s, when an alternative was sought to coding mathematical problems in machine language, Fortran was developed by John Backus and his team at IBM. Given the predominance of vectors and matrices in mathematical problems, it is not surprising that the Formula Translator—Fortran—embodied support for these in the form of arrays. The single- and multidimensional arrays of Fortran are implementations of the mathematical abstractions of vectors, matrices, and tensors (three-dimensional matrices), and serve as the model for similar implementations in Algol, PL/1, Basic, Pascal, and even Modula-2 and Ada.

3.2.1 Vectors and One-Dimensional Arrays

A *vector* of N components is a set of N values which is ordered in the sense that each value is assigned a specific position in the set. For example, the vector V1 = $\langle 3,5,-1 \rangle$ is different from the vector V2 = $\langle 5,-1,3 \rangle$: they both contain the same set of values, but the values appear in different orders.

If we number the components of a vector $1,2,\ldots,N$, then it makes sense to talk about the ith component of this vector. For example, if $i = 2$, then the ith component of vector above V2 is -1. In everyday mathematical notation, we would express a component by means of a *subscript*, as, for example, v_i. In programming, however, we do not usually have the ability to write actual subscripts, so brackets or parentheses are generally used, as in V2[i] and V2(i), respectively. In this discussion we shall use brackets exclusively, since they are the form used by Modula-2.

In everyday programming we implement vectors through the use of one-dimensional arrays. The high-level language we are using generally allows us to indicate to the compiler what size array we need, usually by means of a *declaration* like

      INTEGER A(10) (in Fortran)

or

      A: ARRAY[1 . . 10] OF INTEGER (in Modula-2)

Both of these declarations indicate that the compiler is to set aside space for ten integer values, and that the valid range of subscripts used with the array variable to denote those values is $1,2,\ldots,10$.

Having instructed the compiler to *create* the array (without storing any values in it), we can now carry out two operations on array elements: we can *store* a value, as in the Fortran assignment A(2) = 3, or we can *retrieve* a previously stored value, as in the Modula-2 assignment Y := A[2]. Note that the use of a constant as subscript is arbitrary: we could just as well have stored or retrieved using a variable, as in A(I) = 3, as long as we made certain that at the time the statement was executed I had a value in the range 1 through 10.

Sec. 3.2 Classical Representation of Vectors and Matrices

The point is that we have mentioned two separate operations—a storage operation and a retrieval operation—even though most programming languages permit the same syntax to be used on either side of the assignment operator to reference elements in the array. It is just that these two abstract operations have been implemented in programming in a way which is syntactically convenient and intuitively comfortable.

It is important to realize that the *type* of a vector's elements need not be numerical, although integers and reals are in fact the types most frequently seen in engineering problems and programming texts. We shall have frequent occasion, however, to think of vectors of *records* of one sort or another, like student records. Many examples use numbers as the elements of a vector because they are easily written and because support for vectors in early programming languages was motivated by numerical problems; bear in mind, though, that nearly everything we say in this chapter generalizes very nicely to many other types.

Now, what sort of machine instructions would a compiler have to produce in order to support subscripting? When an array is declared, space is reserved for as many elements as are requested by the declaration. But how much space? That depends on the *type* of the thing to be stored in each element. For example, in many 16-bit computers (like the PDP-11), an integer in Fortran is deemed to occupy one 16-bit word, but a real (or floating-point) number occupies two words, or 32 bits. So an array of 100 floating-point numbers, declared, say, by REAL T(100) in Fortran-66, will require 200 words of PDP-11 memory. On the other hand, in machines like the IBM System/370, where space is allocated in eight-bit bytes rather than 16-bit words, the same array requires 400 bytes, because floating-point numbers in that machine occupy four bytes each. Similarly, one could envision an array of 200-character student records with name, address, course grades, etc., and thus 200 bytes per element might be necessary. A total of 20,000 bytes would then be required for the entire array.

Once the space is allocated for an array, the compiler must generate certain instructions which "understand" the relationship between a subscript reference like T[i] and the internal storage on the particular computer involved. This relationship is usually called the *storage mapping*, or sometimes the *storage mapping function*. Letting add(T) be the machine address of the first storage unit (byte, word, etc.) of the array, and NUNITS be the number of storage units per array element, what is the storage mapping for an array whose lowest subscript value is 1 (as in Fortran-66)? To determine this, we need a formula which tells us how many elements to skip over in order to reach the *i*th one. Clearly, we need to skip over $i - 1$ of them. So the address of T[i] is in fact

$$\text{add}(T) + (i - 1) \times \text{NUNITS}$$

To verify the formula, consider a four-byte real as in the System/370. Here, addressing is done byte by byte, so NUNITS = 4. Thus, the address of T[1] is in fact just add(T) because the second term drops out, the address of T[2] is add(T) + 4, and so on.

The array-declaration scheme in Fortran-66 is really a special case, since the lowest subscript value is required to be 1. In the more general situation, as implemented in most other languages, including Fortran-77 and Modula-2, the lowest value may be any integer, the only restriction being that it must be less than the highest subscript value. These two values are usually called the *range* of the subscripts.

Under the more general scheme, what would the storage mapping look like? Call the lowest subscript value First, and the highest Last. Then, if we declare

$$T: ARRAY[First .. Last] \text{ OF REAL}$$

we will need (Last − First + 1) × NUNITS of space in order for T[First] to map to add(T). And to get to an arbitrary T[i], we will need to skip over $i - $ First elements. To see this, suppose that First = 3 and Last = 10, as in

$$T: ARRAY[3 .. 10] \text{ OF INTEGER}$$

Clearly, there will be eight elements required, each NUNITS long; so to get to, say, T[5], we need to skip elements 3 and 4, which is 5 − 3 = 2 elements. And to get to the first element, T[3], we skip no elements (3 − 3 = 0). So our storage mapping becomes

$$T[i] \text{ maps to add}(T) + (i - \text{First}) \times \text{NUNITS}$$

Notice that this is perfectly consistent with the special case of Fortran-66, where First = 1 *always*. The formula works even if First and/or Last are negative. For example, let us find the storage mapping for NUNITS = 4 and the declaration

$$T: ARRAY[-5 .. 7] \text{ OF REAL}$$

The array will require 7 − (−5) + 1 four-byte elements, or 52 bytes, and T[i] maps to add(T) + (i − (−5)) × 4, or add(T) + (i + 5) × 4. So T[−5] maps to add(T), T[0] maps to add(T) + 20 (we've skipped over five elements), and so on. This arrangement is shown in Figure 3–1.

To conclude this subsection, notice that there is nothing sacred about mapping the *lowest-subscripted* element to add(T). Indeed, in some computers, e.g., the Hewlett-Packard HP-3000, the hardware design is such that add(T) maps most conveniently to the zeroth element, so that elements with negative and positive subscripts are deemed to lie below and above the zero point, respectively. In these computers, even if there is no zeroth element, add(T) is mapped to where it would be located if there were one. For uniformity, we shall retain the convention that the lowest-subscripted element maps to add(T).

3.2.2 Matrices and Two-Dimensional Arrays

We define an $R \times C$ *matrix* M as a rectangular array of elements having R *rows* and C *columns*, and then refer to any particular element in the matrix by using two subscripts r and c in an expression M[r,c]. Note that if we were to view M

Sec. 3.2 Classical Representation of Vectors and Matrices

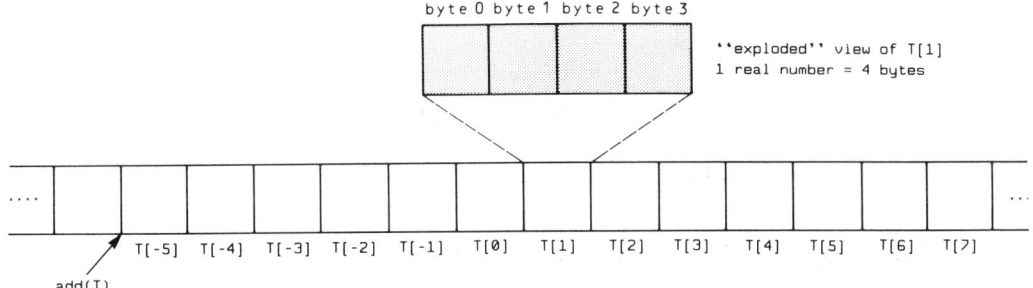

Figure 3-1 Storage allocation in linear memory for T:ARRAY[−5..7] OF REAL

pictorially, as in Figure 3-2, the rows would be oriented horizontally, the columns vertically, and the subscript reference would give the *row* subscript first. (Once again, there is nothing sacred about this view; it is just a convention.)

As in the one-dimensional case, we have an abstraction Matrix, implemented in most programming languages by whatever feature allows us to declare two-dimensional arrays and to store and retrieve elements in them in the form M[r,c]. To see how compilers effect the implementation, recall that in most computers memory is logically organized in a linear fashion, with the addresses of the storage

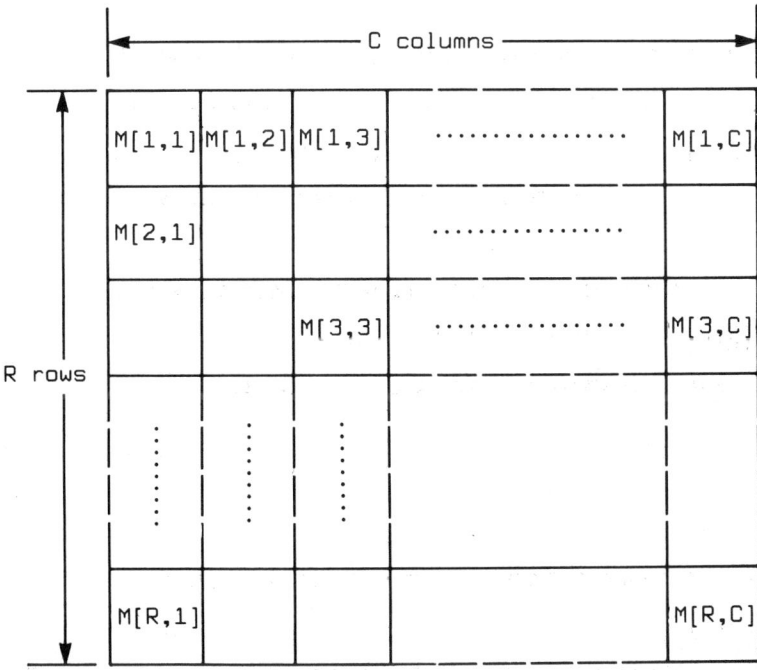

Figure 3-2 Abstract (programmer's) view of an R × C array, or matrix

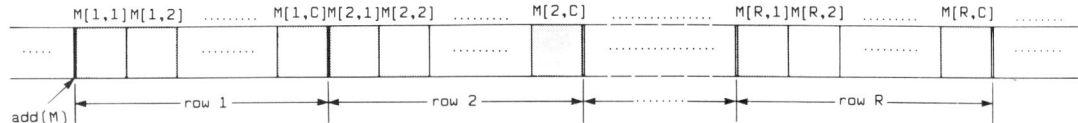

Figure 3-3 Row-major implementation of an R × C array

units (words or bytes) running in a single increasing sequence. So a structure with two dimensions has to be mapped onto a structure with only one dimension. To comply with this condition, most programming languages implement the abstraction Matrix in a form called *row-major,* where the two-dimensional array is stored *row by row* in linear memory, as shown in Figure 3-3. That this is not the only storage scheme is evidenced by the *column-major* scheme in Fortran, in which a two-dimensional array is stored *column by column,* shown for the same array in Figure 3-4.

Now what is the storage mapping function for a two-dimensional array stored in row-major form? As before, let us begin with the familiar case where the rows and columns are numbered 1 . . R and 1 . . C, respectively. Since the array is stored row by row, to reach the rth row we need to skip over $r - 1$ rows; then, to reach the cth element in the rth row, we need to skip over $c - 1$ elements. Since each row has C elements, and since each element requires NUNITS of storage, the mapping function is

$$M[r,c] \text{ maps to } add(M) + (r - 1) \times C \times \text{NUNITS} + (c - 1) \times \text{NUNITS}$$

Letting NUNITS = 4 as above (for, say, a four-byte real in the System/370), we see that for the 5 × 6 array in Figure 3-5, 120 bytes of storage are needed, and

$$M[1,1] \text{ maps to } add(M) + 0$$

$$M[5,6] \text{ maps to } add(M) + 4 \times 6 \times 4 + 5 \times 4 = add(M) + 116$$

and

$$M[3,2] \text{ maps to } add(M) + 2 \times 2 \times 4 + 1 \times 4 = add(M) + 20$$

As in the one-dimensional case, we can generalize this idea to permit subscripts to have an arbitrary integer range, as in Pascal, Modula-2, or Fortran-77. The Modula-2 declaration would be

M: ARRAY[FirstR . . LastR],[FirstC . . LastC] OF REAL

The details of the row-major storage mapping function are left as an exercise,

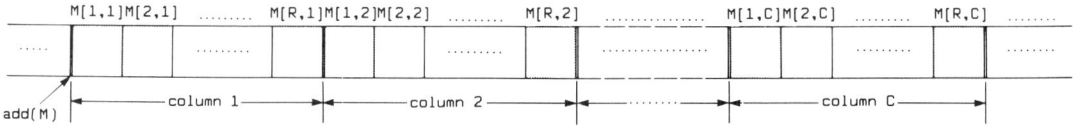

Figure 3-4 Column-major implementation of an R × C array

Sec. 3.2   Classical Representation of Vectors and Matrices   61

```
VAR M: ARRAY [1..5],[1..6] OF REAL
```

(a) Declaration.

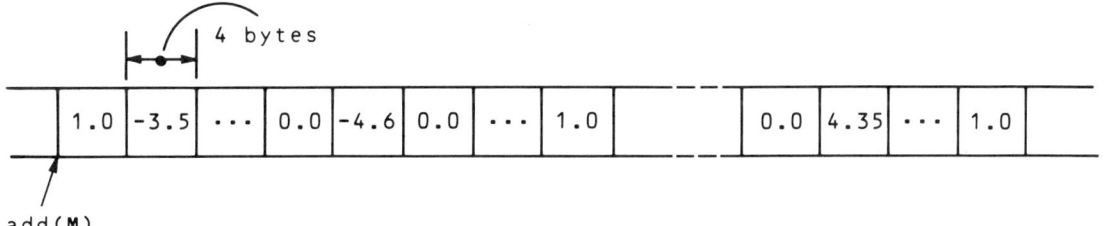

(b) Abstract (programmer's view).

(c) **Row-major implementation in linear memory** (note real numbers are stored as four-byte floating point).

Figure 3-5 Abstraction and implementation of a 5×6 array of reals

as is the question of developing a storage mapping for two-dimensional arrays implemented in column-major form, as in Fortran.

3.2.3 Higher-Dimensional Structures

There is often a need in programming problems to work with arrays of higher dimension than two, and most programming languages support a feature to permit a fairly large number of separate ranges of subscripts. The facility is implemented as a generalization of the two-dimensional case. For example, for three-dimensional arrays, the third dimension is conventionally called a *plane,* and the new subscript is conventionally added *before* the one for a row. So a Modula-2 declaration

A: ARRAY[1 . . 4],[1 . . 5],[1 . . 6] OF REAL

would be interpreted as an array with four planes, each having five rows and six

M:ARRAY[-1..1],[0..3],[5,6] OF ...

(a) Declaration.

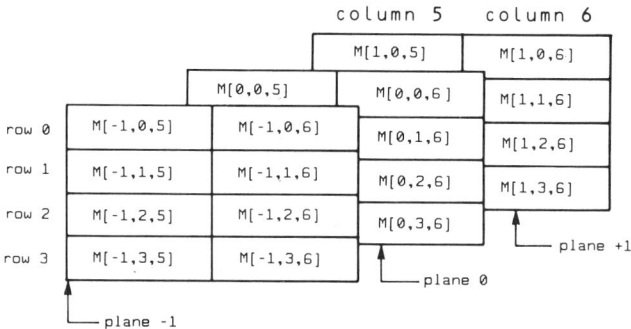

(b) Abstract view.

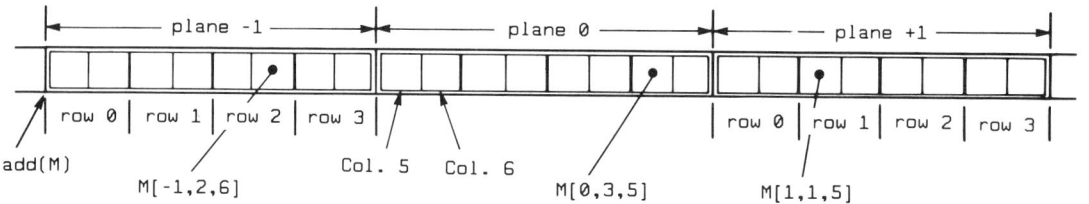

(c) Row-major implementation.

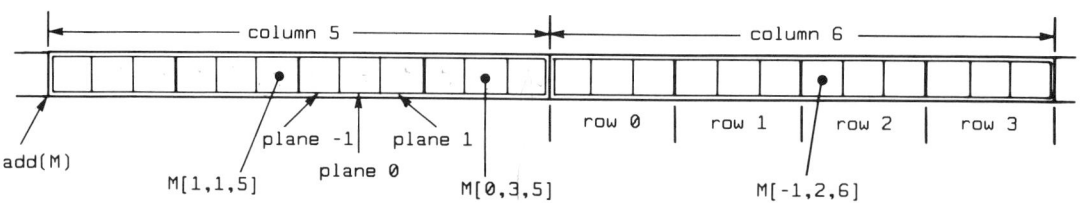

(d) Column-major implementation.

Figure 3–6 Abstraction and implementation of a three-dimensional array

columns. A reference A[p,r,c] would then be interpreted as referring to that element at the intersection of the *p*th plane, *r*th row, and *c*th column.

As in the two-dimensional case, this abstract structure is then mapped onto linear storage in either row-major or column-major form. In row-major form, we reach the element A[p,r,c] by skipping over $p - 1$ planes to reach the *p*th plane, then $r - 1$ rows to reach the *r*th row, and finally $c - 1$ elements (each in a column) to reach the *c*th element. In column-major form, we first skip over $c - 1$ columns to reach the *c*th column, then $r - 1$ rows to reach the *r*th row, and finally $p - 1$ elements (each in a plane) to reach the *p*th element. These schemes are illustrated in Figure 3-6, for the case of an array A[-1..1],[0..3],[5..6]

Obtaining storage mapping functions for this case and for higher dimensions in general is left as a set of exercises. The idea is that in any row-major scheme, of whatever dimension, the *leftmost* subscript varies most slowly, whereas in any column-major scheme the *rightmost* subscript varies most slowly.

3.3 VECTORS, MATRICES, AND MODULA-2 ARRAYS

Having discussed the *storage* of vectors and matrices in the previous section, let us take up the question of *operations* on these structures. In the process, we shall discover a feature of Modula-2 called *open array parameters*, a convenient method of dealing with one-dimensional arrays in a general fashion.

3.3.1 Vector Arithmetic and Open Array Parameters

To represent a vector in Modula-2, we can use a normal one-dimensional array. For example, letting ValueType be INTEGER, we declare

VAR A: ARRAY[- 5 .. 2] OF INTEGER

Let us write a procedure to find the maximum value stored in this array. The algorithm is simple enough: just write a loop that goes through the array, saving the largest value so far. But we would like to generalize the algorithm, putting it in a subprogram which would work correctly for any vector, *regardless of its subscript range*. How, then, do we inform the procedure what this range is? One possibility is to pass the lower and upper bounds as additional parameters. This is cumbersome, however, and Modula-2 provides a better way, called *open array parameters*. Using these, we could declare

PROCEDURE MaxValue (VAR Result: INTEGER;

V:ARRAY OF INTEGER);

Notice that the open array parameter ARRAY OF INTEGER is written without any subscript range.

```
PROCEDURE MaxValue (VAR Result: ValueType;
                        V:          ARRAY OF ValueType);
   VAR
      CurrentMax: ValueType;
      R:          INTEGER;

   BEGIN

      CurrentMax := V[0];

      FOR R := 1 TO HIGH(V) DO
         IF V[R] > CurrentMax THEN
            CurrentMax := V[R];
         END;
      END;

      Result := CurrentMax;

   END MaxValue;
```

Figure 3-7 Function to find the maximum value in a vector

Given this scheme, how does the procedure discover the actual range of integers? In fact, it doesn't. When an open array parameter is used as a *formal* parameter, the subscript range of the *actual* parameter is *mapped* onto the range [0 .. HIGH(V)], where HIGH is a standard Modula-2 function which returns one less than the number of elements in the array. Since the array in our original declaration has eight elements (range [-5 .. 2]), HIGH(V), called inside the procedure, would return 7 if A were the actual parameter. Figure 3-7 shows the MaxValue procedure written in this style; Figure 3-8 illustrates a main program that uses the procedure.

With the help of some definitions, this simple example points the way to procedures capable of carrying out some interesting vector-arithmetic operations. Given two vectors U and V with dimensions [Rmin .. Rmax], whose components are all of type ValueType, and given a scalar K of type ValueType, we have the following operations.

The *vector sum* of U and V is a *vector* T with components of type ValueType and with dimensions [Rmin .. Rmax] such that, for each r between Rmin and Rmax inclusive,

$$T[r] := U[r] + V[r]$$

i.e., the components of the two vectors are added pairwise.

The *inner product*, sometimes called the *scalar product* or *dot product*, of U and V is a *scalar* of type ValueType whose value is the sum of all the pairwise products

$$U[r] \times V[r]$$

taken over all the components.

Sec. 3.3 Vectors, Matrices, and Modula-2 Arrays

```
MODULE TestMaxValue;

   FROM InOut IMPORT Read, WriteLn, WriteString, WriteInt;

   TYPE
      ValueType = INTEGER;
      Vector    = ARRAY [-5..2] OF ValueType;

   VAR
      A : Vector;
      X : ValueType;
      Q : CHAR;         (* wait for keypress before quitting *)


   (* Declare MaxValue as a local procedure here *)


   BEGIN      (* Main body of Program *)
      A[-5] :=   3;
      A[-4] := -17;
      A[-3] :=  20;
      A[-2] :=   0;
      A[-1] :=   4;
      A[ 0] :=  18;
      A[ 1] :=  27;
      A[ 2] := -32;

      MaxValue (X, A);
      WriteString ('Maximum value is ');
      WriteInt (X, 3);
      WriteLn;
      Read(Q);   (* wait for keypress before terminating *)

   END TestMaxValue.
```

Figure 3-8 Program which uses MaxValue

The *sum of a vector V with a scalar K* is a *vector* T, of the same type and dimensions as V, whose components have values

$$T[r] := K + V[r]$$

for every r between Rmin and Rmax;

The *product of a vector V by a scalar K* is a vector T, of the same type and dimensions as V, whose components have values

$$T[r] := K \times V[r]$$

for every r between Rmin and Rmax.

Figure 3-9 shows examples of these calculations for two six-element vectors.

Given the preceding definitions, we seek to design a Modula-2 module for vector arithmetic, defining appropriate types and then writing appropriate procedures to implement the operations, just as we did for the Fraction and Text modules earlier. In doing so, it is convenient to use functional notation for the arithmetic

```
        Given
   U = [1, -3, 2, 0, -1, 4]
   V = [0, -2, 1, -3, 2, 0]
   K = 2
        then
U + V = [1, -5, 3, -3, 1, 4]
U * V = 0 + 6 + 2 + 0 - 2 + 0 = 6
K + V = [2, 0, 3, -1, 4, 2]
K * V = [0, -4, 2, -6, 4, 0]
```

Figure 3–9 Some vector operations

operations, so that expressions can be written as in ordinary vector algebra. Consequently, since we shall represent vectors as arrays, which cannot be returned as function-procedure results, we shall need a solution using pointers, similar to that for Fractions. We present such a solution in the design section of the chapter; for now, let us keep things simple by writing our vector operations as procedures with a VAR parameter to hold the result.

Let us write the code for two of the vector arithmetic functions. First we add a scalar to a vector. This procedure is shown in Figure 3–10; the resulting vector is returned to the caller as a VAR parameter.

Now we write the inner-product program. In order for this to make sense mathematically, the vectors need to be *conformable*, i.e., have the same dimensions. This requirement suggests that our program should check the dimensions. As before, we can indicate to a client program that the input vectors are not conformable by exporting a Boolean exception variable and setting it to TRUE if the dimensions don't match, and FALSE otherwise.

Exception handling and reporting is an interesting and controversial subject in programming. Exactly which mechanism to use for error checking and recovery is part of the controversy; in fact, some languages have special constructs just for identifying and handling errors (e.g., "on-conditions" in PL/1, and "exceptions"

```
PROCEDURE AddScalarToVector (VAR Result: ARRAY OF ValueType;
                             V:          ARRAY OF ValueType;
                             Scalar:     ValueType);

VAR
   CurrentMax: ValueType;
   R:          INTEGER;

BEGIN

   FOR R := 0 TO HIGH(V) DO
      Result[R] := V[R] + Scalar;
   END;

END AddScalarToVector;
```

Figure 3–10 Procedure to add a scalar to a vector

Sec. 3.3   Vectors, Matrices, and Modula-2 Arrays

```
PROCEDURE InnerProduct (VAR Result: ValueType;
                        V,W:        ARRAY OF ValueType;
                        VAR ConformabilityError: BOOLEAN);

   VAR
      CurrentSum: ValueType;
      R:          INTEGER;

   BEGIN
      CurrentSum := Zero;
      ConformabilityError := FALSE;

   (* Check for conformability *)

      IF HIGH(V) <> HIGH(W) THEN
         ConformabilityError := TRUE;
         RETURN;
      END;

   (* If conformable, then compute inner product *)

      FOR R := 0 TO HIGH(V) DO
         CurrentSum := CurrentSum + V[R] * W[R];
      END;

      Result := CurrentSum;

   END InnerProduct;
```

Figure 3-11 Inner product of two vectors

in Ada). Modula-2 has no special construct for exceptions, so we use the device of exported Boolean variables.

Another controversial issue is whether error recovery is the responsibility of the subprogram or its client. Since this book is oriented to the design and construction of *general-use* software components (libraries and modules), we take the position that, in general, errors that cannot be corrected in an obvious way by a library program must be passed back to the calling program, since the subprogram generally cannot guess how the writer of the client program wants to treat errors. In the present example, since the inner-product operation is mathematically meaningless if the vector sizes do not match, and since there is nothing the subprogram can do to correct the mismatch, all we can do is signal the situation to the client program and assume that the writer of the client provided code to handle the error more knowledgeably than we could.

The code for the inner-product program appears in Figure 3-11. Note that the conformability checking and looping is done with respect to the "mapped" subscripts 0 and HIGH. This means that a vector with range [1 . . 6], for example, is conformable with one with range [−2 . . 3]. Such a situation is perfectly acceptable; a mathematician would argue that the actual subscript range is arbitrary, and only the *number* of elements in the vector is important.

The result of the inner-product operation is a scalar. Why, then, is the program not written as a function procedure, since functions can return scalars?

The answer is that the vector could contain, say, fractions instead of integers. And even though, mathematically, a single fraction is a scalar, in programming it is a record structure and so cannot be easily passed from a function. The design section shows how to use functional notation for these operations.

It is left as an exercise to complete the code for the remaining operations.

3.3.2 Matrix Arithmetic

A number of arithmetic operations are mathematically defined for matrices, which are usually represented in programming as two-dimensional arrays. In this section we shall consider the transpose of a matrix, the addition of a scalar to a matrix, multiplication of a matrix by a scalar, and the sum and product of two matrices.

The matrix sum, and addition and multiplication of matrices with scalars, are similar to their counterparts in the vector case. In the matrix-addition case, each component in the result is the sum of the corresponding components in the arguments. As in the vector situation, conformability dictates that the two matrices have the same dimensions.

For the transpose of a matrix and the product of two matrices, we need to consider how to write general Modula-2 procedures for matrices. As in the vector case, we would like these procedures to be able to accept matrix parameters with arbitrary subscript ranges. But how will the procedures discover these ranges? Unfortunately, the open array parameter capability of Modula-2 applies only to one-dimensional arrays, making it inconvenient to write general-purpose subprograms that work with matrices. In the design section we shall develop a satisfactory approach to the solution of this problem; for now, we simply assume that (1) all matrices are declared in the client program to be of the same maximum size, (2) an actual matrix resides in just a part of the overall matrix, and (3) its subscript ranges are passed as parameters to the procedure.

In this case, a procedure like MatrixAdd would require *15* parameters: three matrix parameters and four subscript-range values for each. To reduce the length of the parameter list, and therefore the complexity of the procedure, we can store each set of four subscript limits in a record, say of type MatrixDimensions. Then, an invocation of the procedure would require only six parameters. Figure 3–12 shows a type declaration; Figure 3–13 gives the Modula-2 code for ScalarPlusMatrix. Try writing ScalarTimesMatrix and MatrixAdd as an exercise.

We can now flesh out the last two operations. The *transpose* of a matrix M

```
TYPE MatrixDimensions =
   RECORD
      FirstRow : INTEGER;
      LastRow  : INTEGER;
      FirstCol : INTEGER;
      LastCol  : INTEGER;
   END;
```

Figure 3–12 Record type for matrix dimensions

Sec. 3.3 Vectors, Matrices, and Modula-2 Arrays

```
        PROCEDURE ScalarPlusMatrix (VAR Result: Matrix;
                                    M:          Matrix;
                                    Mdim:       MatrixDimensions;
                                    Scalar:     ValueType);
     VAR
        Row,Col : INTEGER;

     BEGIN

        FOR Row := Mdim.FirstRow TO Mdim.LastRow DO
           FOR Col := Mdim.FirstCol TO Mdim.LastCol DO
              Result[Row,Col] := M[Row,Col] + Scalar;
           END;
        END;

     END ScalarPlusMatrix;
```

Figure 3-13 Procedure to add a scalar to a matrix

is a matrix whose second dimension is the same as M's first dimension, and whose first dimension is the same as M's second dimension. Figure 3-14 shows the transpose of a square matrix; note that all the rows become columns and all the columns become rows.

Finally, consider matrix multiplication. Conformability of M and N for multiplication means that the *second* dimension of M must be the same as the *first* dimension of N. The product is a matrix P which has a *first* dimension equal to the *first* dimension of M and a *second* dimension equal to the *second* dimension of N. Thus, if M is $[1..5,-3..0]$ and N is $[-3..0,6..8]$ then MatrixMult(P,M,N) has dimensions $[1..5,6..8]$. Note that order counts: MatrixMult

$$\begin{bmatrix} 1 & -3 & 2 & 0 \\ 6 & -1 & 4 & 5 \\ -2 & 0 & 0 & 7 \end{bmatrix}$$

(a) A 3 x 4 matrix M.

$$\begin{bmatrix} 1 & 6 & -2 \\ -3 & -1 & 0 \\ 2 & 4 & 0 \\ 0 & 5 & 7 \end{bmatrix}$$

(b) M^1 = Transpose(M).

Figure 3-14 A matrix and its transpose

```
(* assume MatdimM and MatdimN are of type MatrixDimensions and
   give the dimensions of M and N, respectively *)

FOR r := MatdimM.FirstRow TO MatdimM.LastRow DO

   FOR c := MatdimN.FirstCol TO MatdimN.LastCol DO

   (* Find inner product of row of M
      with column of N *)

      P[r,c] := Zero;
         FOR k := MatdimM.FirstCol TO MatdimM.LastCol DO
            P[r,c] := P[r,c] + M[r,k] * N[k,c];
         END;

   END;

END;
```

Figure 3–15 Code fragment for matrix multiplication

(P,M,N) is *not* the same as MatrixMult(P,N,M); in fact, M and N cannot be conformable for both operations unless they're square.

Figure 3–15 gives a code fragment for MatrixMult. A component P[r,c] has a value that is found by taking the inner product of the rth row of M and the cth column of N. Go through the code and make sure you understand the computation; then check your understanding against the example in Figure 3–16. As an exercise, complete the procedures for MatrixTranspose and MatrixMult.

$$\begin{bmatrix} 1 & 2 \\ -1 & 4 \\ 0 & 3 \end{bmatrix}$$

M
(3 X 2)

$$\begin{bmatrix} 3 & -1 & 2 & 0 \\ 1 & -2 & 0 & 4 \end{bmatrix}$$

N
(2 X 4)

$$\begin{bmatrix} 5 & -5 & 2 & 8 \\ 1 & -7 & -2 & 16 \\ 3 & -6 & 0 & 12 \end{bmatrix}$$

M X N
(3 X 4)

Figure 3–16 Example of matrix multiplication

3.4 DENSELY PACKED STRUCTURES WITH MANY "ZERO" ELEMENTS

In this section and the next, we consider how to handle vectors and matrices with the special property that many of their elements are "zero." We write "zero" in quotation marks to indicate that we are using the term as an abstraction for whatever value is appropriate given the type of the element in question. If we are dealing with integers or reals, 0 (the number) is perfectly appropriate. If the elements are Booleans, "zero" is interpreted as FALSE. If the elements are strings of characters, then "zero" means the null or empty string, i.e., the string with no characters in it. On the other hand, if we are dealing with records which we store or retrieve by some key (such as student records stored by student ID number), then perhaps "zero" simply means that the record is missing or that no student exists with that number.

From here on, we shall just write zero (no quotation marks) or 0, and will usually show numerical examples, with the understanding that the ideas of course generalize to many other types.

Structures with a high proportion of zeroes—sometimes as high as 95 percent—are common in real-world applications, and the total number of elements they contain is often very large, sometimes too large to fit in main memory. It is thus important to think about how to store such structures economically. The idea is to devise a way to store only those elements which are not zero, keeping track of which elements they are either by some nice storage-mapping function or by some bookkeeping scheme.

There are two cases to consider: (1) structures in which the nonzero elements are *concentrated* in a predictable part of the structure, and (2) structures in which nonzero elements are *randomly* (and, we shall assume, uniformly) distributed throughout. We discuss these, in turn, in this and the next section, respectively.

3.4.1 Lower Triangular Matrices

A common structure in many computational problems representing physical systems is a square matrix (the number of rows equals the number of columns) in which all the elements above the main diagonal are zero, and thus all nonzero elements are concentrated on and below the diagonal. This structure is called a *lower triangular* matrix and is so common, in fact, that one is led to wonder why "scientific" programming languages like Fortran and PL/1 do not support it as a built-in type.

If the number of rows and columns of a lower triangular matrix is very large, we can reduce the memory required to store it by just under 50 percent by keeping only the nonzero elements, since the zeroes are predictably located. Specifically, in a matrix

$$M: ARRAY[1 .. R],[1 .. R] \text{ OF Element}$$

with elements designated M[r,c], any element located such that $c > r$ is zero, while all the others are (most likely) nonzero. This configuration is illustrated in Figure 3–17.

We can obtain a storage structure for a lower triangular matrix by remembering how two-dimensional arrays are mapped onto linear storage. Recall that we developed a row-major and a column-major arrangement. Let us consider the matrix M from Figure 3–17 and map it onto a one-dimensional array M' in row-major fashion: first all the nonzero elements in row 1 (there is exactly one of these, M[1,1]), then all those in row 2 (there are two of these, M[2,1] and M[2,2]), then all those in row 3, and so on until row R (all R elements in this row are nonzero) is located. Figure 3–18 illustrates this mapping.

Under this scheme, the total number of elements stored in M' is $1 + 2 + \ldots + R = (R + 1)(R/2)$ elements. Also, to find an arbitrary element M[r,c] in M', consider the following. If $c > r$, we know that M[r,c] was zero and so was not stored in M'. Otherwise, we can find the beginning of row r by skipping over $r - 1$ rows, in other words, $1 + 2 + \ldots + r - 1 = r(r - 1)/2$ elements. Then

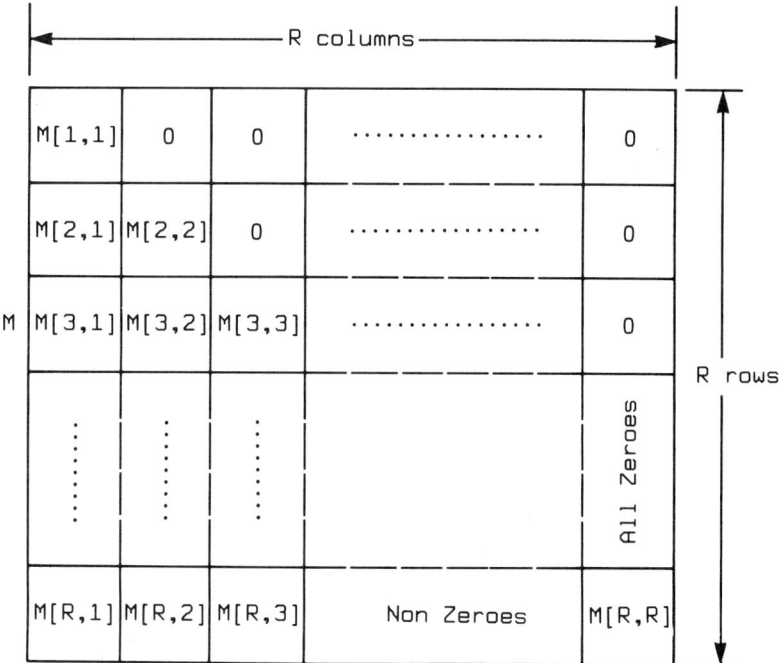

Figure 3–17 Lower-triangular matrix M (zeroes "above" diagonal)

Sec. 3.4 Densely Packed Structures with Many "Zero" Elements

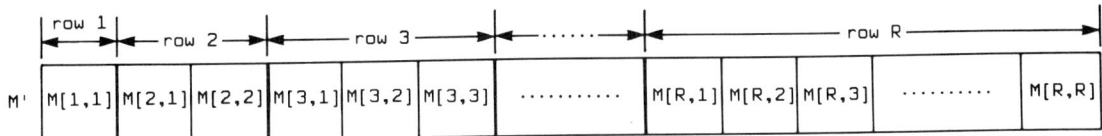

Figure 3–18 Row major implementation, M', of lower-triangular matrix M

we find the right element M[r,c] by skipping over $c - 1$ more elements. If we number the elements in M' starting at 1, the storage mapping for $c \leq r$ is

$$M[r,c] \text{ maps to } M'[1 + r \times (r - 1)/2 + c - 1]$$

We can check this by calculating the end points and typical points in between:

$$M[1,1] \text{ maps to } M'[1 + 1 \times (1 - 1)/2 + 1 - 1] = M'[1]$$

$$M[2,1] \text{ maps to } M'[1 + 2 \times (2 - 1)/2 + 1 - 1] = M'[2]$$

$$M[3,3] \text{ maps to } M'[1 + 3 \times (3 - 1)/2 + 3 - 1] = M'[6]$$

$$M[R,R] \text{ maps to } M'[1 + R \times (R - 1)/2 + R - 1] = M'[(R + 1) \times R/2]$$

Sometimes it is convenient to represent a physical system in the form of an *upper triangular* matrix. The storage structure for this matrix is similar to that just given; we leave it as an exercise to show that the storage mapping function for an upper triangular matrix is "cleaner" if the mapping is column major rather than row major.

3.4.2 Symmetric Matrices

Many physical systems can be presented as a *symmetric* matrix, i.e., a square matrix M such that $M[r,c] = M[c,r]$ for every r and c. In such a situation, we can economize on storage in a way very similar to that used in a triangular matrix: we use the storage structure for the lower triangular matrix, but a slightly different storage mapping function.

To see how this works, consider that, since, for $c > r$, we are in the upper region of the matrix (above the diagonal), we can deduce the value of M[r,c] merely by looking at its "mirror element" M[c,r] below the diagonal. So we just use the storage mapping function from the previous section if $c \leq r$, and for $c > r$ we use that mapping with r and c interchanged. The situation is shown in Figure 3–19.

3.4.3 Band Matrices

In many computational problems the underlying physical system can be modelled as a *band* matrix, in which the nonzero elements are concentrated on the diagonal and within a fixed distance above and below it. The total width of this band of nonzero elements is called the bandwidth.

	1	2	3	4
1	1	3	-2	7
2	3	-5	2	5
3	-2	2	0	0
4	7	5	0	4

(a) Symmetric Matrix M.

1	2	3	4	5	6	7	8	9	10
1	3	-5	-2	2	0	7	5	0	4

(b) Mapping onto one-dimensional array M'.

$$M[r,c] = \begin{cases} M'(1 + r*(r-1)/2 + c-1) & \text{if } r \geq c \\ M'(1 + c*(c-1)/2 + r-1) & \text{if } r < c \end{cases}$$

(c) Storage mapping function.

Figure 3-19 Economical storage of symmetric matrix

Figure 3-20 shows some band matrices. We leave it as an exercise to obtain storage mappings for them. In doing so, think in terms of a "diagonal-major" structure, in which each diagonal is stored as a contiguous block in the one-dimensional array. This might lead to a "cleaner" storage mapping function.

3.5 SPARSE VECTORS AND MATRICES

The second case of vectors and matrices with many zeroes is the one where the zeroes are distributed unpredictably but (we assume) uniformly throughout the structure. We call such structures *sparse* vectors or matrices.

The study of sparse vectors and matrices is useful for two reasons. First, sparse systems of *numerical* values occur very frequently in mathematical programming problems such as arise in engineering, physics, and economics. In recent years, the rise in popularity of microcomputer spreadsheet programs has led to a new concern for sparse structures, since a large spreadsheet is indeed a sparse system, and sparse matrix techniques have been widely used by authors of such programs. Second, many other common structures, such as files of data which

Sec. 3.5   Sparse Vectors and Matrices   75

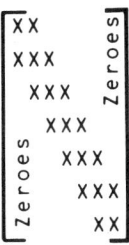

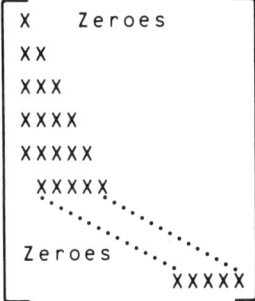

Figure 3-20 Band matrices

need to be structured according to one or more keys, can be viewed as sparse structures. The methods for handling such files are generalizations of the methods we shall consider shortly.

3.5.1 Sparse Vectors

Consider the case of a vector of, say, 1,000 integers where at any given time a large number of values—perhaps 75 percent or more—are zero. Instead of storing this vector as a 1,000-element array, let us try to save some space by doing a little bookkeeping. We can implement each *nonzero* element of the vector as a *record* containing the *index* and the *value* of the element as its two fields, and then store the entire vector as an array of these records. The size of this array is determined by the number of nonzero elements. Modula-2 declarations for this kind of structure are given in Figure 3-21; Figure 3-22 shows an example in which only seven elements are nonzero.

The problem with this structure is that it is too static: in most real applications, the number of nonzero elements is not fixed for the duration of the program, nonzero elements sometimes becoming zero and zero elements sometimes becoming nonzero. So our structure really needs to have some "empty space" built in, to accommodate changing values.

Let us assume that we can make a pretty good guess at the *maximum* number of nonzero elements, and let MaxNonZero be that number. We can also keep track of the *actual* number of nonzero elements using a "header" containing

```
CONST  MaxElements = 1000;
       MaxNonZero  = 15;

TYPE   VectorRange = [1..MaxElements];
       ArrayRange  = [1..MaxNonZero];
       ValueType   = INTEGER;   (* or whatever other type *)

TYPE   SparseVectorElement =

           RECORD
               index: VectorRange;
               value: ValueType;
           END;

TYPE   SparseVector =

           ARRAY ArrayRange OF
               SparseVectorElement;
```

Figure 3–21 Type declarations for sparse vector (number of non-zero elements is fixed)

ActualNonZero as a value. All of this of course depends on our being pretty sure that ActualNonZero will never be more than MaxNonZero; but in many real situations we can judge well enough to make a good choice of MaxNonZero which neither wastes too much space nor overflows. A declaration of this revised structure is shown in Figure 3–23; Figure 3–24 shows the example of Figure 3–22 in the revised structure, where we choose MaxNonZero to be 15.

Since we have assumed that elements are frequently changing value, it makes sense to consider the performance of operations that make this happen. An element changing from zero to nonzero means that we have to add a new item to our structure; a change from nonzero to zero means that we have to eliminate that item to make space for a possible new nonzero one later on. Furthermore, in any

V

	Index	Value
1	3	49
2	9	19
3	22	92
4	43	-123
5	52	-27
6	178	34
7	241	-7

Figure 3–22 Sparse vector, stored as array of ordered pairs

Sec. 3.5 Sparse Vectors and Matrices

```
CONST MaxElements = 1000;
      MaxNonZero   = 15;

TYPE  VectorRange = [1..MaxElements];
      ArrayRange  = [1..MaxNonZero];
      NonZeroes   = [0..MaxNonZero];
      ValueType   = INTEGER;  (* or whatever other type *)

TYPE  SparseVectorElement =

          RECORD
              index: VectorRange;
              value: ValueType;
          END;

TYPE  SparseVectorArray =

          ARRAY ArrayRange OF
              SparseVectorElement;

TYPE  SparseVector =

          RECORD
              ActualNonZero: NonZeroes;
              Store:         SparseVectorArray;
          END;
```

Figure 3–23 Declarations for sparse vector (number of non-zero elements can vary up to a maximum of MaxNonZero)

interesting application, we shall need to retrieve the value of the element with a given index, and occasionally to print out the entire sparse vector in order by index.

In the two preceding examples, we showed the vector elements stored in sorted form—that is, with their indices in numerical order. We might also imagine storing them without regard to the order of the indices.

All this looks similar to something we have seen before. At the end of Chapter 2 we considered maintaining a dynamically changing table using an array, and we calculated the "big Os" for four important operations: update, search, delete, and report. If we think of the index of a vector element as a key, and the number associated with that element as a value, then our sparse-vector problem just becomes a special case of this general table-maintenance problem. Moreover, whether we choose the ordered-array or unordered-array scheme, the sparse-vector store and retrieve operations can be written in terms of the search, update, and delete operations for table handling:

TO RETRIEVE A SPARSE VECTOR ELEMENT BY INDEX:

1. Search the table for an element with the desired index.
2. If the search is successful, return the associated value.
3. If the search is unsuccessful, return 0.

```
              MaxNonZero=15
              ActualNonZero=7
    V
         Index      Value
    1  |   3    |    49   |
    2  |   9    |    19   |
    3  |  22    |    92   |
    4  |  43    |  -123   |
    5  |  52    |   -27   |
    6  | 178    |    34   |
    7  | 241    |    -7   |
    8  |        |         |
    9  |        |         |
   10  |        |         |
   11  |        |         |
   12  |        |         |
   13  |        |         |
   14  |        |         |
   15  |        |         |
```

Figure 3-24 Sparse vector with "extra space"

TO STORE A SPARSE VECTOR ELEMENT BY INDEX:

1. If the element value is 0, delete it from the table (we assume that delete does nothing if the element is not in the table).
2. If the element value is nonzero, search the table for an element with the desired index.
3. If the search is successful, change the value part of the element found.
4. If the search is unsuccessful, update the table with the new element.

You can write programs to perform these operations as an exercise. Note how thinking of a sparse vector as just a special case of a dynamic table has greatly simplified our problem. However, we need to consider what happens to performance if we *oversimplify*. For example, in the store operation, if the table search returns only a value from the table, and not a location, how can we just change the value part? One way is to do a search followed by an update; however, this doubles the time required to perform the store operation.

Sec. 3.5   Sparse Vectors and Matrices

This oversimplification leads us to conclude that if really good performance is desired, either a fairly large set of table operations must be provided—not just the four we developed in Chapter 2—or we must be content to think in generalities only at the conceptual level and write all our operations anew for each application. This is a good example of the tradeoffs we often face between performance and abstraction.

3.5.2 Linking the Elements Together

We can save a fair amount of time in adding and deleting elements while still maintaining the ability to report—i.e., to traverse the array—in linear time by storing the elements *physically* in the order in which they arrive, but keeping them *logically* in order of index. How is this done? Let us add another field, called Next, to the record describing each element. Then, since our elements are stored in an array V with indices in the range 1 . . MaxNonZero, we shall declare Next to have values in the same range. Figure 3–25 shows declarations for this.

Assume the array starts out empty, with no nonzero elements. As each element arrives, we store it in the next empty physical position in V. But then we connect it *logically* to the previous elements by using Next. For each element

```
CONST MaxElements = 1000;
      MaxNonZero  = 15;

TYPE  VectorRange = [1..MaxElements];
      ArrayRange  = [1..MaxNonZero];
      NonZeroes   = [0..MaxNonZero];
      ValueType   = INTEGER;   (* or whatever other type *)

TYPE  SparseVectorElement =

         RECORD
            index: VectorRange;
            value: ValueType;
            next:  NonZeroes;
         END;

TYPE  SparseVectorArray =

         ARRAY ArrayRange OF
            SparseVectorElement;

TYPE  SparseVector =

         RECORD
            ActualNonZero: NonZeroes;
            First:         NonZeroes;
            Store:         SparseVectorArray;
         END;
```

Figure 3–25 Declarations for sparse vector (number of non-zero elements is variable)

```
MaxElements=1000
MaxNonZero=15
First=2
ActualNonZero=4
```

V	Index	Value	Next
1	52	-27	3
2	9	19	4
3	178	34	0
4	22	92	1
5			
6			
7			
8			
9			
10			
11			
12			
13			
14			
15			

Figure 3-26 Sparse vector stored with "Next" links

of V, Next indicates the position of V containing the element with the next higher index. A new field, First, added to the header will indicate the position of V containing the element with lowest index.

Figure 3-26 shows an array V similar to that in the previous examples, with four elements already in place, having arrived in the index order 52,9,178,22. Notice that we use a Next value of zero to indicate that no more elements follow, i.e., that the element with highest index has been reached.

Now let the element with index 43 arrive. Figure 3-27 shows V with the new element added physically to V[5] and logically connected in its right place. Finding the right place is done by a sequential search, but instead of sequencing according to the *physical* order, as we would ordinarily do, we search in *logical* order, following the "daisy chain" of Next indicators until we find an element with higher index than that of the new arrival, and then reconnecting the indicators accordingly.

Sec. 3.5 Sparse Vectors and Matrices

```
MaxElements=1000
MaxNonZero=15
First=2
ActualNonZero=5
```

V

	Index	Value	Next
1	52	-27	3
2	9	19	4
3	178	34	0
4	22	92	5
5	43	-123	1
6			
7			
8			
9			
10			
11			
12			
13			
14			
15			

Figure 3–27 Vector of Figure 3–26, with V[43] = −123

Note that the performance of the insertion of the new element is again linear: reconnecting the Next indicators and placing the new arrival physically are both done in constant time, while finding the right logical position, being a sequential search, performs in linear time.

In Figure 3–28, we have shown how V would look with new arrivals 3 (new lowest index) and 241 (new highest index) inserted. As an exercise, try to figure out the logic of a routine to carry out such an insertion; we will not give any program text yet, since we'll take up the subject in a much more thorough way in the next chapter. In the meantime, let us briefly look at the delete and report operations. To delete a nonzero element that goes to zero, we need merely search in logical order until we find the right element, and then connect the next indicator of the logically *preceding* element to "jump around" the one we want to delete. The vector V with the element with index 52 deleted is shown in Figure 3–29.

```
MaxElements=1000
MaxNonZero=15
First=6
ActualNonZero=7
```

V	Index	Value	Next
1	52	-27	3
2	9	19	4
3	178	34	7
4	22	92	5
5	43	-123	1
6	3	49	2
7	241	-7	0
8			
9			
10			
11			
12			
13			
14			
15			

Figure 3-28 Insertion of V[3] and V[241]

We leave it to the next chapter to consider how to reuse the space formerly occupied by the deleted element.

As regards the report operation, traversing the sparse vector is easy: we just start at the position of V indicated by First, and then follow the daisy chain, printing out all the elements in logical order.

The structure we have just designed is a special case of a *linked list*. Linked lists are commonly used for the implementation of many abstractions in computing, and warrant a chapter of their own. The preceding discussion has merely motivated the need for a linked structure by considering how sparse vectors might be implemented in a space-saving way.

We can generalize the sparse-vector idea to handle two-dimensional sparse structures: sparse matrices. We shall not go into detail on this yet; rather, we'll wait until the next chapter, after you're more familiar with linked lists.

```
                MaxElements=1000
                MaxNonZero=15
                First=6
     V          ActualNonZero=6
```

	Index	Value	Next
1	(52)	(-27)	(3)
2	9	19	4
3	178	34	7
4	22	92	5
5	43	-123	3
6	3	49	2
7	241	-7	0
8			
9			
10			
11			
12			
13			
14			
15			

Figure 3-29 Removal of V[52] (note the link from V[43] to V[178], "jumping around" V[52]

3.6 DESIGN: VECTORS, MATRICES, AND FUNCTIONAL NOTATION

Let us return to the case of "normal" (nonsparse) vectors and matrices and use Modula-2 constructs to create a module capable of doing vector and matrix arithmetic in a general way. In Section 3.3 we wrote operations for vectors and matrices. We used the open array parameter idea for vectors, but discovered that this Modula-2 capability is limited to the one-dimensional case and thus cannot be used for matrices. Our simple solution was to insist that all matrices be declared to be of the same maximum size and that the actual subscript ranges be passed to procedures in a record called MatrixDimensions.

A serious difficulty with this approach is that it wastes an enormous amount of space as the price of generality: for example, to use the procedures first with a 3×3 matrix and then, later in the same client, with a 10×10 matrix, requires that *all* matrices be declared at least 10 × 10. Another difficulty, albeit less painful, is that procedures must be written with VAR result parameters. Thus, vector and matrix operations cannot be written to directly model the mathematical notation of the application, in which the sum C of two matrices A and B is easily written as $C = A + B$. It turns out that in solving the first problem, we can solve the second as well.

3.6.1 A Dynamic Storage Scheme for Vectors and Matrices

Our method for handling vectors and matrices relies on the pointer and dynamic allocation method used in chapter 1, and also the notion of a storage mapping function introduced in section 3.2

Let us create a module Vectors which exports an opaque type Vector and a number of operations. Figure 3-30 shows the definition module for this library; note that ValueType is simply given as a new name for INTEGER. Minimal though this change may be in terms of coding, it allows us to write all vector operations in terms of ValueType, the ramifications of which, as we shall shortly see, are profound.

The constructor CreateVector(V,First,Last) creates a "blank" vector with subscript range [First .. Last]; the selectors VectorFirst(V) and VectorLast(V) return these subscript limits to the client program. DestroyVector destroys the vector and reclaims the storage it occupies. The constructor VectorStore(V,r,x) stores a value x in the rth component of V; the selector VectorRetrieve(V,r) returns the value stored in the rth component of V.

This new module serves as a complete set of primitive operations to manipulate vectors, providing client programs with a generality similar to that given by the Ada and PL/1 languages themselves. But how shall we implement it? Recall from chapter 1 that an opaque type is almost always a pointer. In Figure 3-31

```
DEFINITION MODULE Vectors;

    TYPE Vector;

    TYPE ValueType = INTEGER;
    TYPE Index     = INTEGER;

    VAR IndexError: BOOLEAN;

    PROCEDURE CreateVector    (VAR V: Vector; First, Last: Index);
    PROCEDURE VectorFirst     (V: Vector) : Index;
    PROCEDURE VectorLast      (V: Vector) : Index;
    PROCEDURE DestroyVector   (VAR V: Vector);
    PROCEDURE VectorStore     (V: Vector; r: Index; x: ValueType);
    PROCEDURE VectorRetrieve  (V: Vector; r: Index) : ValueType;

END Vectors.
```

Figure 3-30 Definition module for Vectors

Sec. 3.6 Design: Vectors, Matrices, and Functional Notation

```
TYPE VectorDimensions =

  RECORD
    First: Index;
    Last:  Index;
  END;

TYPE VectorRec =

  RECORD
    Dims: VectorDimensions;
    Storage: POINTER TO ValueType;
  END;

TYPE Vector = POINTER TO VectorRec;
```

Figure 3-31 Full type declaration for vector as opaque type

we give the full type declaration of the vector type. Note that it is a record containing the First and Last values, and a pointer to the actual storage to be occupied by the vector. Figure 3–32 shows a diagram of the storage allocated for a Vector; the actual allocation of this space from the heap is effected by the constructor CreateVector (Figure 3–33). *With this instrumentation, we can now "size" a vector to be precisely as large as we need it to be.*

Figure 3–34 gives the code for VectorStore, which uses concepts from section 3.2 and the Modula-2 procedure INC, which allows us to modify the value of a pointer variable by a given number of bytes. We leave it as an exercise to complete the implementation module by writing the remaining procedures, as well as to show how the module might be imported by a module that does vector arithmetic. In furtherance of the latter, Figure 3–35, in which VectorAdd is written as a function procedure, offers a clue. Figure 3–36 gives, as an example, a main program in which VectorAdd is called.

3.6.2 A General Module for Matrices

The concepts just outlined can be used to design a module to handle matrices in the same general fashion. Here the idea of a storage mapping is even more important. In Figure 3–37 we show the full type declaration for a Matrix; this could be exported opaquely from a module Matrices. Figure 3–38 gives the code for CreateMatrix, and Figure 3–39 gives the code for MatrixStore. These are

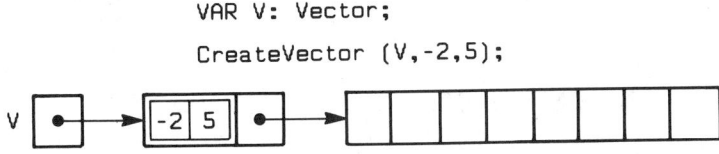

Figure 3-32 Storage allocation for a vector

```
PROCEDURE CreateVector   (VAR V: Vector; First, Last: Index);
BEGIN

   IndexError := FALSE;
   IF Last-First <= 0 THEN
      IndexError := TRUE;
      RETURN;
   END;

   Allocate(V, SIZE(VectorRec));
   V^.Dims.First := First;
   V^.Dims.Last  := Last;
   Allocate(V^.Storage,(Last-First+1)*SIZE(ValueType));

END CreateVector;
```

Figure 3–33 Procedure to create a "blank" vector

analogous to their counterparts in the Vectors module and use a row-major storage mapping. As an exercise, you might try encapsulating these operations in a module and then writing a client module to do arithmetic on these dynamic matrices.

```
PROCEDURE VectorStore(V: Vector; r: Index; x: ValueType);

VAR
   P:      POINTER TO ValueType;
   A:      ADDRESS;   (* address arithmetic to find offset *)
   Offset: LONGCARD; (* some systems may require CARDINAL *)

BEGIN

   IndexError := FALSE;
   IF (r < V^.Dims.First) OR (r > V^.Dims.Last) THEN
      IndexError := TRUE;
      RETURN;
   END;

   Offset := (r-V^.Dims.First) * SIZE(ValueType);

   A   := V^.Storage;
   INC(A,Offset);            (* find where to store x; *)
   P   := A;                 (* address arithmetic OK  *)
   P^  := x;                 (* and put it there       *)

END VectorStore;
```

Figure 3–34 Procedure to store an element into a vector

3.7 STYLE GUIDE: LANGUAGES WITHOUT RECORD TYPES

Suppose you need to code and debug programs in a language like Fortran or Basic, both of which are sometimes called "data structure poor" and do not permit the direct use of record types? You should then view arrays of records as an *abstrac-*

```
PROCEDURE VectorAdd(U,V: Vector) : Vector;

VAR
   Result: Vector;
   R:      INTEGER;
   X:      ValueType;

BEGIN
   ConformabilityError := FALSE;
   IF (VectorFirst(U) <> VectorFirst(V)) OR
      (VectorLast(U)  <> VectorLast(V))
   THEN
      ConformabilityError := TRUE;
      RETURN(Result);
   END;

   CreateVector(Result,VectorFirst(U),VectorLast(U));

   FOR R := VectorFirst(U) TO VectorLast(U) DO

      X := VectorRetrieve(U,R) + VectorRetrieve(V,R);
      VectorStore(Result,R,X);
      (* i.e. Temp[R] := U[R] + V[R] *)

   END;

   RETURN(Result);

END VectorAdd;
```

Figure 3–35 VectorAdd, now written as a function procedure

tion, useful for program *design*, and *implement* them in a way acceptable to your compiler or interpreter. This means that each field of the record (or in some cases groups of fields) needs to be declared as a separate array of the appropriate type, since the language permits only scalar types to be values in an array. In Figure 3–40 we show how this might be done by giving Modula-2, Fortran-77, and Basic declarations for the abstraction consisting of a 100-record array of records containing some student information. A program which *uses* these records should be implemented as directly as possible from the abstract design, using, for example, a *single* index variable to select a given record from the array. This leads to a clear and readable result.

The example shows, once again, the benefits of thinking in terms of an abstract design which expresses the solution of a problem in a natural way, and then implementing the design using whatever features are available in the given coding language.

3.8 SUMMARY

This chapter has covered a number of important issues pertaining to vectors and matrices, which are mathematical structures; to arrays, which are programming structures; and to the relation between them. Arrays of one or more dimensions

```
MODULE TestDynamicVectorAdd;

    FROM InOut IMPORT Read, WriteLn, WriteString, WriteInt;

    FROM Vectors IMPORT ValueType, Vector, CreateVector,
                    VectorFirst, VectorLast, VectorStore,
                    VectorRetrieve;

    VAR
       A, B, C : Vector;
       Q:         CHAR;


    PROCEDURE WriteVector(V: Vector);
       VAR
          R: INTEGER;
       BEGIN
          FOR R := VectorFirst(V) TO VectorLast(V) DO
             WriteInt(VectorRetrieve(V,R),4);
          END;
          WriteLn;
       END WriteVector;


    (* code for VectorAdd goes right here *)


    BEGIN       (* Main body of Program *)

       CreateVector(A,-5,2);     CreateVector(B,-5,2);

       VectorStore(A,  -5,   3);  VectorStore(B,  -5,  17);
       VectorStore(A,  -4,-17);   VectorStore(B,  -4,  24);
       VectorStore(A,  -3,  20);  VectorStore(B,  -3,   5);
       VectorStore(A,  -2,   0);  VectorStore(B,  -2,   0);
       VectorStore(A,  -1,   4);  VectorStore(B,  -1,   3);
       VectorStore(A,   0,  18);  VectorStore(B,   0,   6);
       VectorStore(A,   1,  27);  VectorStore(B,   1, -15);
       VectorStore(A,   2, -32);  VectorStore(B , 2, -32);

       WriteVector(A);
       WriteVector(B);

       C := VectorAdd(A,B);

       WriteVector(C);
       WriteLn;
       Read(Q);

    END TestDynamicVectorAdd.
```

Figure 3-36 A program demonstrating dynamic vector operations

are allocated to one-dimensional storage on most computers; the formula showing the relationship between a subscripted array reference in a high-level language and the physical storage to which it is allocated is called the *storage mapping function*. One way to store multidimensional arrays is in *row-major* form, in which the rightmost subscript varies fastest; another is in *column-major* form, in which the leftmost subscript varies fastest.

Sec. 3.8   Summary

```
TYPE MatrixDimensions =

  RECORD
    FirstRow: Index;
    LastRow:  Index;
    FirstCol: Index;
    LastCol:  Index;
  END;

TYPE MatrixRec =

  RECORD
    Dims: MatrixDimensions;
    Storage: POINTER TO ValueType;
  END;

TYPE Matrix = POINTER TO MatrixRec;
```

Figure 3-37 Full type declaration for matrix as opaque type

Several different kinds of special matrices were presented, including those with a large number of zero elements distributed in a predictable fashion. Upper triangular, lower triangular, symmetric, and band matrices can be stored more efficiently if their zero values are not stored; the storage mapping function is then responsible for carrying out the subscripted reference for efficient storage. Since high-level languages do not provide built-in facilities for handling these special matrices, building the right package and storage mapping function is up to the programmer.

Several different strategies for storing sparse vectors—those with a large number of zeroes in unpredictable locations—were also considered. Two of these were very similar to the ordered-array and unordered-array table-handling schemes discussed in chapter 2. A third scheme involves linking together the nonzero

```
PROCEDURE CreateMatrix (VAR M:            Matrix;
                  FirstRow, LastRow: Index;
                  FirstCol, LastCol: Index);

  BEGIN

    IndexError := FALSE;
    IF (LastRow-FirstRow <= 0) OR (LastCol-FirstCol <= 0) THEN
      IndexError := TRUE;
      RETURN;
    END;

    Allocate(M, SIZE(MatrixRec));
    M^.Dims.FirstRow := FirstRow;
    M^.Dims.LastRow  := LastRow;
    M^.Dims.FirstCol := FirstCol;
    M^.Dims.LastCol  := LastCol;

    Allocate(M^.Storage,
       (LastRow-FirstRow+1)*(LastCol-FirstCol+1)*SIZE(ValueType));

  END CreateMatrix;
```

Figure 3-38 Procedure to create a "blank" matrix

```
PROCEDURE MatrixStore (M:    Matrix;
                       r,c:  Index;
                       x:    ValueType);

VAR
   P:              POINTER TO ValueType;
   A:              ADDRESS;
   Offset:         LONGCARD;
   FirstR, LastR:  Index;
   FirstC, LastC:  Index;

BEGIN
   FirstR := M^.Dims.FirstRow;
   LastR  := M^.Dims.LastRow;
   FirstC := M^.Dims.FirstCol;
   LastC  := M^.Dims.LastCol;

   IndexError := FALSE;
   IF (r < FirstR) OR (r > LastR) OR
      (c < FirstC) OR (c > LastC) THEN
      IndexError := TRUE;
      RETURN;
   END;

   Offset := SIZE(ValueType) *
             ((r-FirstR) * (LastC-FirstC+1) + (c-FirstC));

   A    := M^.Storage;
   INC(A,Offset);              (* find where to store x *)
   P    := A;
   P^   := x;                  (* and put it there      *)

END MatrixStore;
```

Figure 3-39 Procedure to store a value into a matrix

elements so that their logical order does not in general match their physical order. This is a special case of a *linked-list* method, a more general study of which follows in chapter 4.

The chapter also covered the array-handling facilities of Modula-2, specifically the *open-array parameter*, which allows a one-dimensional array to be passed to subprograms without having to pass the array size explicitly. In the design section, some techniques were presented for implementing vectors and matrices in a general way using dynamic storage allocation and functional notation.

Finally, the style guide presented some hints for simulating record structures in Fortran and Basic, languages that do not provide a direct facility for coding such structures. Even in the absence of such a facility, there is considerable programming merit to thinking of records and arrays of records as abstract data types.

Chap. 3   Exercises

```
TYPE HWRange    = [0..10];
TYPE TestRange  = [0..100];
TYPE NameString = ARRAY[0..19] OF CHAR;

TYPE StudentRecord =

  RECORD
    Name:           NameString;
    HoweworkGrades: ARRAY[1..5] OF HWRange;
    TestScores:     ARRAY[1..2] OF TestRange;
    TermAverage:    REAL;
  END;
```

(a) Modula-2

```
CHARACTER NAMES  (1:100)     * 20
INTEGER   HWGRDE (1:100,1:5)
INTEGER   TESTS  (1:100,1:2)
REAL      AVERGE (1:100)
```

(b) Fortran-77

```
DIM N$(1:100)        :REM "$" declares string array
DIM G%(1:100,1:5)    :REM "%" declares integer array
DIM T%(1:100,1:2)
DIM A (1:100,1:2)    :REM real by default
```

(c) Basic

Figure 3-40   Declaration of "arrays of records" in three languages

CHAPTER 3   EXERCISES

1. Obtain a detailed storage mapping function for a two-dimensional array stored in row-major form.
2. Obtain a detailed storage mapping function for a two-dimensional array stored in column-major form.
3. Obtain a detailed storage mapping function for a three-dimensional array stored in row-major form.
4. Obtain a detailed storage mapping function for a three-dimensional array stored in column-major form.
5. Obtain a general storage mapping function for an array of D dimensions stored in row-major form.
6. Obtain a general storage mapping function for an array of D dimensions stored in column-major form.
7. Language M (perhaps Modula-2) stores its multi-dimensional arrays in row-major form;

language F (perhaps Fortran) stores them in column-major form. Suppose a program in language M creates a three-dimensional array, then needs to pass it to a subroutine written in language F. To save time in passing arrays, most subroutine linkage arrangements just pass the address of the array, thus the same physical copy of the array is used by both programs. A reference to, say, T[1,5,4] in the language M program refers to a different physical location than the same reference in the language F program. What has to be done to make the two languages communicate better? Write whatever programs you need.

8. Write a procedure ScalarTimesVector, which finds and returns the product of a scalar K by a vector V, and which is suitable for inclusion in the module described in Section 3.3. Given that, mathematically, K × V and V × K give the same result, explain how you could allow a client program to write either expression.

9. Write a VectorAdd procedure, which finds and returns the sum of its two vector arguments, and which is suitable for inclusion in the module described in Section 3.3. Do not forget to check for conformability!

10. Write a procedure ScalarTimesMatrix, which finds and returns the product of a scalar K by a matrix M, and which is suitable for inclusion in the module described in Section 3.3. Given that, mathematically, K × M and M × K give the same result, explain how you could allow a client program to write either expression.

11. Write a MatrixAdd procedure, which finds and returns the sum of its two matrix arguments, and which is suitable for inclusion in the module described in Section 3.3. Do not forget to check for conformability!

12. Write a procedure MatrixTranspose, which finds and returns the transpose of a matrix, and which is suitable for inclusion in the module described in Section 3.3.

13. Obtain a column-major storage mapping function for a lower triangular matrix and explain why the row-major form is simpler.

14. Obtain both row-major and column-major storage mapping functions for an upper triangular matrix and explain why the column-major form is simpler.

15. Obtain storage mapping functions for the band matrices given in section 3.4.3. There are many different possible storage mappings, thus many correct answers.

16. Try to write a procedure to insert an element into a sparse vector with "next" links (this may be difficult, but chapter 4 will clear up a lot of your troubles).

17. Write an implementation module for Vectors which uses opaque types and functional notation as described in section 3.6.1. Your module should contain only a minimum number of operations (basic constructors and selectors, no arithmetic).

18. Write a library module for vector arithmetic, which imports the module from problem 17.

19. Write an implementation module for Matrices using opaque types and functional notation, as described in section 3.6.2.

20. Write a matrix arithmetic module which imports the module developed in problem 19.

Chapter 4
LINKED LISTS, POINTERS, AND CURSORS

4.1 GOAL OF THE CHAPTER

In this chapter we shall examine how linked storage is manipulated in languages in which the compiler provides built-in facilities for such storage. We give an ADT for linked lists, and several ways of implementing this ADT using Modula-2's *recursive data structures*, which are something like recursive programs: the type definition for the structure has, embedded within it, a reference to another structure of the same type. Thus, in a sense, the type definition "calls itself."

The chapter contains two design sections. In the first, we reconsider the subject of sparse vectors and matrices, showing how these can be implemented in a general linked structure and giving the linked version of the arithmetic operations introduced in Section 3.5. In the second, we improve the text-handler package introduced in chapter 1 in order to limit the inconvenience to client programs caused by the fact that strings have a fixed maximum length. We employ linked-list organization to allow strings to have essentially unlimited length with very little wasted storage.

There are also two style guides in this chapter. The first considers the question of "cursor-oriented" linked allocation, as an aid to the many people who need to write programs using linked lists, but who are constrained to use languages like Fortran and Basic, which lack built-in support for linked lists. The second presents a strategy for reclaiming the storage used by linked structures in a way that minimizes the need for client-program concern about such storage. The principles discussed are also applicable to cursor-oriented memory management.

4.2 LINKED STRUCTURES

The last few examples in section 3.5 showed some of the advantages of storing information in a structure with *links* or *pointers* so that the *logical* sequence of the items did not have to correspond with their *physical* sequence in an array structure. In the scheme presented, the links were positive integers taken from the range of the index of the array. This is often the way linked structures must be handled, especially in languages like Fortran, Basic, and Cobol, where no better mechanism is provided by the language design.

On the other hand, linked structures are useful in general as an implementation mechanism for many of the abstract structures to be taken up later in this book. It is precisely for this reason that modern languages like PL/1, C, Pascal, Modula-2, and Ada provide a better way to work with linked structures. Accordingly, we shall consider linked structures in relation to these languages first in the next section.

4.3 POINTERS AND DYNAMIC MEMORY MANAGEMENT

In languages providing built-in support for linked structures, the compiler associates a special storage area with an object program which it leaves initially unassigned to any program variable. This area is usually called the *heap* or the *dynamic storage pool*. A facility in the run-time support for the language is given responsibility for the allocation of sections of storage from the heap *at execution time*. We shall call this facility the *heap manager*.

Special kinds of variables called *pointer variables* are provided in these languages for referencing space allocated dynamically from the heap. In chapter 3 we discussed the use of these variables in Modula-2 to facilitate the use of functional notation and opaque export; here we shall use pointer variables in Modula-2 to link together a number of objects whose space is allocated from the heap. The syntax differs, but the idea is the same, in the other languages that support linked structures.

Recall from chapter 3 that a pointer is just an abstraction for a machine address. Each language has a way of declaring variables that can hold pointers. In Modula-2, the mechanism is through the use of pointer types. Suppose we had already defined a record type called RecType, something like

TYPE RecType = RECORD

. . . fields . . .

END;

Then a type definition such as

TYPE RecPointer = POINTER TO RecType;

gives us the ability to declare variables of type RecPointer, that is, variables that

Sec. 4.3   Pointers and Dynamic Memory Management

can hold pointers to things for type RecType. For example, a declaration

VAR P1, P2, P3: RecPointer;

brings into existence three such variables.

Notice that declaring these variables does *not* cause any records to be allocated: only enough space to hold the address of (a pointer to) a record is given to each variable. How, then, do the records themselves come into being? Recall the procedure Allocate discussed earlier. The statement

Allocate(P1,SIZE(RecType));

causes the heap manager to search the heap, looking for a block of space large enough to hold a record of type RecType. When such a block is found, its address is stored in the variable P1.

Suppose we now write

Allocate(P2,SIZE(RecType));

What happens? A *second* block of space is allocated from the heap, and its address is stored in P2. Now suppose that we have found a way to store values in all of the fields of the record pointed to by P1, and then write

P3 := P1;

Then P3 is made to point to the same record P1 points to. (N. B.: An assignment statement using an access variable copies only the pointer to (the address of) the structure involved, not the structure itself.)

If we now write, a second time,

Allocate(P1,SIZE(RecType));

then space for yet a third record is found in the heap, its address is stored in P1, and P3 is left pointing to the "old" record. If we then write, again,

Allocate(P1,SIZE(RecType));

the record previously created is left with nothing pointing to it, and is thus *inaccessible*.

Some languages (e.g., Snobol, Lisp, and Ada) provide that space previously in use but currently inaccessible will be returned to the heap whenever the heap manager "thinks" the space in the heap is running out. A system procedure usually called, picturesquely, a *garbage collector* will wander through the heap, looking for inaccessible blocks of space and "recycling" them for future use. In other languages, including Pascal, C, and Modula-2, the inaccessible space is not usually recycled automatically; rather, the programmer must do it explicitly. We shall return to this subject later in the chapter.

Figure 4-1 shows diagrammatically how dynamic allocation works. The cloudlike shape represents the heap; pointers are represented by arrows.

It is important to remember that using pointer variables is always a bit risky in any language. In Modula-2, one source of trouble is the fact that pointer

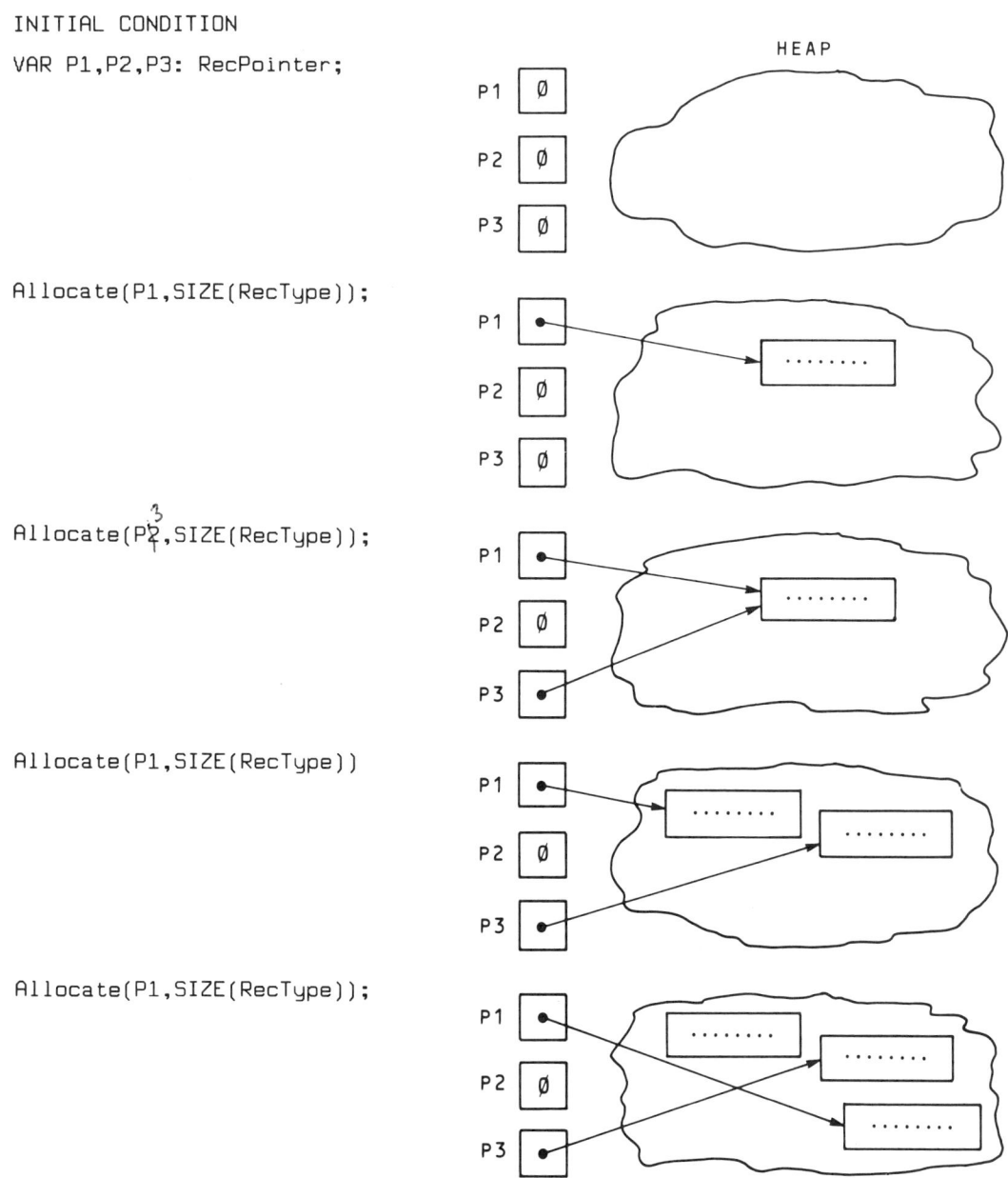

Figure 4–1 Run-time allocation from the heap (statements are executed in sequence)

variables initially have "garbage" or utterly useless values, so great care must be taken that they are assigned predictable values as soon as possible. The special value NIL is used to mean "not pointing to anything right now," or "empty." This is a known value, quite different from the garbage held by a pointer before it is given a value. In our diagrams we shall use the Greek letter phi (ϕ) as an abbreviation of NIL.

4.4 USING DYNAMIC ALLOCATION FOR LINKED STRUCTURES

Dynamic allocation and pointers can be used to great advantage in working with the kind of linked structures that were introduced in section 3.5. The secret is to define a type which we shall call a *node*. (Some authors prefer *cell* instead.) A node contains an information part and a pointer part. We make the pointer part be of type "pointer to another node," and then proceed to link nodes together into a chain or linked list.

Figure 4–2 shows the Modula-2 type definitions for a node. Notice that the pointer type is mentioned first so that it can be used as a field name within the record type. Clearly, InfoType can be any type at all. For definiteness in the next few examples, we shall assume that InfoType has been defined somewhere as a character type. Also, it will become obvious later on why we have called the node a *one-way list node*.

The type OneWayListNode is a special case of *recursive types*, (sometimes called "self-referencing" types) which are so called because each node's pointer part points to another node of the same type. In this respect, recursive types are similar to recursive programs: they are defined, in some sense, "in terms of themselves."

4.4.1 Creating a One-Way Linked List

Suppose we declare two variables N1,N2: NodePointer in a program and then write:

> Allocate(N1,SIZE(OneWayListNode));
>
> N1^.Info := 'E';
>
> N1^.Next := NIL;

```
TYPE InfoType = CHAR; (* for example *)

TYPE NodePointer = POINTER TO OneWayListNode;

TYPE OneWayListNode =
   RECORD
      Info: InfoType;
      Next: NodePointer;
   END;
```

Figure 4–2 Type declarations for linked-list node

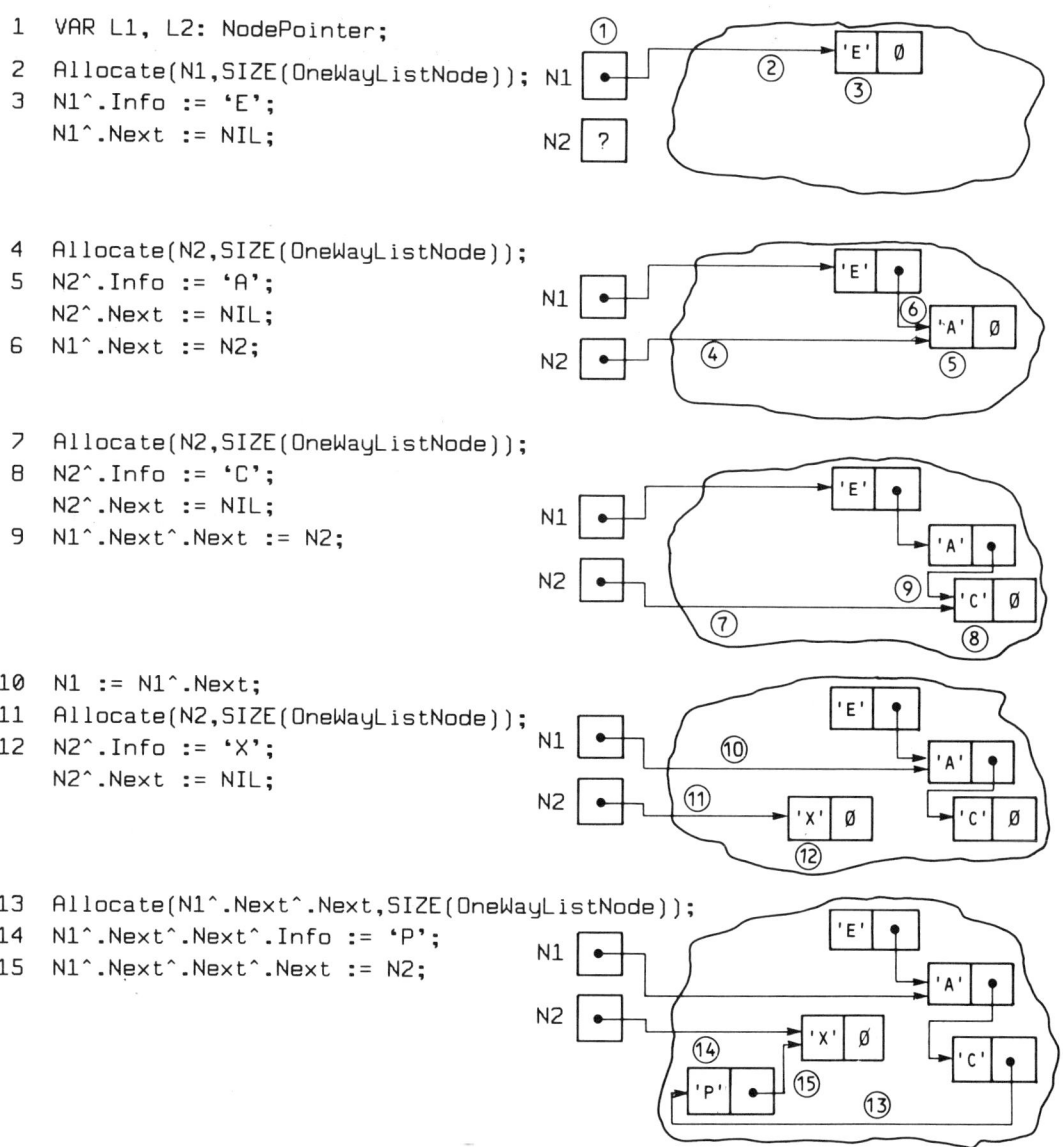

```
1   VAR L1, L2: NodePointer;
2   Allocate(N1,SIZE(OneWayListNode));
3   N1^.Info := 'E';
    N1^.Next := NIL;

4   Allocate(N2,SIZE(OneWayListNode));
5   N2^.Info := 'A';
    N2^.Next := NIL;
6   N1^.Next := N2;

7   Allocate(N2,SIZE(OneWayListNode));
8   N2^.Info := 'C';
    N2^.Next := NIL;
9   N1^.Next^.Next := N2;

10  N1 := N1^.Next;
11  Allocate(N2,SIZE(OneWayListNode));
12  N2^.Info := 'X';
    N2^.Next := NIL;

13  Allocate(N1^.Next^.Next,SIZE(OneWayListNode));
14  N1^.Next^.Next^.Info := 'P';
15  N1^.Next^.Next^.Next := N2;
```

Figure 4–3 Building one-way lists

where the dereferencing expression N1^.Info means "the Info field of whatever N1 points to." Then these statements set up a chain or list with one node in it. The variable N1 is known as the *head* of the list; since it is outside the heap (that is, it is in our program as a declared variable), it serves as our "doorway" into the dynamically allocated list.

Sec. 4.4   Using Dynamic Allocation for Linked Structures

Now suppose we write

Allocate(N2,SIZE(OneWayListNode));

N2^.Info := 'A';

N2^.Next := NIL;

N1^.Next := N2;

Then we shall have *linked* or connected the two nodes together into a chain or list. Figure 4-3 shows the results of a sequence of such list operations. Follow each operation through, and make certain you understand the dynamic changes going on in the heap. Notice that the node containing 'E' becomes inaccessible, since, even though its pointer is left pointing at something, nothing points to it.

4.4.2 An ADT for One-Way Linked Lists

Now let us develop an ADT which encapsulates the low-level details. What are the appropriate operations on one-way lists? To create an ADT that will be useful for many applications, it makes sense not to make any assumptions about the type of thing stored in the Info part of a list node. Accordingly, the ADT we describe

```
TYPE List = POINTER TO ListHeader;

TYPE ListHeader =

   RECORD
      Head : NodePointer;
   END;


PROCEDURE ListInit(VAR L: List);

   BEGIN
      Allocate(L,SIZE(ListHeader));
      L^.Head := NIL;
   END ListInit;

PROCEDURE MakeNode(Item: InfoType) : NodePointer;

   VAR
      p : NodePointer;

   BEGIN
      Allocate(p,SIZE(OneWayListNode));
      p^.Info := Item;
      p^.Next := NIL;
      RETURN p;
   END MakeNode;


PROCEDURE InfoPart(N: NodePointer) : InfoType;

   BEGIN
      RETURN N^.Info;
   END InfoPart;
```

Figure 4-4 Some declarations for one-way lists

here will deal mostly with keeping track of the connections in the list, working with list nodes in which the Info part has been prestored.

Instead of starting a list just by declaring a variable of type NodePointer, let us define a type List as a pointer to a record containing just one field, namely a NodePointer and then create empty lists by declaring variables of this type. The type List is a pointer to a header record instead of the record itself, so that we will be able to make List an opaque type (recall that Modula-2 requires an opaque type to be a pointer). Different applications may require additional header information, and so the header is a record to permit us just to add more fields later. Figure 4–4 shows all the relevant type definitions, as well as three functions: a constructor ListInit, which allocates the header record and sets to NIL the head pointer; a constructor MakeNode, which accepts an object of type InfoType and returns a pointer to a node with that object stored in it; and a selector InfoPart. Figure 4–5 shows the declaration of L1 and L2 as Lists and a few calls of the functions, together with diagrams showing the results of these declarations and calls. Note that we show the record for a ListHeader as a box within a box, to distinguish it visually from a scalar.

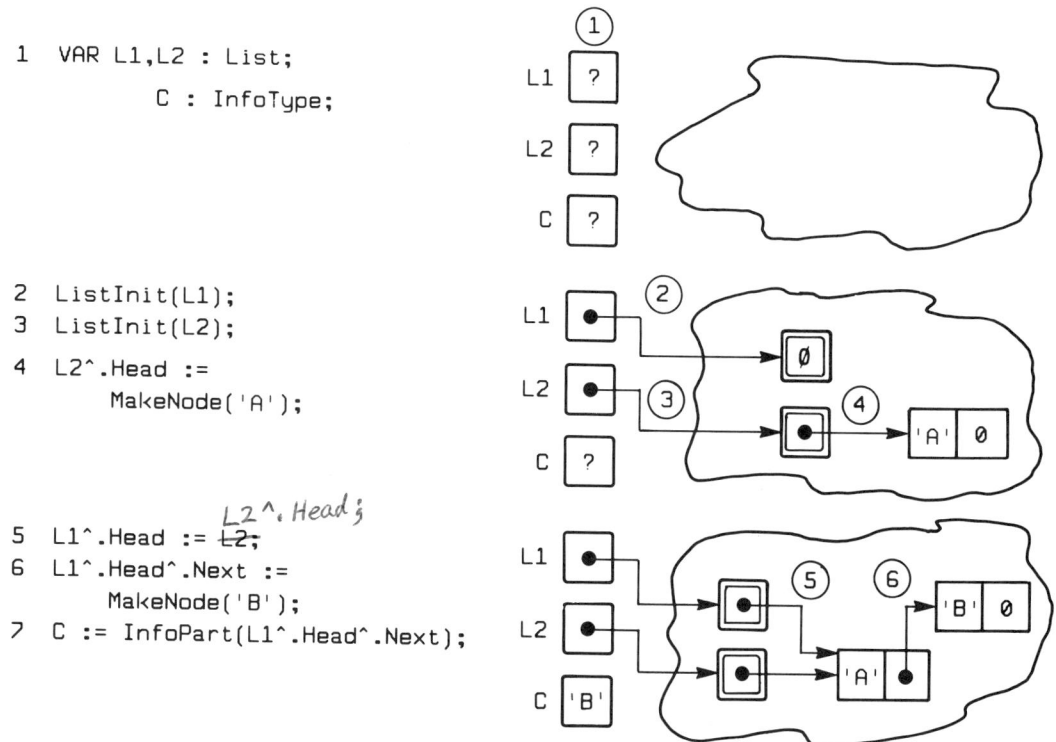

```
1  VAR L1,L2 : List;
        C : InfoType;

2  ListInit(L1);
3  ListInit(L2);
4  L2^.Head :=
       MakeNode('A');

5  L1^.Head := L2^.Head;
6  L1^.Head^.Next :=
       MakeNode('B');
7  C := InfoPart(L1^.Head^.Next);
```

Figure 4–5 Using list constructors and selectors

Sec. 4.4   Using Dynamic Allocation for Linked Structures

```
PROCEDURE AddToFront(VAR L: List; N: NodePointer);
    BEGIN
        N^.Next := L^.Head;
        L^.Head := N;
    END AddToFront;
```

Figure 4-6 Procedure to add a node to the front of a list

An important list operation connects a node to the *front* of the list. If L is a List and N a pointer to a list node, then the procedure in Figure 4-6 carries out this operation. Notice that L must be declared as a VAR parameter, because we need to modify its contents. Figure 4-7 shows the results of a number of calls to ListInit, MakeNode, and AddToFront.

Another important operation connects a node to the *rear* of a list. Here, since the variable L points to the *front* of the list, the only way we can find the end is to search the whole list until we find a node whose link field is NIL.

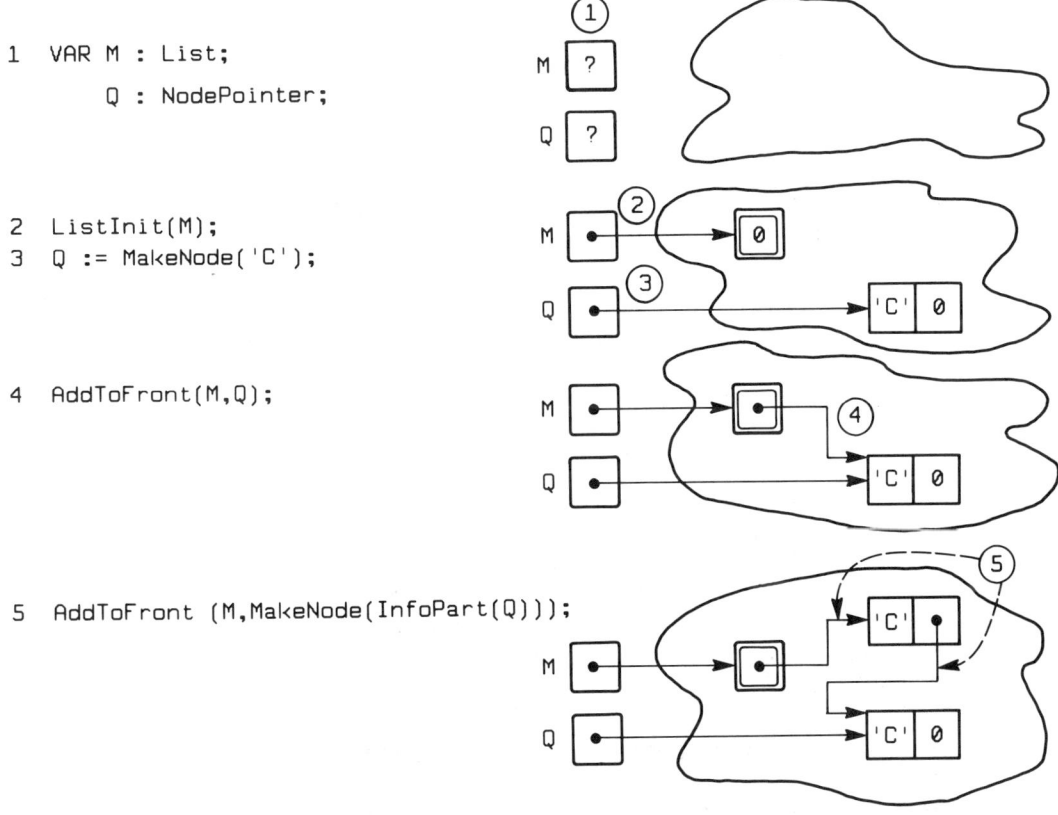

```
1  VAR M : List;
       Q : NodePointer;

2  ListInit(M);
3  Q := MakeNode('C');

4  AddToFront(M,Q);

5  AddToFront (M,MakeNode(InfoPart(Q)));
```

Figure 4-7 Using AddToFront and other list routines

```
PROCEDURE AddToRear(VAR L: List; N: NodePointer);

VAR
    Current:  NodePointer; (* Current 'walks through' list *)
    Previous: NodePointer; (* Previous trails behind Current *)

BEGIN
    (* first handle the special case of an empty list *)
    IF L^.Head = NIL THEN
        AddToFront(L,N);
        RETURN;
    END;

    Current := L^.Head;
    WHILE Current <> NIL DO
        Previous := Current;
        Current  := Current^.Next;
    END;

    (* at this point Previous points to the last node *)

    Previous^.Next := N;
    N^.Next := NIL;

END AddToRear;
```

Figure 4–8 Procedure to add a node to the rear of a list

Moreover, a special case is required if the list L is empty. The AddToRear procedure is shown in Figure 4–8; pay careful attention to the way the loop operates, particularly the statement Current := Current^.Next. Try "walking through" the code for this procedure as you follow the diagrams in Figure 4–9. For simplicity we have eliminated the "cloud" indicating the heap.

Two more useful operations are the function procedures Successor(L,N), which returns a pointer to the node which *follows* N in the list L, and Predecessor(L,N), which returns a pointer to the node which *precedes* N in L. Again, because we have no immediate way of finding the predecessor node, this function must search the list in the manner of AddToRear. Note that the predecessor of the first node in the list is deemed to be NIL, as is the successor of the last node in the list. The Successor and Predecessor functions are shown in Figure 4–10.

Given these operations, Figure 4–11 shows a definition module for an ADT that deals with one-way lists. The module exports the opaque types NodePointer and List and shows a function Front, which returns a pointer to the first node in a list, and three more procedures: Delete, InsertBefore, and InsertAfter. It is left as an exercise to complete the implementation module. Note that Delete returns the deleted node to the heap by a call to Deallocate, which is imported from System.

4.4.3 One-Way Lists with Head and Tail Pointers

The procedure AddToRear requires a search through the entire list to find the end. This takes a time proportional to the existing length of the list. We can speed the process up in a classic example of a time/space tradeoff: we just change

Sec. 4.4 Using Dynamic Allocation for Linked Structures

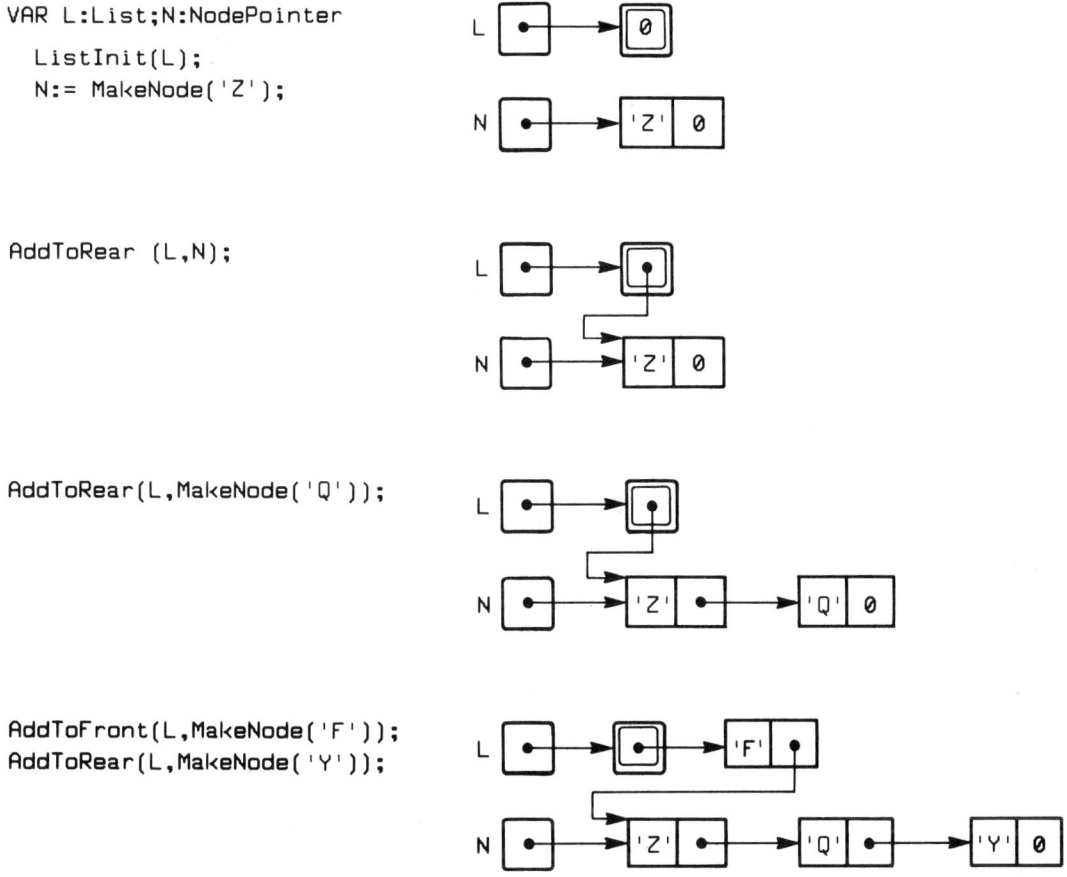

```
VAR L:List;N:NodePointer
  ListInit(L);
  N:= MakeNode('Z');
```

```
AddToRear (L,N);
```

```
AddToRear(L,MakeNode('Q'));
```

```
AddToFront(L,MakeNode('F'));
AddToRear(L,MakeNode('Y'));
```

Figure 4-9 Using AddToRear

our definition of ListHeader so that the record contains *two* pointers, one to the first or head node in the list, the other to the last or tail node. This uses a bit of extra space, of course, but is well worth it, because now a node can immediately be added to the end of the list without a search—that is, in *constant* time. Figure 4-12 gives the procedure AddToRear modified to take the two-pointer list head into account; we leave it as an exercise to discover whether modifications are required to any other operations in the ADT.

The ADT for one-way linked lists gives a rather minimal set of operations. Using the fact that new lists are created just by declaring variables, and that Modula-2 (like other languages supporting pointers) allows two pointer values to be tested for equality or inequality, we can write many other operations in terms of those in the minimal set. No knowledge of the internal structure of lists or list nodes is required. Some of these other operations are finding the number of nodes in a list; making a copy of a list; attaching one list onto the end of another; and

```
PROCEDURE Successor(L: List; N: NodePointer) : NodePointer;

BEGIN
    RETURN N^.Next;
END Successor;


PROCEDURE Predecessor(L:List; N: NodePointer) : NodePointer;

VAR
    Current: NodePointer;      (* Current starts at beginning *)
    Previous: NodePointer;     (* Previous trails by 1 node *)
BEGIN

    Current := L^.Head;
    IF Current = N THEN        (* N is first node in list *)
        RETURN NIL;
    END;

    WHILE (Current <> NIL) AND (Current <> N) DO
        Previous := Current;   (* Previous points to Current's
                                              predecessor *)
        Current := Current^.Next;
    END;

    IF Current = NIL THEN      (* N is not in the list! *)
        RETURN NIL;
    ELSE
        RETURN Previous;
    END;

END Predecessor;
```

Figure 4-10 Successor and Predecessor

destroying an entire list. As an example, Figure 4-13 gives a procedure which returns a copy of a list L. Writing the other operations mentioned is left as an exercise.

4.4.4 Building an Ordered List

The linked-list ADT operations were designed to have no knowledge of the structure of the InfoPart of the nodes. Some applications, however, build a list in such a way that this kind of knowledge is required. For instance, if the list is in ascending order based on a field in the InfoPart, the program has to be able to find the field on which the list is sorted and know what "less than" and "greater than" mean for the type of that field. Let us call the field in question the *key* field and the rest of the InfoPart the *value* field, and define a list node accordingly, as in Figure 4-14. We shall assume (1) that the selector functions Key(N) and Value(N) are available to return the fields of a node pointed to by N, and (2) that a function KeyCompare is available (by importation from a module in which InfoType is defined) which returns a value from the enumeration Relation = {Less, Equal,Greater}.

```
DEFINITION MODULE OneWayLists;

    TYPE InfoType = CHAR;

    TYPE NodePointer;

    TYPE List;

    PROCEDURE ListInit(VAR L: List);
    PROCEDURE MakeNode(Item: InfoType): NodePointer;
        (* Constructors *)

    PROCEDURE InfoPart(N: NodePointer): InfoType;
    PROCEDURE Front(L: List):           NodePointer;
        (* Selectors *)

    PROCEDURE Empty(L: List):           BOOLEAN;
        (* true iff L is empty *)

    PROCEDURE Successor   (L: List; N: NodePointer): NodePointer;
    PROCEDURE Predecessor(L: List; N: NodePointer): NodePointer;
        (* Successor, Predecessor *)

    PROCEDURE AddToFront  (VAR L: List; N: NodePointer);
    PROCEDURE AddToRear   (VAR L: List; N: NodePointer);
    PROCEDURE InsertBefore(VAR L: List; N1,N2: NodePointer);
    PROCEDURE InsertAfter (VAR L: List; N1,N2: NodePointer);
        (* Various update routines *)

    PROCEDURE Delete      (VAR L: List; N: NodePointer);
        (* Remove node pointed to by N from L *)

END OneWayLists.
```

Figure 4–11 Definition module for one-way lists

```
TYPE List = POINTER TO ListHeader;

TYPE ListHeader =
    RECORD
       Head : NodePointer;
       Tail : NodePointer;
    END;

PROCEDURE AddToRear(VAR L: List; N: NodePointer);

    BEGIN
       IF L^.Head = NIL THEN
          AddToFront(L,N);
       ELSE
          L^.Tail^.Next := N;
          L^.Tail := N;
       END;
    END AddToRear;
```

Figure 4–12 AddToRear procedure for list with head and tail pointers

```
PROCEDURE CopyList(L: List) : List;

   VAR
      NewList: List;
      p:       NodePointer;
   BEGIN
      ListInit(NewList);

      p := Front(L);
      WHILE p <> NIL DO
         AddToRear(NewList,MakeNode(InfoPart(p)));
         p := Successor(L,p);
      END;

      RETURN NewList;

   END CopyList;
```

Figure 4-13 Copying a one-way list

Now let us write a procedure InsertInOrder which inserts a node pointed to by N into a list L, in ascending order according to the key field. The insertion process has four separate cases:

1. An inserted node is the first one to be added to an empty list.
2. The inserted node's key is less than those of all others in the list, so that the node goes at the beginning of a nonempty list.
3. The key is greater than all the others, so that the node goes at the end of the list.

```
TYPE KeyType   = CHAR;

TYPE ValueType = INTEGER;

TYPE InfoType =

   RECORD
      KeyField: KeyType;
      ValueField: ValueType;
   END;


TYPE NodePointer = POINTER TO OneWayListNode;

TYPE OneWayListNode =

   RECORD
      Info: InfoType;
      Next: NodePointer;
   END;


TYPE List = POINTER TO ListHeader;

TYPE ListHeader =

   RECORD
      Head : NodePointer;
      Tail : NodePointer;
   END;
```

Figure 4-14 Type declarations for ordered one-way list

```
PROCEDURE InsertInOrder(VAR L: List; N: NodePointer);

VAR
   Current:  NodePointer;
   Previous: NodePointer;   (* trailing pointer *)
   found:    BOOLEAN;
   k:        KeyType;

BEGIN
   Current := Front(L);
   Previous := Current;
   found := FALSE;
   k := Key(N);

   IF Empty(L) THEN                                    (* case (1) *)
      AddToFront(L, N);

   ELSIF KeyCompare(k,Key(Current)) = Less THEN        (* case (2) *)
      AddToFront(L, N);

   ELSE
      WHILE (Current <> NIL) AND (NOT found) DO
         IF KeyCompare(k,Key(Current)) = Less THEN
            found := TRUE;
         ELSE
            Previous := Current;
            Current := Successor(L,Current);
         END;
      END;

      IF Current = NIL THEN                            (* case (3) *)
         AddToRear(L,N);
      ELSE
         InsertAfter(L,N,Previous);                    (* case (4) *)
      END;

   END;

END InsertInOrder;
```

Figure 4–15 InsertInOrder for one-way list

4. The key lies between two others, so that the node goes somewhere in the middle of the list.

The procedure InsertInOrder is shown in Figure 4–15. Notice that it uses operations from the linked-list ADT and needs no deeper knowledge of the linked-list organization. Notice also how the four cases are handled, and the fact that a so-called trailing pointer is used to keep track of the predecessor of a node, allowing us to avoid the linear searching required to use the Predecessor function. As an exercise, trace the actions of the procedure in the examples shown in Figure 4–16. A similar procedure can be written to delete the node containing a given key from a list.

The algorithms for InsertInOrder and Delete can be greatly simplified if we use a "dummy" or "wasted" node at the beginning of a list. A MakeEmptyList routine for ordered lists can set up the dummy node. Then, a list which is *logically*

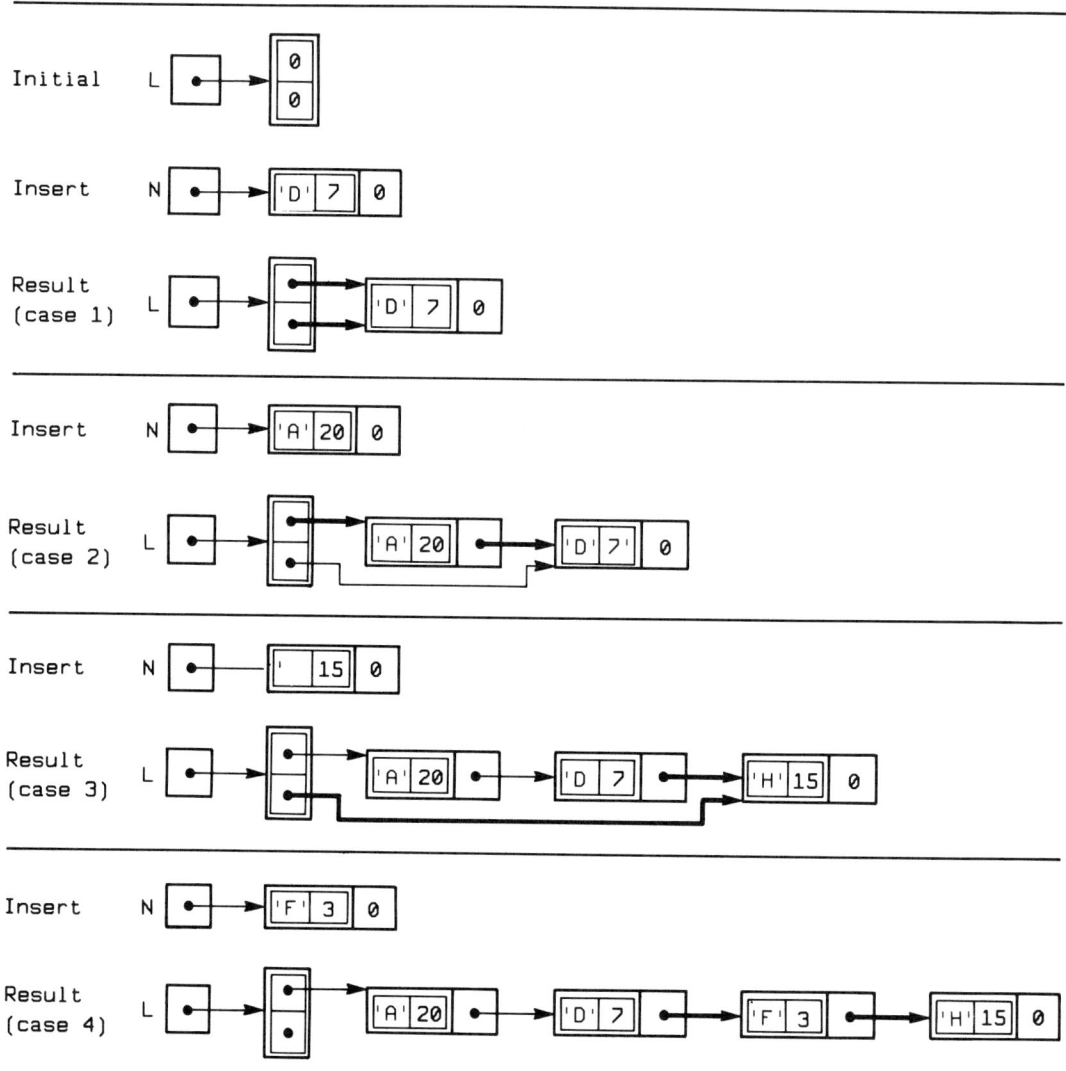

Figure 4-16 Building an ordered list (changed pointers are indicated by heavier lines)

empty actually has one node in it *physically*. The procedure MakeEmptyList is shown in Figure 4-17, and a sample initialized list is given in Figure 4-18. The "dummy node" technique allows the four cases of the InsertInOrder procedure to be collapsed into two, as Figure 4-19 shows. "Walk through" this new procedure to see how it works.

Sec. 4.4 Using Dynamic Allocation for Linked Structures

```
PROCEDURE EmptyNode() : NodePointer;

    VAR
        p: NodePointer;

    BEGIN
        Allocate(p,SIZE(OneWayListNode));
        p^.Next := NIL;
        RETURN p;
    END EmptyNode;

PROCEDURE MakeEmptyList (VAR L: List);

    VAR
        Temp: NodePointer;

    BEGIN
        ListInit(L);
        AddToFront(L,EmptyNode());
    END MakeEmptyList;
```

Figure 4–17 Initializing a list with a dummy node

As another exercise, you can write a deletion procedure for the "dummy node" implementation, as well as a search function. It makes sense to define the latter so that it returns a pointer, rather than a value, to the node containing a given key. Then it can easily return NIL if the desired node is not in the list. Thinking about these operations reveals that an ordered list can really be viewed as just another implementation of a dynamic table as developed in chapter 2.

4.4.5 Two-Way Linked Lists

Finding the predecessor of a node in a one-way list requires a linear search. Again, we can trade time and space by adding, *to each node*, a pointer to its predecessor. Once again, this will consume a significant amount of space, but since it allows finding the predecessor in *constant* time, it might be a price worth paying if the application requires finding predecessor nodes frequently.

Figure 4–20 shows a type definition for a node with forward and backward pointers, and Figure 4–21 shows the code for Predecessor and Delete. It is left

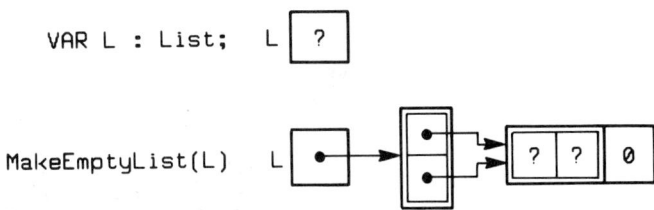

Figure 4–18 Initialized one-way list with dummy node

```
PROCEDURE InsertInOrder(VAR L: List; N: NodePointer);

VAR
   Current:  NodePointer;
   Previous: NodePointer; (* trailing pointer *)
   found:    BOOLEAN;
   k:        KeyType;

BEGIN
   Current := Successor(L,Front(L)); (* remember, dummy node! *)
   Previous := Front(L);
   found := FALSE;
   k := Key(N);

   WHILE (Current <> NIL) AND (NOT found) DO
      IF k < Key(Current) THEN
         found := TRUE;
      ELSE
         Previous := Current;
         Current := Successor(L,Current);
      END;
   END;

   IF Current = NIL THEN                (* cases (1,3) *)
      AddToRear(L,N);
   ELSE
      InsertAfter(L,N,Previous);        (* cases (2,4) *)
   END;

END InsertInOrder;
```

Figure 4-19 InsertInOrder for one-way list with dummy node

as an exercise to figure out the necessary changes to other ADT operations to accommodate the new structure. Note that, as in the one-way list, we assume that the Next pointer of the last node in a list and the Previous pointer of the first node in a list are both NIL.

```
TYPE NodePointer = POINTER TO TwoWayListNode;

TYPE TwoWayListNode =

   RECORD
      Info:     InfoType;
      Next:     NodePointer;
      Previous: NodePointer;
   END;

TYPE List = POINTER TO ListHeader;

TYPE ListHeader =

   RECORD
      Head : NodePointer;
      Tail : NodePointer;
   END;
```

Figure 4-20 Type declarations for two-way list

Sec. 4.5 Design One: Sparse Vectors and Matrices Revisited

```
PROCEDURE Predecessor(L: List; N: NodePointer) : NodePointer;

BEGIN
   RETURN N^.Previous;
END Predecessor;


PROCEDURE Delete(VAR L: List; N: NodePointer);

BEGIN
   IF L^.Head = L^.Tail THEN      (* list empty after deletion *)
      L^.Head := NIL;
      L^.Tail := NIL;
   ELSIF N = L^.Head THEN         (* deleting first node *)
      L^.Head := N^.Next;
      N^.Next^.Previous := NIL;
   ELSIF N = L^.Tail THEN         (* deleting last node *)
      L^.Tail := N^.Previous;
      N^.Previous^.Next := NIL;
   ELSE                           (* deleting intermediate node *)
      N^.Next^.Previous := N^.Previous; (* skip going backward *)
      N^.Previous^.Next := N^.Next;     (* skip going forward *)
   END;

   Deallocate(N);

END Delete;
```

Figure 4–21 Predecessor and Delete operations for two-way lists

4.5 DESIGN ONE: SPARSE VECTORS AND MATRICES REVISITED

Armed with an understanding of linked lists, let us consider how sparse vectors should be represented as lists. Let a vector be a one-way ordered list as in the previous section. Then a node in the list represents an ordered pair ⟨index,value⟩, and the *key* is a vector index. Figure 4–22 shows an opaque type Sparse-Vector, which points to a header record VectorRec. This latter contains information on the minimum and maximum indices allowable for the vector, as well as a list header for the one-way list representing the vector itself. MaxNonZero = Last − First + 1 denotes the maximum number of nonzero nodes in the vector, and will do so throughout the discussion that follows. You should draw a diagram of a typical vector to understand the entire structure.

4.5.1 Vector Arithmetic

In chapter 3 we presented arithmetic packages for vectors and matrices. Here we show the implementation of two of the operations presented there—multiplication by a scalar and vector addition—for the case of sparse vectors. The other operations are left as an exercise.

In Figure 4–23 we give the code for a function VectorRetrieve(V,i) which

```
TYPE KeyType   =   INTEGER;
TYPE ValueType =   INTEGER;   (* or whatever *)
CONST Zero =       0;         (* of type ValueType *)

TYPE VectorDimensions =

   RECORD
      First: KeyType;
      Last:  KeyType;
   END;

TYPE InfoType =

   RECORD
      KeyField:   KeyType;
      ValueField: ValueType;
   END;

TYPE NodePointer = POINTER TO OneWayListNode;

TYPE OneWayListNode =

   RECORD
      Info: InfoType;
      Next: NodePointer;
   END;

TYPE List = POINTER TO ListHeader;

TYPE ListHeader =

   RECORD
      Head : NodePointer;
      Tail : NodePointer;
   END;

TYPE VectorRec =

   RECORD
      Dims: VectorDimensions;
      Storage: List;
   END;

TYPE SparseVector = POINTER TO VectorRec;
```

Figure 4-22 Declarations for sparse vector, implemented as one-way list

searches a list for a node containing a given key and returns the value of the key if it is there and 0 otherwise. The operation corresponds to the Retrieve operation sketched out in section 3.5. We leave as an exercise the procedure VectorStore(V,i,x), which stores the value x in the location V with index i. This procedure, also sketched in section 3.5, is a bit more involved than it looks at first.

Figure 4-24 shows a function for multiplying a sparse vector by a scalar. Notice that it looks very similar to the corresponding function in section 3.5 that worked with nonsparse vectors, except that the array references are replaced by calls to VectorStore and VectorRetrieve. Figure 4-25 shows a vector addition function built on the same principle.

Sec. 4.5 Design One: Sparse Vectors and Matrices Revisited

```
PROCEDURE VectorRetrieve(V: SparseVector; i: KeyType): ValueType;

VAR
   Current: NodePointer;
   S:       List;
BEGIN
   S := V^.Storage;
   Current := Successor(S,Front(S));
      (* Dummy node is used *)

   WHILE Current <> NIL DO
      CASE KeyCompare(Key(Current),i) OF
         Equal:
            RETURN Value(Current) |

         Greater:
            RETURN Zero |            (* it wasn't there *)

         Less:
            Current := Successor(S,Current);
      END;

   END;

   RETURN Zero;                      (* we got to the end *)

END VectorRetrieve;
```

Figure 4-23 Retrieve operation for sparse vector

Let us look at the performance of vector addition using this scheme. Suppose that each vector has ActualNonZero nonzero components (i.e., the list has ActualNonZero nodes not counting the dummy). Then each VectorRetrieve operation has performance $O(\text{ActualNonZero})$, since on the average half the list must be searched. Now, since the vector is defined for MaxNonZero components, MaxNonZero calls to VectorRetrieve must be executed for each of the two vectors, and—in the worst case—MaxNonZero insertions must be done. Since each insertion also has performance $O(\text{ActualNonZero})$, the overall addition operation

```
PROCEDURE ScalarTimesVector
            (V: SparseVector; X: ValueType): SparseVector;

VAR
   Result: SparseVector;
   k:      INTEGER;

BEGIN
   CreateVector(Result,VectorFirst(V),VectorLast(V));

   FOR k := VectorFirst(Result) TO VectorLast(Result) DO
      VectorStore(Result,k,(VectorRetrieve(V,k) * X));
   END;

   RETURN Result;

END ScalarTimesVector;
```

Figure 4-24 Multiplication of a sparse vector by a scalar (the naive way)

```
    PROCEDURE VectorAdd(U,V: SparseVector) : SparseVector;

    VAR
       Result: SparseVector;
       R:      INTEGER;
       X:      ValueType;

    BEGIN

       ConformabilityError := FALSE;
       IF (VectorFirst(U) <> VectorFirst(V)) OR
          (VectorLast(U)  <> VectorLast(V))
       THEN
          ConformabilityError := TRUE;
          RETURN(Result);
       END;

       CreateVector(Result,VectorFirst(U),VectorLast(U));

       FOR R := VectorFirst(U) TO VectorLast(U) DO

          X := VectorRetrieve(U,R) + VectorRetrieve(V,R);
          VectorStore(Result,R,X);
          (* i.e. Temp[R] := U[R] + V[R] *)

       END;

       RETURN(Result);

    END VectorAdd;
```

Figure 4-25 Addition of sparse vectors (the naive way)

has performance $O(\text{MaxNonZero} \times \text{ActualNonZero})$. But, from chapter 3, addition of nonsparse vectors has performance $O(\text{MaxNonZero})$, since the two arrays are traversed only once. Thus, we have paid a price in performance in return for the economy of space achieved by storing vectors in sparse form.

In actuality, the tradeoff is between abstraction and performance: the slow performance came from our unwillingness to let the addition function know the details of a sparse vector. Instead, we required the addition function to call VectorStore and VectorRetrieve. If, on the other hand, we are willing to let addition know that sparse vectors are stored as *lists*, we can speed it up considerably. Then, only one pass through the two lists is required: since the lists are stored in order by index, we use a very simple merge algorithm.

To see the algorithm in operation, begin at the beginning of both vectors U and V. If the index of the first node of U is less than that of V, we know that the corresponding element of V is zero. So we just add the element of U onto the end of the sum vector W. Then we find the successor node in U and try again. On the other hand, if the V index is less, then we add the node from V onto the end of the W list, moving to *its* successor. If the two indices are ever equal, then both U and V have nonzero values in that position, and we add the values together before adding a node for the index in question onto the end of W. In any case, eventually, we reach the end of one of the vectors, say, U. At that point, we just copy the rest of V onto W, since all the remaining components of U are zero.

Figure 4-26 shows the function to carry out the addition just described. It

Sec. 4.5 Design One: Sparse Vectors and Matrices Revisited

```
PROCEDURE VectorAdd(U,V: SparseVector) : SparseVector;

   VAR
      W:       SparseVector;
      k1, k2: KeyType;
      p1, p2: NodePointer;
      UL, VL: List;

   BEGIN

      CreateVector(W,VectorFirst(U),VectorLast(U));

      (* initialize *)
      UL := U^.Storage;
      VL := V^.Storage;

      p1 := Successor(UL,Front(UL));
      p2 := Successor(VL,Front(VL));

      (* merge until end of one or the other is reached *)
      WHILE (p1 <> NIL) AND (p2 <> NIL) DO
         k1 := Key(p1);
         k2 := Key(p2);

         CASE KeyCompare(K1,K2) OF
            Less:              (* corresponding V element = 0 *)
               AddToRear(W^.Storage,MakeVectorElement(k1,Value(p1)));
               p1 := Successor(UL,p1) |
            Greater:           (* corresponding U element = 0 *)
               AddToRear(W^.Storage,MakeVectorElement(k2,Value(p2)));
               p2 := Successor(VL,p2) |
            Equal:             (* indices are equal            *)
               AddToRear(W^.Storage,
                  MakeVectorElement(k2,Value(p1) + Value(p2)));
               p1 := Successor(UL,p1);
               p2 := Successor(VL,p2);
         END;

      END;

      (* if U is empty, copy tail of V *)
      IF p1 = NIL THEN
         WHILE p2 <> NIL DO
            k2 := Key(p2);
            AddToRear(W^.Storage,MakeVectorElement(k2,Value(p2)));
            p2 := Successor(VL,p2);
         END;

      (* if V is empty, copy tail of U *)
      ELSE
         WHILE p1 <> NIL DO
            k1 := Key(p1);
            AddToRear(W^.Storage,MakeVectorElement(k1,Value(p1)));
            p1 := Successor(UL,p1);
         END;

      END;

      RETURN W;

   END VectorAdd;
```

Figure 4–26 A faster but more complex sparse-vector addition routine

is somewhat more complicated than the naive version, but it is a lot faster since its performance is $O(\text{ActualNonZero})$ (Why?). Note that if MaxNonZero is much larger than ActualNonZero—a good assumption, of course, in the case of sparse vectors—then the speedup is really significant.

We shall return to merge algorithms in the discussion on sorting external files.

4.5.2 Sparse Matrices

The linked-list implementation of sparse vectors can be extended to two dimensions to handle sparse matrices. First consider the classical array case, as we did for vectors. Let Rmax = LastRow − FirstRow + 1 and Cmax = LastCol − FirstCol + 1. The matrix sum of two matrices M and N, each with Rmax rows and Cmax columns, is a matrix P such that for each r and c, $P(r,c) = M(r,c) + N(r,c)$. This sum operation has performance $O(\text{Rmax} \times \text{Cmax})$.

Now recall that for two matrices M and N, if the number of rows, Rmax2, of N is the same as the number of columns, Cmax1, of M, then M and N are multiplication-conformable and their matrix product is a matrix P with Rmax1 rows and Cmax2 columns. A fragment of a procedure for finding the matrix product is shown in Figure 4–27. The performance of the procedure is $O(\text{Rmax1} \times \text{Cmax2} \times \text{Cmax1})$, i.e., cubic.

Finding the product of matrices is an important application, so if our sparse matrix implementation is to be realistic, we shall need to be able to scan rows and columns with equal ease. We thus define a node of a sparse matrix as having row and column indices and row and column pointers. Type definitions for such a node are found in Figure 4–28, as is the definition of a sparse matrix.

The "header" of a sparse matrix contains two arrays, one with Rmax elements, the other with Cmax elements, to serve as heads of the respective row and column lists. Each node is thus on two lists, a row list and a column list. To scan a row we follow its row list; to scan a column, we follow its column list.

Filling in the details of a package for sparse-matrix arithmetic is left as an exercise. As a hint to figuring out the structure of a matrix, an example is given in Figure 4–29 of the nonsparse and sparse forms of a 5 × 5 matrix of integers.

The linked-list implementation of sparse matrices is often called *cross lists* or

```
...
FOR R := FirstRow1 TO LastRow1 DO
   FOR C := FirstCol2 TO LastCol2 DO

      M3[R,C] := 0;
      FOR K := FirstCol1 TO LastCol1 DO
         M3[R,C] := M1[R,K] * M2[K,C] + M3[R,C];
      END;

   END;
END;
...
```

Figure 4–27 Fragment for matrix product

```
CONST FirstRow = ...;
CONST LastRow  = ...;
CONST FirstCol = ...;
CONST LastCol  = ...;

TYPE RowIndexType = [FirstRow..LastRow];
TYPE ColIndexType = [FirstCol..LastCol];

TYPE InfoType =

   RECORD
      RowIndex : RowIndexType;
      ColIndex : ColIndexType;
      value    : ValueType;
   END;

TYPE NodePointer = POINTER TO SparseMatrixNode;

TYPE SparseMatrixNode =

   RECORD
      Info: InfoType;
      NextRow : NodePointer;
      NextCol : NodePointer;
   END;

TYPE CrossListHeader =

   RECORD
      Head: NodePointer;
      Tail: NodePointer;
   END;

TYPE MatrixRec =

   RECORD
      RowBody: ARRAY RowIndexType OF CrossListHeader;
      ColBody: ARRAY ColIndexType OF CrossListHeader;
   END;

TYPE SparseMatrix = POINTER TO MatrixRec;
```

Figure 4–28 Declaration for sparse matrix, implemented as a cross-list structure

orthogonal lists. It is a special case of a more general situation in which each node has N indices or keys and N pointers, and the values may be, for example, information records of some sort. In the general situation, the structure is often called a *multilist* structure and appears in discussions of data base organization.

4.6 DESIGN TWO: TEXT HANDLING REVISITED

In this section we return to the module for text handling designed in chapter 1. In that design, a Text object consisted of a 32-character array and a current-length field. Such a design is severely limited for general text-handling work by its

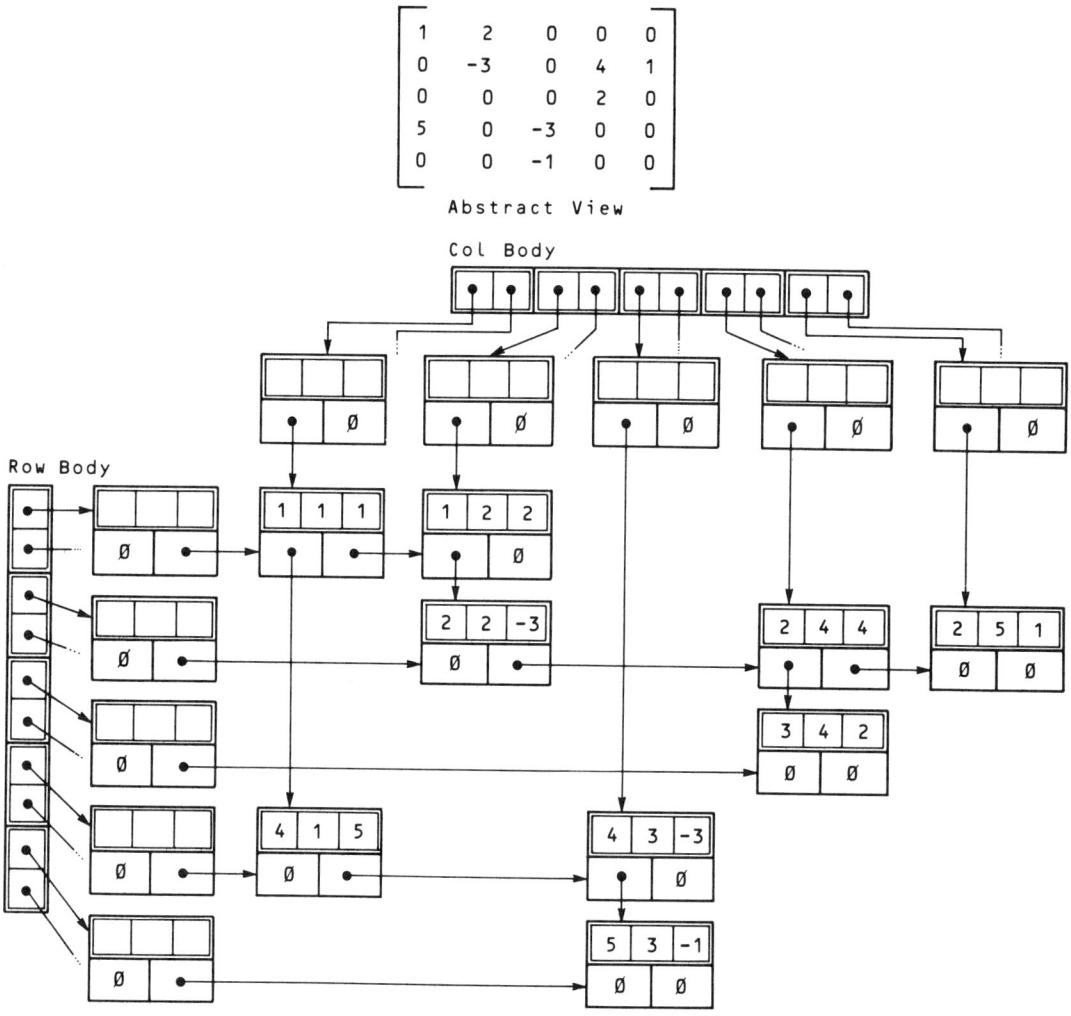

Figure 4-29 A sparse matrix and its cross-list implementation (tail pointers are omitted to avoid clutter)

requirement that every text object contain exactly 32 characters: if many of the objects have only a few actual characters, a lot of space is wasted; and if a concatenation is done of two Text objects with a total length greater than 32, an overflow condition results.

We can solve both problems with an alternative design, namely, letting a Text object consist of a linked list of smaller objects—e.g., objects with a length of eight characters. Then, even if the actual length of a string is not a multiple of

Sec. 4.7 Style Guide One: Simulating Dynamic Allocation

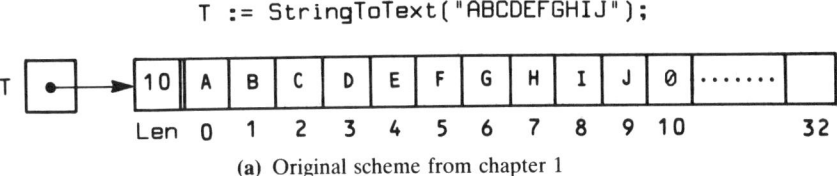

(a) Original scheme from chapter 1

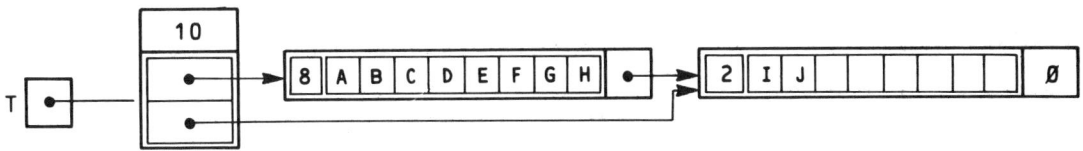

(b) New scheme using one-way list

Figure 4-30 Old and new structures for Text objects

eight, only a little space is wasted at the end, almost certainly less than in the other design. A comparison of how a particular 10-character string would be stored in both schemes is shown in Figure 4-30.

In fact, we can reuse most of our work from the package in chapter 1 if we just change Maximum to, say, 8, and change the name of the Text object to ShortText, to avoid confusion with our new Text object. Figure 4-31 shows the definition module for the revised library module, called ShortTexts, and Figure 4-32 shows the type definitions for the new scheme, which assumes that both ShortTexts and OneWayLists are available.

In Figure 4-33 we show the code for StringToText in this new implementation; Figure 4-34 gives a function TextCompare that works for the new Text structure. Concatenating one Text object to another is now merely a matter of copying the first object to a resulting list and then copying the second Text object, character by character, onto the end of the result. Copying character by character is necessary in order to fill in the empty space at the end of the first list. Doing the latter makes substring searching and comparison easier.

A diagram showing the concatenation of two of the newly defined Text objects is given in Figure 4-35; writing the function and completing the module is left as an exercise.

4.7 STYLE GUIDE ONE: SIMULATING DYNAMIC ALLOCATION

Chapter 3 discussed storing sparse vectors in the form of arrays holding the vector elements, and integers in the range of the array subscripts to represent links or pointers. In this section we show how a sparse vector system (or any other system

```
DEFINITION MODULE ShortTexts;

    (* For brevity, not all operations from Texts are repeated *)
    CONST ShortTextMaximum = 8;
    TYPE  STRING = ARRAY[0..ShortTextMaximum] OF CHAR;

    TYPE  ShortText;

    TYPE  Relation = (Less, Equal, Greater);

    PROCEDURE StringToShortText(S: ARRAY OF CHAR) : ShortText;
    PROCEDURE CharToShortText(C: CHAR) : ShortText;
    PROCEDURE NullShortText(): ShortText;
    (* basic constructors *)

    PROCEDURE ShortTextLength (T: ShortText) : INTEGER;
    PROCEDURE ShortTextToString(VAR S: ARRAY OF CHAR; T: ShortText);
    (* basic selectors *)

    PROCEDURE ShortTextEmpty (T: ShortText) : BOOLEAN;
    (*       predicate         *)

    PROCEDURE ShortTextHead(T: ShortText) : CHAR;
    PROCEDURE ShortTextTail(T: ShortText) : ShortText
    PROCEDURE ShortTextCopy(VAR DEST: ShortText; SOURCE: ShortText);

    PROCEDURE ShortTextConcat (T1, T2: ShortText) : ShortText;
    (*       concatenation          *)

    PROCEDURE ShortTextCompare(T1, T2: ShortText) : Relation;
    (*       lexical comparison      *)


END ShortTexts.
```

Figure 4-31 Defintion module for ShortTexts

```
TYPE InfoType = ShortText;

TYPE NodePointer = POINTER TO OneWayListNode;

TYPE OneWayListNode =

    RECORD
      Info: InfoType;
      Next: NodePointer;
    END;

TYPE List = POINTER TO ListHeader;

TYPE ListHeader =

    RECORD
       Head : NodePointer;
       Tail : NodePointer;
    END;


TYPE Text = POINTER TO TextRecord;

TYPE TextRecord =

  RECORD
     Len: INTEGER;
     Val: List;
  END;
```

Figure 4-32 Declarations for Text objects constructed as one-way lists

Sec. 4.7 Style Guide One: Simulating Dynamic Allocation

```
PROCEDURE StringToText(S: ARRAY OF CHAR): Text;

VAR
   T:           Text;
   L:           List;
   ShortCount:  INTEGER;    (* count chars in current block *)
   LongCount:   INTEGER;    (* total chars in string *)
   ShortS:      ARRAY[0..ShortTextMaximum] OF CHAR;
   MoreBlocks:  BOOLEAN;

BEGIN
   Allocate(T,SIZE(TextRecord));
   LongCount := 0;
   MoreBlocks := TRUE;
   ListInit(L);

   WHILE MoreBlocks DO (* copy a block into ShortS *)

      FOR ShortCount := 0 TO ShortTextMaximum DO
         ShortS[ShortCount] := 0C;
      END;

      ShortCount := 0;
      LOOP
         ShortS[ShortCount] := S[LongCount];
         IF (S[LongCount] = 0C) THEN
            MoreBlocks := FALSE;
            EXIT;
         ELSIF (LongCount = HIGH(S)) THEN
            LongCount := LongCount + 1;
            MoreBlocks := FALSE;
            EXIT;
         ELSIF (ShortCount = ShortTextMaximum) THEN
            EXIT;
         ELSE
            ShortCount := ShortCount + 1;
            LongCount := LongCount + 1;
         END;
      END;

      AddToRear(L, MakeNode(StringToShortText(ShortS)));
   END;

   T^.Len := LongCount;
   T^.Val := L;
   RETURN T;

END StringToText;
```

Figure 4-33 New constructor for Text

using linked lists) might be set up in a language—Fortran, for example—that supports neither pointers nor record types.

We suppose that our implementation language has no heap or dynamic storage allocation built in and no record types either, and attempt to *simulate* these by using subscripts to represent pointers and a separate one-dimensional array to represent each field of a record type. Our basic routines for handling list and vector elements can then easily be recoded to suit the new implementation.

We assume that the application package requires storage of *many* vectors at

```
PROCEDURE TextCompare(T1,T2: Text) : Relation;

VAR
  N1, N2: NodePointer;
  R:      Relation;

BEGIN
  N1 := Front(T1^.Val);
  N2 := Front(T2^.Val);

  WHILE (N1 <> NIL) AND (N2 <> NIL) DO
    R := ShortTextCompare(InfoPart(N1),InfoPart(N2));

    CASE R OF
      Less:
        RETURN Less      |
      Greater:
        RETURN Greater   |
      Equal:
        N1 := Successor(T1^.Val,N1);
        N2 := Successor(T2^.Val,N2);
    END;

  END;

  IF (N1 = NIL) AND (N2 = NIL) THEN    (* Equal Length *)
    RETURN Equal;
  ELSIF (N1 = NIL) THEN                (* N2 has more blocks *)
    RETURN Less;
  ELSE                                 (* N1 has more blocks *)
    RETURN Greater;
  END;

END TextCompare;
```

Figure 4–34 New Text comparison operation

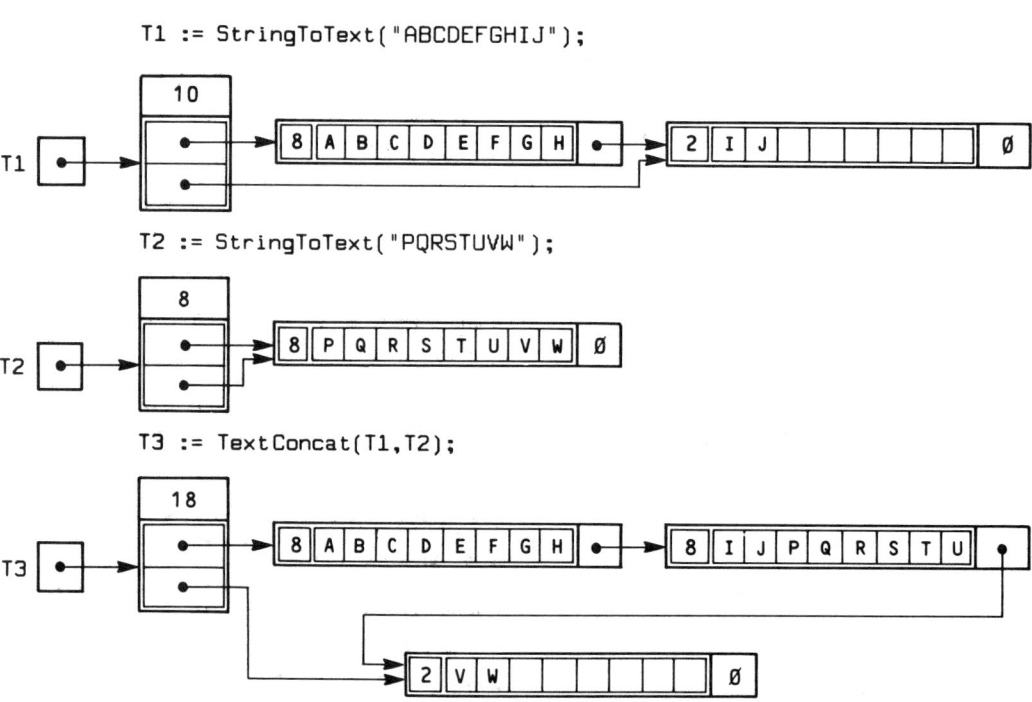

Figure 4–35 Concatenation of Text objects

Sec. 4.7 Style Guide One: Simulating Dynamic Allocation

```
CONST Null         = 0;
CONST MaxElements  = 1000;
CONST MaxNonZero   = 40;

TYPE  IndexType    = [1..MaxElements];
TYPE  ValueType    = INTEGER;             (* or whatever *)
TYPE  NodePointer  = [Null..MaxNonZero];
CONST Zero = 0;

CONST FirstRow     = 1; (* indices for first and last row vals *)
CONST LastRow      = 2;
CONST Head         = 3; (* indices for head and tail pointers *)
CONST Tail         = 4;

TYPE SparseVector = ARRAY[1..4] OF INTEGER;
(* carry all info as 1 variable *)

VAR  Avail:    NodePointer;               (* Head of LAVS *)

(* The next three arrays jointly comprise the "storage pool" *)

VAR  Indices: ARRAY [1..MaxNonZero] OF IndexType;
VAR  Values:  ARRAY [1..MaxNonZero] OF ValueType;
VAR  Links:   ARRAY [1..MaxNonZero] OF NodePointer;
```

Figure 4-36 Declarations for simulated heap implementation of sparse vectors

one time, on which operations may be performed. Then let the vector range be 1..1000 as before, and suppose that no more than 500 vector *elements* will be nonzero at any one time, independently of the number of *vectors* currently active. Recalling the terminology of chapter 3, and using the type NodePointer now to designate an integer link, we show in Figure 4-36 the declarations for a number of structures. (Be sure you understand why there are three arrays and why the dimensions and types are what they are.) According to this scheme, a sparse vector is just a one-dimensional array of two elements: a head pointer and a tail pointer, as shown in the figure.

To handle allocation of nodes in this simulated heap structure, we declare another vector, called Avail, which will let us know the location of the next available node in the array, and initialize the whole array by calling a routine StoragePoolInit, which just sets the link of each node to point at the next physical node. The Modula-2 code for StoragePoolInit is given in Figure 4-37, and a diagram of the initialized space is shown in Figure 4-38.

```
PROCEDURE StoragePoolInit;
   VAR
      p: NodePointer;
   BEGIN
      FOR p := 1 TO MaxNonZero-1 DO
         Links[p]   := p + 1;
      END;

      Links[MaxNonZero] := Null;
      Avail := 1;
   END StoragePoolInit;
```

Figure 4-37 Storage pool initialization for array-based heap

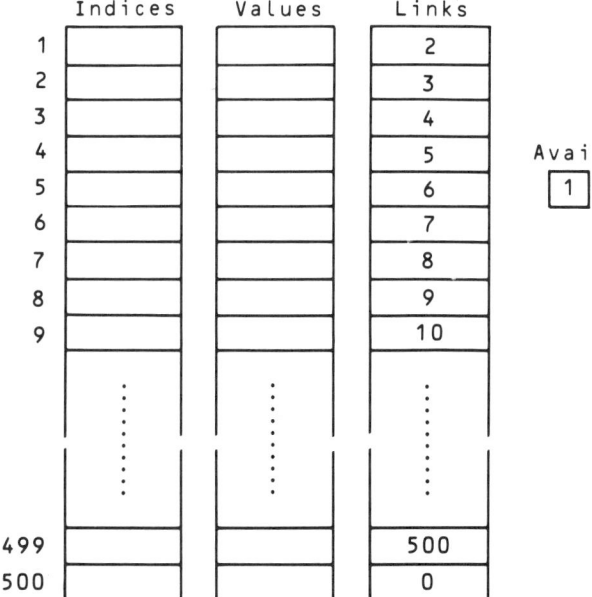

Figure 4-38 Initialized storage pool

This scheme is a miniature version of the Allocate operation in Modula-2 and its equivalents in other languages. As new nodes are required to store values in any of the vectors, they are allocated from this list of available space, or LAVS, as it is usually called. If a node is deleted, it is just returned to LAVS by adding it to the front. Returning it serves to keep track of all the "garbage" as we go along, so no separate "garbage collection" is needed. This is analogous to the Modula-2 Deallocate operation.

In this way, a number of vectors can be stored in the same pool of space and can grow and shrink as required. The link structure maintains the logical order of each list; in turn, all the lists share the same physical heap space. The only time it is necessary to refuse to add a new item to a vector is when all nodes in the list space are simultaneously occupied and allocated to vectors.

The name "cursor" is sometimes given to a pointer which is simulated by a value in an array in order to distinguish it from the "genuine" pointers available in Modula-2 and other such languages.

Figure 4-39 gives the code for a few more routines that are equivalent to the ones given earlier for built-in dynamic allocation. These routines are coded in Modula-2, of course, but they could be translated very straightforwardly into, say, Fortran, if the need ever arose.

In Figure 4-40 a number of statements are shown that assign zero and nonzero values to vector elements as though they were stored in a "normal," nondynamic array structure. Figure 4-41 shows a storage pool after the execution of the entire

Sec. 4.7 Style Guide One: Simulating Dynamic Allocation

```
PROCEDURE Index(N: NodePointer) : IndexType;

   BEGIN
      RETURN Indices[N];
   END Index;


PROCEDURE Successor(V: SparseVector; N: NodePointer) : NodePointer;

   BEGIN
      RETURN Links[N];
   END Successor;


PROCEDURE MakeNode(K: IndexType; V: ValueType) : NodePointer;

   VAR
      p : NodePointer;

   BEGIN
      StorageError := FALSE;
      p := Avail;

      IF p = Null THEN            (* no storage left! *)
         StorageError := TRUE;
         RETURN Null;
      ELSE
         Avail := Links[Avail];   (* remove node from LAVS *)
         Indices[p] := K;
         Values[p]  := V;
         Links[p]   := Null;
         RETURN p;
      END;

   END MakeNode;


PROCEDURE CreateVector(VAR V: SparseVector; First,Last: IndexType);

   BEGIN
      IndexError := FALSE;

      IF Last-First <= 0 THEN
         IndexError := TRUE;
      ELSE
         V[FirstRow] := First;
         V[LastRow]  := Last;
         V[Head] := EmptyNode();  (* dummy node! *)
         V[Tail] := V[Head];
      END;

   END CreateVector;

PROCEDURE AddToFront(VAR V: SparseVector; N: NodePointer);

   BEGIN
      IF V[Head] = Null THEN   (* list was empty *)
         V[Head] := N;
         V[Tail] := N;

      ELSE
         Links[N] := V[Head];
         V[Head]  := N;
      END;

   END AddToFront;
```

Figure 4-39 Some routines for sparse vector manipulation

```
PROCEDURE VectorRetrieve(V: SparseVector; i: IndexType): ValueType;

VAR
   Current: NodePointer;

BEGIN
   Current := VectorFront(V);

   WHILE Current <> Null DO
      CASE KeyCompare(Index(Current),i) OF
         Equal:
            RETURN Value(Current) |
         Greater:
            RETURN Zero |              (* it wasn't there *)
         Less:
            Current := Successor(V,Current);
      END;
   END;
   RETURN Zero;                        (* we got to the end *)

END VectorRetrieve;
```

Figure 4-39 (Con't)

```
CreateVector(V1, 1, 50);
VectorStore (V1, 5, -3);

CreateVector(V2, 1, 50);
VectorStore (V2,17,  2);
VectorStore (V1, 2,  4);
VectorStore (V1, 5, -3);
VectorStore (v2, 7,  5)
CreateVector(V3, 1, 50);
VectorStore (V3,18, -5);
VectorStore (V1,10, -2);
VectorStore (V1,20,  4);
VectorStore (V1,10,  0);
VectorStore (V2,10,  4);
VectorStore (V3, 5,  2);
VectorStore (V1,30, -3);
VectorStore (V2, 7,  0);
```

Figure 4-40 Sparse vector operations

sequence of assignment statements. Start with the initialized storage pool of Figure 4-38, and trace each statement's effect in order to fully understand how this simulated dynamic memory scheme works.

The example points up the distinct advantages inherent in thinking of an application first in general design terms appropriate to it—linked lists, pointers, type definitions, and so on—and only afterwards in terms of the code required for it in whatever language is available. With a correct understanding of what linked lists and sparse vectors are all about, conversion to a cursor-oriented implementation is easy, requiring recoding of only a small number of routines and then translating the whole into a less well-equipped implementation language.

Sec. 4.8 Style Guide Two: Fractions Revisited

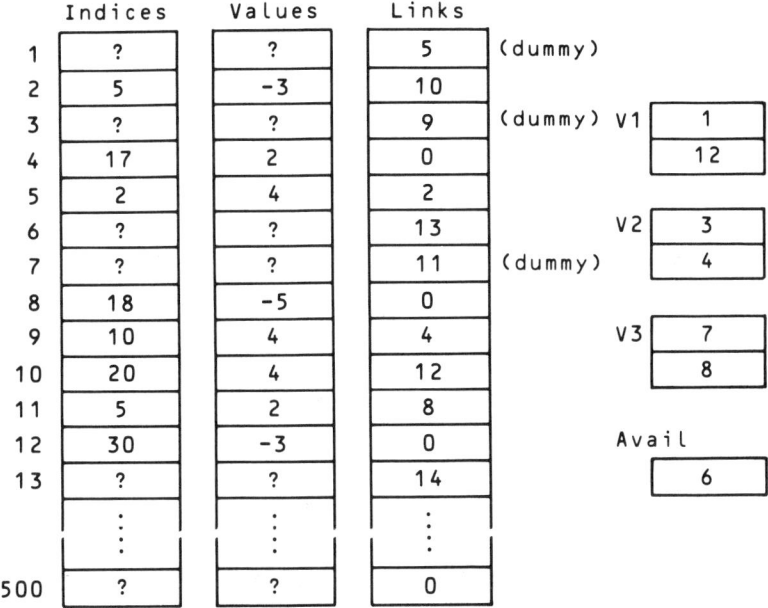

Figure 4-41 Storage pool after executing the operations of Figure 4-40

4.8 STYLE GUIDE TWO: FRACTIONS REVISITED

Earlier, it was mentioned that in some language systems space allocated from the heap which later becomes inaccessible can be recycled by a "garbage collector" module. This automatic recycling of inaccessible space usually relieves the programmer of having to think about storage reclamation. However, other common programming languages—Pascal, C, PL/1, and Modula-2, for example—lack garbage collectors. (There are interesting reasons for this, but they are beyond our scope.) In these languages, inaccessible space remains inaccessible, and thus unavailable for future use during execution of a program. If a programmer is concerned about running out of heap space while a program is executing, he or she is responsible for doing something explicit about it.

As a vehicle for discussing this situation, let us return to the Fractions module introduced in chapter 1. Recall that our objective there was to build operations using functional notation, so that these operations could be combined in expressions. For convenience, we repeat the relevant type definitions and a few key functions from that module in Figure 4-42.

In the module, the constructor MakeFract calls Allocate each time it is called, allocating a record to store each fraction. If F1 and F2 have been declared as

```
TYPE FractRecord =

    RECORD
        Numerator:   INTEGER;
        Denominator: INTEGER;
    END;


TYPE Fraction = POINTER TO FractRecord;

PROCEDURE MakeFract(N,D: INTEGER) : Fraction;

    VAR
        Z: Fraction;

    BEGIN
        ZeroDenom := FALSE;

        IF D = 0 THEN    (* Set Error Flag,
                            return undefined value *)
            ZeroDenom := TRUE;
            RETURN Z;
        END;

        Allocate(Z,SIZE(FractRecord));
        Z^.Numerator   := N;
        Z^.Denominator := D;
        RETURN Z;

    END MakeFract;


PROCEDURE FractAdd(X,Y: Fraction) : Fraction;

    VAR
        N,D: INTEGER;
        Z: Fraction;

    BEGIN
        N := FractNumer(X) * FractDenom(Y)
           + FractNumer(Y) * FractDenom(X);
        D := FractDenom(X) * FractDenom(Y);
        Z := FractReduce(MakeFract(N,D));
        RETURN Z;
    END FractAdd;
```

Figure 4-42 Repetition of some declarations and operations from Chapter 1

Fractions, then

$$F1 := MakeFract(1,-3);$$

$$F2 := MakeFract(-2,5);$$

will work exactly as intended. Similarly, if F3 is a Fraction, then

$$F3 := FractAdd(F1,F2)$$

will work correctly. But what about

$$F3 := FractMult(FractAdd(F1,F2), FractSub(F1,F2))$$

which is equivalent to F3: = (F1 + F2) × (F1 − F2)? Since FractAdd and FractSub both call MakeFract, two calls to Allocate are made for the results of these operations. But these are only *temporary* or *intermediate* results; they are never assigned to program variables, so they should be reclaimed as soon as they are used as arguments to FractMult. Unfortunately, however, Modula-2's lack of a garbage collector leaves them allocated but inaccessible.

A solution to this problem can be found in the fact that Modula-2 has a built-in procedure Deallocate(P) which, given a pointer P, returns to the heap whatever is pointed to by P. In other words, Deallocate is the inverse of Allocate. A difficulty to be overcome is that our Fraction module must be able to discern whether a given Fraction is temporary, as in the preceding intermediate expressions, or whether it is permanent, like F1, F2, and F3. One way to do this is to add to each record a Boolean field IsTemp which is set to TRUE by MakeFract. This would change the type definitions and the MakeFract function to those shown in Figure 4-43.

Two questions remain: how to designate "permanent" Fractions, and how to dispose of temporary ones. The first question can be answered by noticing that the only time we need a "permanent" Fraction is when we are assigning it to a declared variable. To do this, we need to write a procedure FractAssign(F1,F2) which copies F2 into F1, *replacing the := operation*, and sets IsTemp to FALSE. Writers of client programs must then be careful to use FractAssign instead of :=, the results of which may be unpredictable. We also need a procedure FractInit, which guarantees that a fraction variable always has a NIL value initially (why is this essential?).

```
TYPE FractRecord =

   RECORD
      Numerator:   INTEGER;
      Denominator: INTEGER;
      IsTemp:      BOOLEAN;   (* new field *)
   END;

TYPE Fraction = POINTER TO FractRecord;


PROCEDURE MakeFract(N,D: INTEGER) : Fraction;

   VAR
      Z: Fraction;

   BEGIN
      ZeroDenom := FALSE;
      IF D = 0 THEN  (* Set Error Flag,
                        return undefined value *)
         ZeroDenom := TRUE;
         RETURN Z;
      END;

      Allocate(Z,SIZE(FractRecord));
      Z^.Numerator    := N;
      Z^.Denominator  := D;
      Z^.IsTemp       := TRUE;    (* new assignment *)
      RETURN Z;

   END MakeFract;
```

Figure 4-43 Revised type declarations and MakeFract operation

```
PROCEDURE FractInit(VAR F: Fraction);

BEGIN
    F := NIL;
END FractInit;


PROCEDURE FractAssign(VAR Destination, Source: Fraction);
BEGIN
    IF Destination <> NIL THEN
        Deallocate(Destination);   (* Get rid of current value *)
    END;

    Destination :=
        MakeFract(FractNumer(Source),FractDenom(Source));
    Destination^.IsTemp := FALSE;

END FractAssign;


PROCEDURE FractRelease(VAR F: Fraction);
BEGIN
    IF F^.IsTemp THEN
        Deallocate(F);
    END;
END FractRelease;
```

Figure 4-44 FractAssign and FractRelease

To dispose of temporary structures, we can write a procedure FractRelease(F) which calls Allocate whenever F is temporary. Then, we call FractRelease in each of our arithmetic functions, so that input arguments to these functions can be given back if temporary. In effect, FractRelease is a "conditional Deallocate." FractInit, FractAssign and FractRelease are shown in Figure 4-44.

The revised fractions module may seem a bit messy, and burdensome to client programs as well, but since Modula-2 will not collect garbage automatically, it's about the only way to reclaim storage. As an example of where FractRelease would be used, consider the revised FractAdd given in Figure 4-45. In this scheme, each arithmetic function has to be made responsible for calling Fract-

```
PROCEDURE FractAdd(X,Y: Fraction) : Fraction;

VAR
    N,D: INTEGER;
    Z:   Fraction;

BEGIN
    N := FractNumer(X) * FractDenom(Y)
        + FractNumer(Y) * FractDenom(X);
    D := FractDenom(X) * FractDenom(Y);
    Z := FractReduce(MakeFract(N,D));

    FractRelease(X);    (* new statements *)
    FractRelease(Y);

    RETURN Z;
END FractAdd;
```

Figure 4-45 Revised fraction addition operation

Sec. 4.9 Summary

Release for its arguments. The scheme can also be used in applications such as the cursor-oriented sparse vector system of the previous section. There, of course, there are no built-in Allocate and Deallocate operations, so the simulated ones are used instead.

4.9 SUMMARY

In this chapter we have worked with *linked lists*, emphasizing how they are handled in languages providing built-in support. The example of sparse vectors and matrices has served as an important vehicle for understanding list structures. Note that there is nothing in the linked-list abstraction that limits its use to numerical problems.

A couple of levels of abstraction are operative in considering linked lists. First, there is the abstraction of a sparse vector or other appropriate application structure which is implemented as a linked list. Then, the list in turn is treated as an abstraction, with implementation either as a record-and-pointer structure in languages which permit such structures, or as an array-and-cursor structure in languages which don't.

Performance analysis of vector addition revealed the oft-present tradeoff between *abstraction* and *run-time performance*. Indeed, the entire discussion of sparse vectors was built on the need to economize on space, thus requiring another entire set of tradeoffs.

Finally the style guide sections showed how to deal with linked lists and storage reclamation in languages which fail to provide these things as standard features.

In the coming chapters you will see many cases where linked structures are used in developing implementations of other abstractions.

CHAPTER 4 EXERCISES

1. Implement the list operations Front, Delete, InsertBefore, and InsertAfter for the one-way list module specified in Figure 4–11.
2. Modify the operations of OneWayLists to account for head and tail pointers in the list header records.
3. Write a function to return the number of nodes in a one-way list.
4. Write a procedure which attaches one list to the end of another. Note that this procedure destroys the original lists.
5. Write a function which returns the concatenation of two lists L1 and L2, that is, a list containing copies of all the nodes of L1 followed by copies of all the nodes of L2. Note that this function does *not* destroy either L1 or L2.
6. Write a procedure Delete(L:List; K:KeyType), which deletes from an ordered list L the *first* node containing a given key K. Do this two ways: first for a list without a dummy node, then for a list with a dummy node.

7. Write a procedure DeleteLast(L:List; K:KeyType), which deletes from an ordered list L the *last* node containing a given key K. Do this two ways: first for a list without a dummy node, then for a list with a dummy node.

8. Write a procedure DeleteAll(L:List; K:KeyType), which deletes from an ordered list L all nodes containing a given key K. Do this two ways: first for a list without a dummy node, then for a list with a dummy node.

9. Write a function Search(L:List; K:KeyType) which searches a list L with a dummy node, returning a pointer to the first node containing a given key K.

10. Some writers in the ADT field advocate that every ADT should contain a *destructor* which destroys an object. Write a procedure DestroyList which returns all the nodes in a list to the heap. Do this three ways: one-way list with head pointer only; list with head and tail pointers but no dummy node; list with dummy node. Consider whether, in the last case, the dummy node should be returned to the heap or not, i.e., whether the list should be *physically* or only *logically* destroyed.

11. Starting from the specification in Figure 4–11, develop a module for two-way lists.

12. Develop a module for sparse vectors represented as one-way lists.

13. For the cross-list implementation of sparse matrices, calculate the space requirements for a matrix in this form which has K non-zero values. Assume that each pointer occupies P bytes, each index occupies I bytes, and each value occupies V bytes. Recalling that a "classical" square matrix with N rows and N columns occupies $V \times N^2$ bytes, find the crossover point, or the point where the classical and list methods require equal storage. Calculate the "sparseness ratio" K/N^2 for the realistic case in which $P = I = 2$ and $V = 4$.

14. Write a concatenation operation for two Text objects stored in the list form described in Section 4.6.

15. Complete the Texts module discussed in Section 4.6.

16. In thinking about the text module discussed in Section 4.6, discuss the tradeoffs between the suggested method of concatenating Text objects—copying character-by-character and leaving no empty space at the ends of nodes—and an alternative strategy in which concatenation simply concatenates the nodes together. This uses extra space because both operands may have empty space at the ends, but saves time. How does this alternative design affect other routines in the module such as equality checking and substring searching?

17. Implement a sparse-vector module using the heap-simulation scheme discussed in Section 4.7.

Chapter 5

QUEUES AND STACKS

5.1 GOAL OF THE CHAPTER

Two very common, important, and easy-to-understand structures in computing are the *queue* and the *stack*. They are distinguished from each other and from vectors and lists by the rules by which their elements are accessed for storage and retrieval. In a vector the access is *random* in the sense that we can store a value at an arbitrary location or retrieve a value from an arbitrary location without having to search any other locations. In a list the access is *sequential* in that a sequential search must be done to locate the position of an arbitrary element. In queues and stacks the method of access is *controlled*: first in, first out (FIFO) and last in, first out (LIFO), respectively.

In this chapter we present ADTs for stacks and queues, together with various implementation schemes for each. We consider, in particular, how stacks and queues can be constructed using arrays as well as linked lists, and discuss an important application of stacks: evaluating and translating arithmetic expressions.

5.2 QUEUES AND STACKS

The queue is analogous to the waiting line at a supermarket checkout or bank teller: customers are served one at a time, in the exact order of their arrival. Because of this first-come, first-served serving strategy, the queue is often called a first-in, first-out, or FIFO, structure. This means that customers must always get at the end of the line when they arrive, and the server (checkout clerk or whatever) waits on whomever is first in line. (We assume that all customers are polite and that "breaking in" never occurs.) The customer who has been served then leaves the line, and the line moves up. Obviously, it doesn't make sense for the server to try to serve an empty queue.

The stack, on the other hand, finds its intuitive analogy in the spring-loaded tray stackers often found in self-service restaurants. In such a stacker, only the top tray is visible and may be removed. When a new, clean tray is placed on top of the stack all the others are pushed down, and when a tray is removed all the others move up. Of course, it doesn't make sense to remove the top tray on an empty stacker. The stack is a last-in, first-out, or LIFO, device because, continuing with the restaurant analogy, the last tray put on the stacker is the first one removed.

With this intuitive introduction let us formalize our consideration of queues and stacks.

5.3 QUEUES

Abstractly speaking, a queue has a head and a tail, and, at any given time, a certain length (the number of items awaiting service). Also, items arrive on the queue and are removed from it. (The type of items involved obviously depends on the application, so we shall leave it abstract and unspecified.)

An item joins the queue only at the tail and leaves the queue only at the head, and only the head item can be examined. Thus, the appropriate set of operations on queues are the constructors *QueueInit* (reset a queue to the empty condition), *Enqueue* (put an item on the queue), and *Dequeue* (take an item off the queue); the predicate *QueueIsEmpty* (test whether a queue is empty); and the selector *QueueFirst* (examine the first element on the queue without removing it). This set corresponds to the actions that occur with most real queues, in which the object at the front of the queue is served and then leaves the queue. Figure 5-1

```
DEFINITION MODULE Queues;

TYPE Queue;

TYPE QueueItem = ...;

PROCEDURE QueueInit     (VAR Q: Queue);
PROCEDURE Enqueue       (VAR Q: Queue; E: QueueItem);
PROCEDURE Dequeue       (VAR Q: Queue);
PROCEDURE QueueFirst    (Q: Queue) : QueueItem;

PROCEDURE QueueIsFull (Q: Queue) : BOOLEAN;
PROCEDURE QueueIsEmpty(Q: Queue) : BOOLEAN;

VAR QueueOverflow:  BOOLEAN;
    QueueUnderflow: BOOLEAN;

END Queues.
```

Figure 5-1 Definition module for Queues; QueueItem must be supplied

shows a definition module for the queue ADT. The type of the elements to be stored in the queue will be filled in later. Note the inclusion of the Boolean exception variables *QueueOverflow* and *QueueUnderflow* to signal accidental misuse of the Enqueue and Dequeue operations. We shall examine their use shortly.

5.3.1 Array Implementation of a Queue

One implementation of a queue uses an array with a cursor indicating the tail of the queue. The capacity of the queue is then determined by the length of the array. Initially the tail cursor is set to 0. A new arrival is inserted into the array at the location indicated by the cursor, and then the cursor is incremented to indicate the next available location. What happens when an item is removed from the head of the queue? In a supermarket, when a customer is finished at the checkout, the remaining customers move up one position in the queue. The array implementation works in an analogous fashion. The Enqueue and Dequeue operations are demonstrated in Figure 5-2.

We can code the array implementation in Modula-2 by declaring a type *queue* which is a record containing the tail cursor and the queue array as its fields. This is shown in Figure 5-3; Figure 5-4 gives the procedure bodies for the various queue operations. Notice that we have used a general type Queue Item to be stored in the queue, and that the queue has a maximum of MAX items. These two entities would of course have to be declared previously. Notice also how we have treated the QueueUnderflow and QueueOverflow conditions. Unwittingly attempting to remove an item from an empty queue usually indicates a misunderstanding of the abstract notion of a queue, whereas attempting to add an item to a full queue is usually done without the realization that the queue is full (a practical, rather than abstract, consideration). Always keep in mind this kind of defensive programming or "antibugging:" isolating exceptional conditions is a good design discipline, and methods for reporting and handling them can usually be implemented easily in most programming languages.

5.3.2 Circular Array Implementation of a Queue

The array implementation of a queue has a major performance problem associated with it: although an Enqueue operation requires only a constant amount of time for its execution, namely a one-position move of the tail cursor, the Dequeue operation requires a number of moves, and hence an amount of time, proportional to the queue length since the scheme requires moving the entire queue every time an element is removed from the head (as in real-life supermarket queues). Accordingly, instead of requiring the queue to move whenever a Dequeue is done, let us "move the cash register" instead. That is, maintain a cursor to the current *head* of the queue and move it ahead one position when an element is removed. Thus, only a constant amount of time will now be needed to remove an element, since only the head cursor moves, and then only by one position.

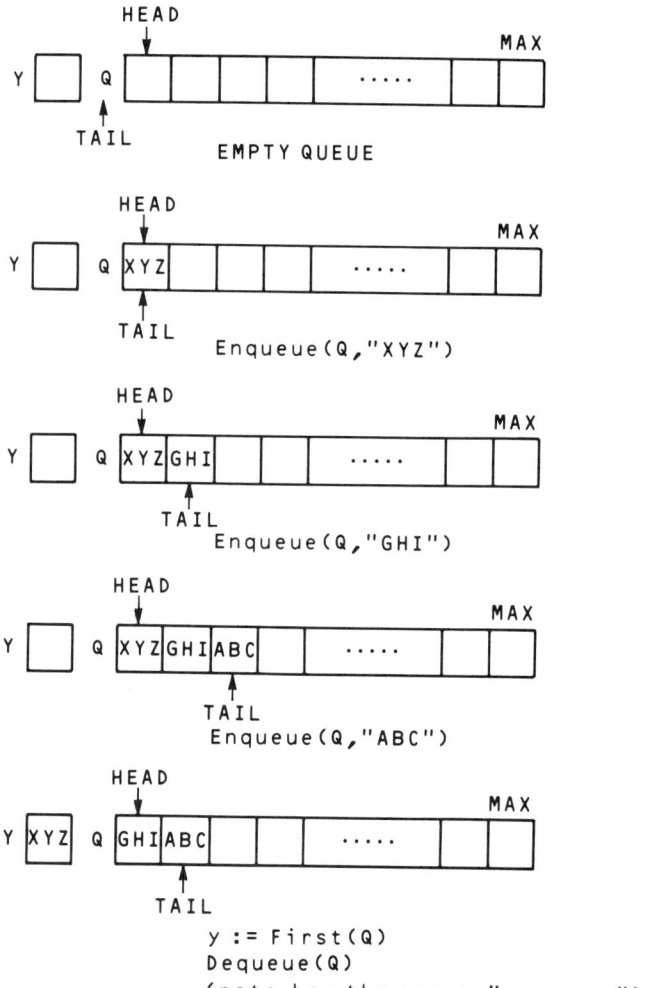

Figure 5-2 Operations on a queue: Enqueue, Dequeue, QueueFirst

```
CONST QueueMaximum = ...

TYPE Queue = POINTER TO QueueRecord;

TYPE QueueRecord =

   RECORD
     Store: ARRAY[1..QueueMaximum] OF QueueItem;
     tail: INTEGER;
   END;
```

Figure 5-3 Declarations for array implementation of a queue

Sec. 5.3 Queues

This scheme works smoothly until the tail cursor reaches the upper limit of the array, at which time it breaks down since no new elements can be enqueued. Note that the queue is probably not full, since elements are presumably being dequeued as well, so there is still space available if only we can discover how to use it. One solution would be to reorganize the queue whenever the tail cursor

```
IMPLEMENTATION MODULE Queues;

    FROM System IMPORT Allocate;

    TYPE Queue = (* FROM PREVIOUS FIGURE *)

    PROCEDURE QueueInit(Q: VAR Queue) ;
        BEGIN
            Allocate(Q,SIZE(QueueRecord));
            QueueUnderflow := TRUE;
            QueueOverflow  := FALSE;
            Q^.tail := 0;
        END QueueInit;

    PROCEDURE Enqueue(Q: VAR Queue; E: QueueItem) ;
        BEGIN
            IF Q^.tail = QueueMaximum THEN
                QueueOverflow := TRUE;
            ELSE
                Q^.tail := Q^.tail + 1;
                Q^.Store[Q^.tail] := E;
            END;
        END Enqueue;

    PROCEDURE Dequeue(Q: VAR Queue) ;
        BEGIN
            IF Q^.tail = 0 THEN
                QueueUnderflow := TRUE;
            ELSE
                FOR i := 2 TO Q^.tail DO
                    Q^.Store[i-1] := Q^.Store[i];
                END;
            END;
        END Dequeue;

    PROCEDURE QueueFirst(Q: Queue): QueueItem ;
        BEGIN
            IF Q^.tail = 0 THEN
                QueueUnderflow := TRUE;
            ELSE
                RETURN Q^.Store[1];
            END;
        END QueueFirst;

    PROCEDURE QueueIsEmpty(Q: Queue): BOOLEAN;
        BEGIN
            RETURN Q^.tail = 0;
        END QueueIsEmpty;

    PROCEDURE QueueIsFull(Q: Queue): BOOLEAN;
        BEGIN
            RETURN Q^.tail = QueueMaximum;
        END QueueIsFull;

END Queues.
```

Figure 5-4 Implementation module for Queues using array

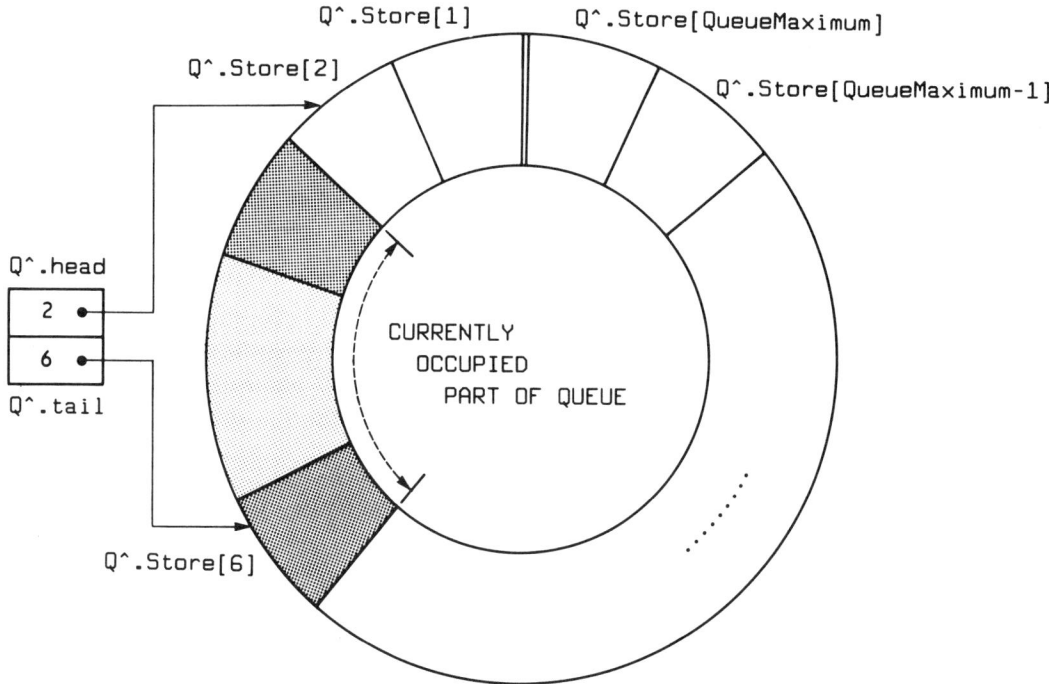

Figure 5-5 Array treated as circular queue

reached the end of the array: simply move the element at the (current) head of the queue to the first array position, and then move the others up behind it. As before, this would require a time proportional to the queue length, but at least the operation would be done much less frequently.

A more elegant and "self-regulating" solution is to treat the array as though the last position were "glued" back to the first position, so that the tail cursor would "wrap around," using empty space at the beginning of the array for new arrivals—space that was vacated by previous departures. This circular arrangement is depicted in Figure 5-5, and a type declaration is given in Figure 5-6.

```
CONST QueueMaximum = ...

TYPE Queue = POINTER TO QueueRecord;

TYPE QueueRecord =
    RECORD
      Store: ARRAY[1..QueueMaximum] OF QueueItem;
      head: INTEGER;
      tail: INTEGER;
    END;
```

Figure 5-6 Type declaration for circular queue

Sec. 5.3 Queues

Since the queue is full whenever the tail cursor "catches up" to the head cursor, a queue-full condition will be evident only if the queue is genuinely full.

An implementation module for this circular scheme is shown in Figure 5–7. The head and tail pointers are initialized to QueueMaximum because of the "wrap-around" and are incremented modulo QueueMaximum. Notice that only QueueMaximum − 1 positions of the array can be used, since otherwise we cannot

```
IMPLEMENTATION MODULE CircularQueues;

   FROM System IMPORT Allocate;

   TYPE Queue = (* FROM PREVIOUS FIGURE *)

   PROCEDURE QueueInit(Q: VAR queue) ;
      BEGIN
         Allocate(Q,SIZE(QueueRecord));
         QueueUnderflow := TRUE;
         QueueOverflow  := FALSE;
         Q^.tail := QueueMaximum;
         Q^.head := QueueMaximum;
      END QueueInit;

   PROCEDURE Enqueue(Q: VAR Queue; E: QueueItem) ;
      BEGIN
         IF (Q^.tail MOD QueueMaximum) + 1 = Q^.head THEN
            QueueOverflow := TRUE;
         ELSE
            Q^.tail := (Q^.tail MOD QueueMaximum) + 1;
            Q^.Store[Q^.tail] := E;
         END;
      END Enqueue;

   PROCEDURE Dequeue(Q: VAR Queue) ;
      BEGIN
         IF Q^.tail = Q^.head THEN
            QueueUnderflow := TRUE;
         ELSE
            Q^.head := (Q^.head MOD QueueMaximum) + 1;
         END;
      END Dequeue;

   PROCEDURE QueueFirst(Q: Queue): QueueItem ;
      BEGIN
         IF Q^.tail = Q^.head THEN
            QueueUnderflow := TRUE;
         ELSE
            RETURN Q^.Store[(Q^.head MOD QueueMaximum) + 1];
         END;
      END QueueFirst;

   PROCEDURE QueueIsEmpty(Q: Queue): BOOLEAN;
      BEGIN
         RETURN Q^.tail=Q^.head;
      END QueueIsEmpty;

   PROCEDURE QueueIsFull(Q: Queue): BOOLEAN;
      BEGIN
         RETURN (Q^.tail MOD QueueMaximum) + 1 = Q^.head;
      END QueueIsFull;

END CircularQueues.
```

Figure 5–7 Implementation module for circular queues

```
TYPE NodePointer = POINTER TO OneWayListNode;

TYPE OneWayListNode =

   RECORD
      Info: QueueItem;
      Next: NodePointer;
   END;

TYPE Queue = POINTER TO QueueRecord;

TYPE QueueRecord =

   RECORD
      Head: NodePointer;
      Tail: NodePointer;
   END;
```

Figure 5-8 Type declaration for queue using linked-list implementation

```
IMPLEMENTATION MODULE Queues;

   FROM System IMPORT Allocate, Deallocate;

   (* types from previous figure *)

   PROCEDURE QueueInit(VAR Q: Queue);

      VAR
         P: NodePointer;

      BEGIN
         Allocate(Q,SIZE(QueueRecord));
         Allocate(P,SIZE(OneWayListNode));
         P^.Next   := NIL;
         Q^.Head   := P;
         Q^.Tail   := P;
         QueueUnderflow := TRUE;

      END QueueInit;

   PROCEDURE QueueIsEmpty(Q: Queue): BOOLEAN;

      BEGIN
         RETURN Q^.Tail=Q^.Head;
      END QueueIsEmpty;

   PROCEDURE Enqueue(VAR Q: Queue; E: QueueItem);

      VAR
         P: NodePointer;

      BEGIN
         Allocate(P,SIZE(OneWayListNode));
         P^.Info := E;
         P^.Next := NIL;
         Q^.Tail := P;
      END Enqueue;

   (* the remaining queue operations are left as an exercise *)

   END Queues.
```

Figure 5-9 Implementation module for queues using linked lists

distinguish between the QueueIsFull and QueueIsEmpty conditions. All operations perform in $O(1)$ time.

5.3.3 Linked-List Implementation of a Queue

Where a programming language supports pointers and dynamic allocation, the linked-list implementation of a queue is useful. As with sparse vectors, we allocate a new list node whenever an item is to be enqueued. We then use whatever deallocation mechanism is available to free the node when an item is dequeued. In this implementation, the header node for the queue contains pointers to the queue head and tail. As before, enqueueing and dequeueing are made easier by the presence of a "dummy" node at the actual head of the queue, set up by the QueueInit constructor.

Two immediate advantages to using a linked-list implementation of a queue are that enqueueing and dequeueing occur in a fixed amount of time that is independent of the queue length, and that a queue has no *a priori* fixed maximum length. (Its only limitation in this regard is the compiler's heap size.) Figure 5-8 shows the necessary type declarations for the linked-list implementation, while Figure 5-9 gives a partial implementation module. Completing the module is an exercise.

5.4 STACKS

In terms of ADTs, an item is inserted in a stack ("pushed") and deleted from it ("popped") only at the top, and only the top item can be examined. So the appropriate operations on stacks are the constructors *StackInit*, *StackPush*, and *StackPop*, and the predicate *StackIsEmpty*. As in the case of queues, we give a selector operation *StackTop* which examines the top item without removing it. Figure 5-10 gives the definition module.

```
DEFINITION MODULE Stacks;

    TYPE Stack;

    TYPE StackItem = ...;

    PROCEDURE StackInit    (VAR S: Stack);
    PROCEDURE StackPush    (VAR S: Stack; E: StackItem);
    PROCEDURE StackPop     (VAR S: Stack);
    PROCEDURE StackTop     (S: Stack) : StackItem;
    PROCEDURE StackIsFull  (S: Stack) : BOOLEAN;
    PROCEDURE StackIsEmpty (S: Stack) : BOOLEAN;

    VAR StackOverflow:  BOOLEAN;
        StackUnderflow: BOOLEAN;

END Stacks.
```

Figure 5-10 Definition module for Stacks

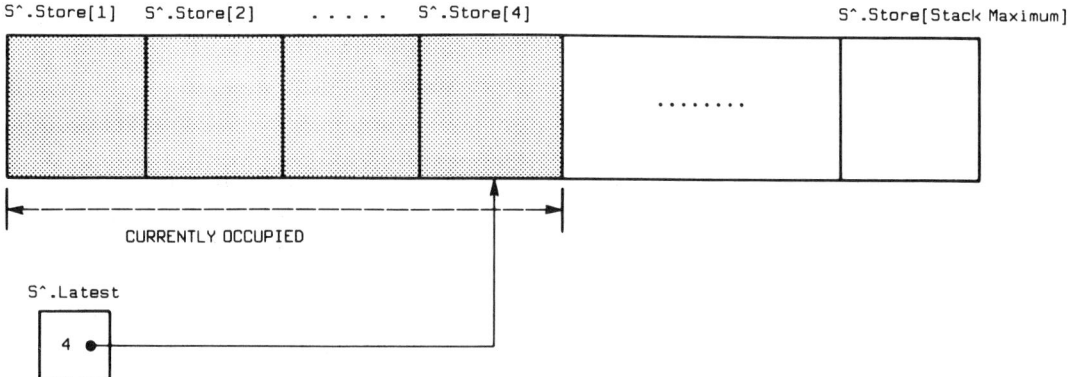

Figure 5-11 Array implementation of stack

Our choice of stack and queue operations merits some discussion. Some authors write StackPop as a procedure in which the top item on the stack is removed *and* returned in an output parameter. Similarly, those authors write Dequeue as a procedure which removes the head item, adjusts the queue, *and* returns the removed item in an output parameter. We think our choice is better because StackTop is a pure selector and StackPop is a pure constructor, and similarly for QueueFirst and Dequeue. These are *primitive* operations on stacks and queues; they only do one thing per invocation. Showing that the other forms can be implemented as combinations of ours is left as an exercise.

5.4.1 Array Implementation of a Stack

In the array implementation of a stack, we shall consider the stack to be a record consisting of a cursor pointing to the current stack top and an array representing the stack itself. The top of the stack keeps moving toward the bottom of the array, avoiding the necessity of moving items when a new item arrives. (Alternatively, we could fill the array from the bottom, which might be more intuitive.) Thus, StackPush and StackPop operations are done in fixed time, independently of the current stack depth.

Figure 5-11 illustrates the array structure, and Figure 5-12 shows an appro-

```
TYPE Stack = POINTER TO StackRecord;

TYPE StackRecord =

    RECORD
       Latest: INTEGER;
       Store: ARRAY[1..StackMaximum] OF StackItem;
    END;
```

Figure 5-12 Type declarations for stack implemented as array

Sec. 5.4 Stacks

priate type declaration. Note again that StackItem is of a type that must be previously declared, and that StackMaximum is a natural number that must be previously declared. As in the case of queues, we have included StackOverflow and StackUnderflow exception flags or "protestors" to isolate error detection and handling. Figure 5–13 gives the implementation module.

```
IMPLEMENTATION MODULE Stacks;

    FROM System IMPORT Allocate;

    CONST StackMaximum = ...
    TYPE Stack = ... (* FROM PREVIOUS FIGURE *)

    PROCEDURE StackInit(VAR S: Stack);
        BEGIN
            Allocate(S,SIZE(StackRecord));
            S^.Latest := 0;
            StackUnderflow := TRUE;
            StackOverflow  := FALSE;
        END StackInit;

    PROCEDURE StackIsEmpty(S: Stack) : BOOLEAN;
        BEGIN
            RETURN S^.Latest<=0;
        END StackIsEmpty;

    PROCEDURE StackIsFull (S: Stack) : BOOLEAN;
        BEGIN
            RETURN S^.Latest>=StackMaximum;
        END StackIsFull;

    PROCEDURE StackPush(VAR S: Stack; E: StackItem);
        BEGIN
            IF StackIsFull(S) THEN
                StackOverflow := TRUE;
            ELSE
                S^.Latest := S^.Latest + 1;
                S^.Store[S^.Latest] := E;
            END;
        END StackPush;

    PROCEDURE StackPop(VAR S: Stack);
        BEGIN
            IF StackIsEmpty(S) THEN
                StackUnderflow := TRUE;
            ELSE
                S^.Latest := S^.Latest - 1;
            END;
        END StackPop;

    PROCEDURE StackTop(S: Stack) : StackItem;
        BEGIN
            IF StackIsEmpty(S) THEN
                StackUnderflow := TRUE;
            ELSE
                RETURN S^.Store[S^.Latest];
            END;
        END StackTop;

END Stacks.
```

Figure 5–13 Implementation module for Stacks, array implementation

5.4.2 Linked-List Implementation of a Stack

A linked-list implementation of a stack is really quite straightforward. Since items are both pushed and popped at the head of the list, both operations are done in fixed time. Given the full linked-list implementation module for queues in the previous section, you should have little difficulty in writing such a module for the linked-list implementation of stacks, so we leave this task as an exercise.

5.5 STACKS, EVALUATION OF EXPRESSIONS, AND POLISH NOTATION

Consider the sort of lengthy computation often carried out on a hand-held calculator given by the sequence of operations

$$(5 \times 2) - (((3 + 4 \times 7) + 8/6) \times 9)$$

Suppose a Brand X calculator allows the user to enter the above expression, parentheses and all, as is, i.e., in *parenthesized* or *infix* notation, whereas a Brand Y calculator requires the user to convert the expression into *reverse Polish notation (RPN)*, also called *postfix* notation, before entering it. In the latter notation the expression would be

$$5\ 2\ \times\ 3\ 4\ 7\ \times\ +\ 8\ 6/\ +\ 9\ \times\ -$$

which looks thoroughly unintelligible. Most people seem to prefer the parenthesized form.

Since the calculator is just a special purpose computer, it follows some algorithm to *evaluate* (find the final result of) the expression, given in one form or the other. The purpose of this section is to introduce the relationship between a parenthesized or infix expression and RPN. You will see how a stack can be used to evaluate an RPN expression and also how to convert "by hand" from the parenthesized form to the RPN form (which is what the Brand X calculator does internally). Design section 1 will give an algorithm for this conversion.

Polish notation got its name from the Polish mathematician Jan Lukasiewicz, who first published an article on it in 1951. Lukasiewicz was more interested in mathematical logic than in computers *per se* (computers weren't very widespread in the early 1950s!), and his notation was developed as a convenient, *parenthesis-free* way to represent logical expressions. Today, of course, Polish notation embraces widespread computer usage.

Polish notation is widely used in interpreters and compilers as an intermediate representational form for statements. (A hand-held calculator is nothing but a kind of interpreter.) The term *reverse* or *postfix* is used to indicate that an operator *follows* its operands, instead of appearing in between them, as in infix notation. There is also a *prefix* or *forward Polish notation* in which the operator precedes its operands. We consider this latter form in an exercise.

Sec. 5.5   Stacks, Evaluation of Expressions, and Polish Notation   145

A bit later on, we shall consider how to convert an infix expression to its corresponding RPN form; for the moment, examine the examples of RPN in Figure 5–14. In each case, brackets are used to indicate the two operands of each operator; they are not part of the RPN. Notice that the numerical quantities in each RPN expression occur *in the same order* as they do in the original infix form. This is always true.

5.5.1 Evaluating RPN Expressions

We evaluate an RPN expression by a left-to-right scan. Since in RPN an operator is preceded by its two operands, we need a way to remember what the operands are until we encounter the applicable operator. This is easy if the expression has only one operator in it—like 3 5 +, for instance. We store the first operand, 3, store the second operand, 5, and then, when the operator arrives, determine that it is indeed + and add the two operands together, getting 8 as the result.

But suppose the RPN has more than one operator, as, for example, in the expression 3 5 + 10 × (the equivalent, in infix notation, of (3 + 5) × 10, or 80). If we scan this expression from left to right, we store 3, then store 5, and then add them, getting 8 as before. But what do we do with the 8? In fact, we need to store it, then store the 10, then discover the × and multiply the 8 by the 10, getting 80.

Evidently, then, we have to save intermediate results as well as input numbers, and then, when we see an operator, apply it to the *last two* things we stored. The expression 3 5 2 × − (equivalent, in infix notation, to 3 − (5 × 2), or −7) makes this even clearer. We need to store the 3, then the 5, and then the 2. Then, when the × is scanned, we multiply the last two numbers stored (2 and 5), and store this intermediate result. Finally, when the − arrives, we have two operands for it: 3 and the intermediate result from the multiplication, 10, so we get −7.

Notice that in evaluating RPN expressions we have been saving values in such a way that the last two values saved become the first two retrieved. This is a perfect application for a stack. Accordingly, let us represent the RPN expression

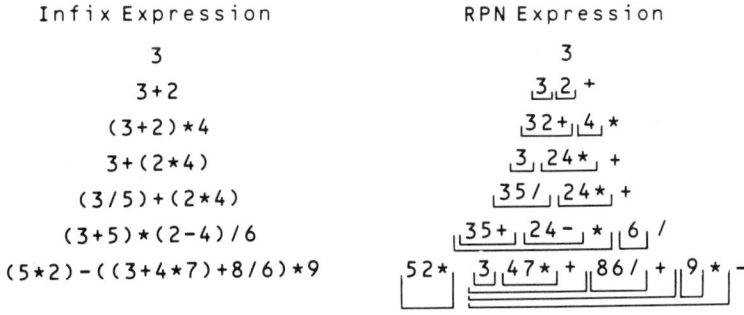

Figure 5–14 Infix expressions and their corresponding RPN forms

```
PROCEDURE EvalExp(X: Text) : INTEGER;

   VAR
      C: CHAR;
      T: Text;
      S: Stack;
      Y: INTEGER;
      Z: INTEGER;
      WeirdChar: BOOLEAN;

   BEGIN
      StackInit(S);
      TextCopy(T,X);
      IF TextEmpty(T) THEN
         RETURN 0;
      END;

      REPEAT
         C := TextHead(T);
         CASE C OF

            '0'..'9' :
               StackPush(S, ORD(C) - ORD("0"))  |

            '+' ,'-', '*', '/' :
               Y := StackTop(S); StackPop(S);
               Z := StackTop(S); StackPop(S);

               CASE C OF
                   '+' : StackPush(S, Z+Y)
                |  '-' : StackPush(S, Z-Y)
                |  '*' : StackPush(S, Z*Y)
                |  '/' : StackPush(S, Z DIV Y)
               END;

            ELSE
               WeirdChar := TRUE;
         END;

         T := TextTail(T);
      UNTIL TextEmpty(T);

      RETURN StackTop(S);

   END EvalExp;
```

Figure 5-15 Function for evaluating a numerical RPN expression (simulation of a Brand Y hand-held calculator)

in the form of a Text object using the text-handler module of Chapter 1, and assume that it is well formed, that is, that it follows the RPN rules for forming an expression. For simplicity, we use single digits to represent numbers. We also use the array-stack module discussed earlier, coded so that a stack can hold integers. A Modula-2 function is shown in Figure 5–15.

Here is how the algorithm works. We scan the text from left to right, removing the first character as we go and checking to see whether it is a number (single digit) or an operator, i.e., $+$, $-$, $\times$, or $/$. If it is a number, we need to convert it to its integer form so that we can do arithmetic with it, and then push it onto the stack. If it is an operator, we remove the top two items from the stack, do the operation, and then push the result back onto the stack. Assuming we started with a legal RPN expression, when all the characters in it have been examined, the final value will be the only value left on top of the stack.

Figure 5–16 shows the evaluation of an RPN expression by this algorithm; try out the algorithm "by hand" on a number of examples to be sure how it works. Note that it works correctly even when the RPN expression is just a single digit.

5.5.2 Converting Manually from Infix to RPN Form

The preceding algorithm works with RPN expressions containing only numeric values. However, most applications need to work with expressions containing variables as well. Consequently, we need a more general understanding of an expression. For our purposes, an arithmetic expression consists of identifiers or variable names, limited to single letters for simplicity; numerical constants, limited to one-digit integers for simplicity; the operators $+$, $-$, $\times$, and $/$, which have their familiar meanings; and parentheses. Also, we consider, at least at first, only fully parenthesized expressions, that is, expressions where parentheses are *always* used to indicate the order in which operations are to be performed.

An RPN expression is either a single variable or constant, or it is two RPN expressions followed by an operator. The last (rightmost) operator in the expression is the "main" operator of the expression, that is, the operator whose operation is performed *last* as the expression is evaluated, to produce the final result of the evaluation. To give a few examples, Figure 5–17 shows the RPN expressions for A, $A - B$, $(A - B) + C$, $A - (B + C)$, and $(A + B) \times (C - D)$. Note how these are constructed, and observe especially how $(A - B) + C$ and $A - (B + C)$ give rise to *different* RPN expressions: in $(A - B) + C$ the $+$ is the main operator, since it is performed *last*; in $A - (B + C)$ the $-$ is the main operator. If numerical values were assigned to A, B, and C, say, 2, 3, and 4, respectively, the result of evaluating $(A - B) + C$ would be 3, whereas the result of evaluating $A - (B + C)$ is -5. As an exercise, try evaluating $(A \times B) - (C + (D/E))$ and $((A - B) + (C/D)) \times E$.

We can now relax the condition that expressions be fully parenthesized. To do so, we replace the rules regarding parentheses with rules concerning the order

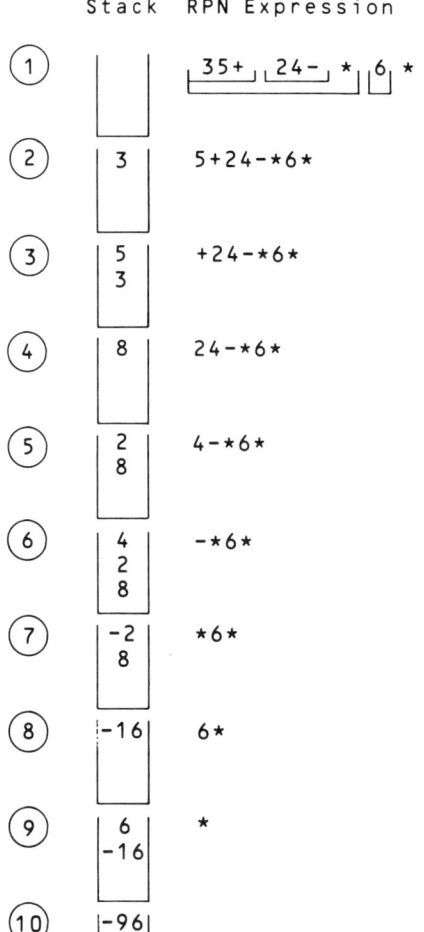

Figure 5-16 Evaluation of an RPN expression. Each "snapshot" is taken just before the leftmost input character is examined.

in which operations will be performed. For example, in the expression $A - B - C$, how do we know whether to evaluate it as though it were $(A - B) - C$, or as though it were $A - (B - C)$? Most programming languages use the rule that a sequence of $+$ and $-$ operations, without parentheses, is evaluated from *left to right*, so that $A - B - C$ is treated as though it were $(A - B) - C$ and $A - B + C$ is done as though it were $(A - B) + C$. We shall use such an *associative* rule, i.e., our addition and subtraction operators *associate* left to right. Figure 5-18 shows unparenthesized and parenthesized infix forms and their corresponding RPN expressions.

Sec. 5.5 Stacks, Evaluation of Expressions, and Polish Notation

Infix Expression	RPN Expression
A	A
A-B	AB-
(A-B)+C	AB-C+
A-(B+C)	ABC+-
(A+B)*(C-D)	AB+CD-*

Figure 5-17 Infix and RPN expressions

Unparenthesized	Assumed Parenthesized Form	RPN
A+B+C	(A+B)+C	AB+C+
A-B-C	(A-B)-C	AB-C-
W-X+Y	(W-X)+Y	WX-Y+
W+X-Y+Z	((W+X)-Y)+Z	WX+Y-Z+

Figure 5-18 Left-to-right associativity of + and −

The same associative rule applies to sequences of × and / operators: these are also evaluated in left-to-right order. Thus, $A/B/C$ is always done as though it were $(A/B)/C$, and $A/B \times C$ as though it were $(A/B) \times C$. Figure 5-19 shows several infix expressions whose only operators are × and / in both parenthesized and unparenthesized forms, together with their corresponding RPN expressions.

Left-to-right associativity is not the only possibility for evaluating the results of operations. Programming languages with a built-in exponentiation operator, often represented as **, usually apply a right-to-left rule for this operator, so that $A ** B ** C$ is treated as if it were $A ** (B ** C)$. An exercise deals with why this rule makes sense. We shall ignore exponentiation and use only left-to-right associativity.

What happens in the case of expressions where all four operators occur? These are usually handled by assigning *priorities* or *precedences* to the different operators. Usually, + and − have the same priority, and × and / have the same priority. For definiteness, let + and − be called priority 2 operators, and × and / be called priority 1 operators. Then, given two adjacent operators, one of priority 1 and the other of priority 2, the priority 1 operator will have its operation performed

Unparenthesized	Assumed Parenthesized Form	RPN
A*B*C	(A*B)*C	AB*C*
K/G/Z	(K/G)/Z	KG/Z/
Q/S*D	(Q/S)*D	QS/D*
P*D/E*K	((P*D)/E)*K	PD*E/K*

Figure 5-19 Left-to-right associativity of * and /

Unparenthesized	Assumed Parenthesized Form	RPN
A+B*C	A+(B*C)	ABC*+
A*B+C	(A*B)+C	AB*C+
A+B*C+D	A+(B*C)+D	ABC*+D+

Figure 5-20 Operator priorities

Original	Assumed Parenthesized Form	RPN
A+B-C+D	((A+B)-C)+D	AB+ C- D+
A-(B+C)*D	A-((B+C)*D)	A BC+ D * -

Figure 5-21 Parenthesized expressions

first. Thus, the expression $A + B \times C$ will be evaluated as though it were parenthesized $A + (B \times C)$; and the expression $A/B - C$ will be evaluated as though it were parenthesized $(A/B) - C$. That is to say, in the first expression $+$ is the main operator, in the second $-$. These expressions and their RPN forms are shown in Figure 5-20.

It is now easy to see how to manually convert an arbitrary infix expression, in which parentheses are *sometimes* used to group subexpressions, into RPN: just add the necessary parentheses (on paper until you have gained enough experience to do it by inspection), and then, follow the preceding rules, produce the RPN from the fully parenthesized version.

To give a couple of examples, first consider $A + B - C + D$. Since adjacent operators of equal priority are handled from left to right, we parenthesize as $((A + B) - C) + D$. Now look at $A - (B + C) \times D$. Here the two adjacent operators of interest are $-$ and $\times$ ($+$ doesn't count because it's inside a subexpression), the latter of which is performed first because it is priority 1. So this expression is handled as though it were $A - ((B + C) \times D)$. The corresponding RPN expressions are shown in Figure 5-21. As an exercise, try putting $A - B \times C/(D - E)$ and $A \times B - (C + D) + E$ in RPN.

5.6 DESIGN ONE: AN INFIX-TO-RPN TRANSLATOR

In this section we shall develop a program to do algorithmically what we did manually in the previous section: translate an expression from infix form to RPN. The program will be a function taking the infix expression as its input and returning the RPN expression as its result.

Consider first an *unparenthesized* expression with all operators of the same priority and left-to-right associativity. In such an expression the operators and operands alternate, and the operands in the RPN appear in the same order as in the original. Assume that the input expression is implemented as a Text object T as defined in the text-handler package discussed earlier. We represent the

Sec. 5.6 Design One: An Infix-to-RPN Translator

	RPN	OP	Input Expression
①			A+B-C+D
②	A		+B-C+D
③	A	+	B-C+D
④	AB	+	-C+D
⑤	AB+	-	C+D
⑥	AB+C	-	+D
⑦	AB+C-	+	D
⑧	AB+C-D	+	
⑨	AB+C-D+		

Figure 5-22 Simple infix-to-RPN translation; each "snapshot" is taken just before the leftmost input character is examined.

resulting RPN expression R as a Text object as well. In translating the input expression, we scan it from left to right. If the first character we see is an operand, we can immediately output it, beginning the RPN string. If it is an operator, we need to remember it until after we have seen both of its operands, which will not be until after the next operator is scanned, at which time we output the saved operator and save the new one. An example of this scanning procedure is shown in Figure 5-22, and a Modula-2 function is given in Figure 5-23.

Now assume that operators of different priorities are allowed, and consider the infix expression $A + B \times C$. Its RPN form is $A\ B\ C \times +$. Consequently, we cannot just output the $+$ when the B is scanned, because the $\times$, having higher priority, must be performed first. So the $+$ must be remembered longer, and we need to tackle the problem a bit more systematically. That is to say, the priority of the incoming operator needs to be checked against the priority of the previous one; then, if the new operator has higher priority, we need to remember *it* as well as the previous one, until we've scanned *its* second operand. Only after its second operand has been scanned and output, can we output the operator.

Observe that in translating the given infix expression, we have had to remember two operators, and the last one remembered is the first one output. This suggests that the computer implementation of what we have done by inspection ought to be a stack, which, after all, is precisely a LIFO device. This is shown in Figure 5-24, where an example is worked through.

Figure 5-25 gives a simple function for determining the priority of an operator in the set $\{+, -, \times, /\}$. A modified version of the infix-to-RPN algorithm which uses priorities is shown in Figure 5-26, where it is assumed that a package implementing stacks of characters is available. In this algorithm, an operator is stacked until one of equal or lower priority comes along, and then it is popped and added to the RPN. The new operator is then pushed onto the stack. The process

```
PROCEDURE RPN(X: Text) : Text;

VAR
    C:          CHAR;
    Op:         CHAR;
    T:          Text;
    Result:     Text;
    WeirdChar: BOOLEAN;
BEGIN
    TextCopy(T,X);
    Result := NullText();
    Op := ' ';
    IF TextEmpty(T) THEN
        RETURN NullText();
    END;

    LOOP
        C := TextHead(T);
        CASE C OF
           'A'..'Z':
               Result := AppendChar(Result,C)    |
           'a'..'z' :
               Result := AppendChar(Result,C)    |
           '0'..'9' :
               Result := AppendChar(Result,C)    |
           '+' , '-' , '*' , '/' :
               IF Op = ' '    THEN
                   Op := C;
               ELSE
                   Result := AppendChar(Result,Op);
                   Op := C;
               END
        ELSE
            WeirdChar := TRUE;
        END;
        T := TextTail(T);
        IF TextEmpty(T) THEN EXIT END;
    END;

    Result := AppendChar(Result, Op);

    RETURN Result;

END RPN;
```

Figure 5-23 Infix-to-RPN translator (simplest case, no priorities)

continues until the input is empty, at which time the stack is emptied of all remaining operators. This function calls an auxiliary one to return the priority of an operator. Try the function on a few made-up examples to understand its operation.

The final modification accommodates parentheses. How this is done becomes clear when it is realized that parentheses really *override* the priority scheme, essentially creating a whole new expression inside them. We can thus allow parentheses by changing our algorithm so that it pushes a *left* parenthesis onto the stack, creating a sort of "false bottom" in the stack. The algorithm then progresses as in the previous case, but when a *right* parenthesis is seen, the stack is emptied as far back as the "false bottom," and then the "false bottom" is discarded.

In Figure 5-27 an example involving parentheses is worked out; we leave it as an exercise to modify the algorithm so that it can handle parentheses.

Sec. 5.6 Design One: An Infix-to-RPN Translator

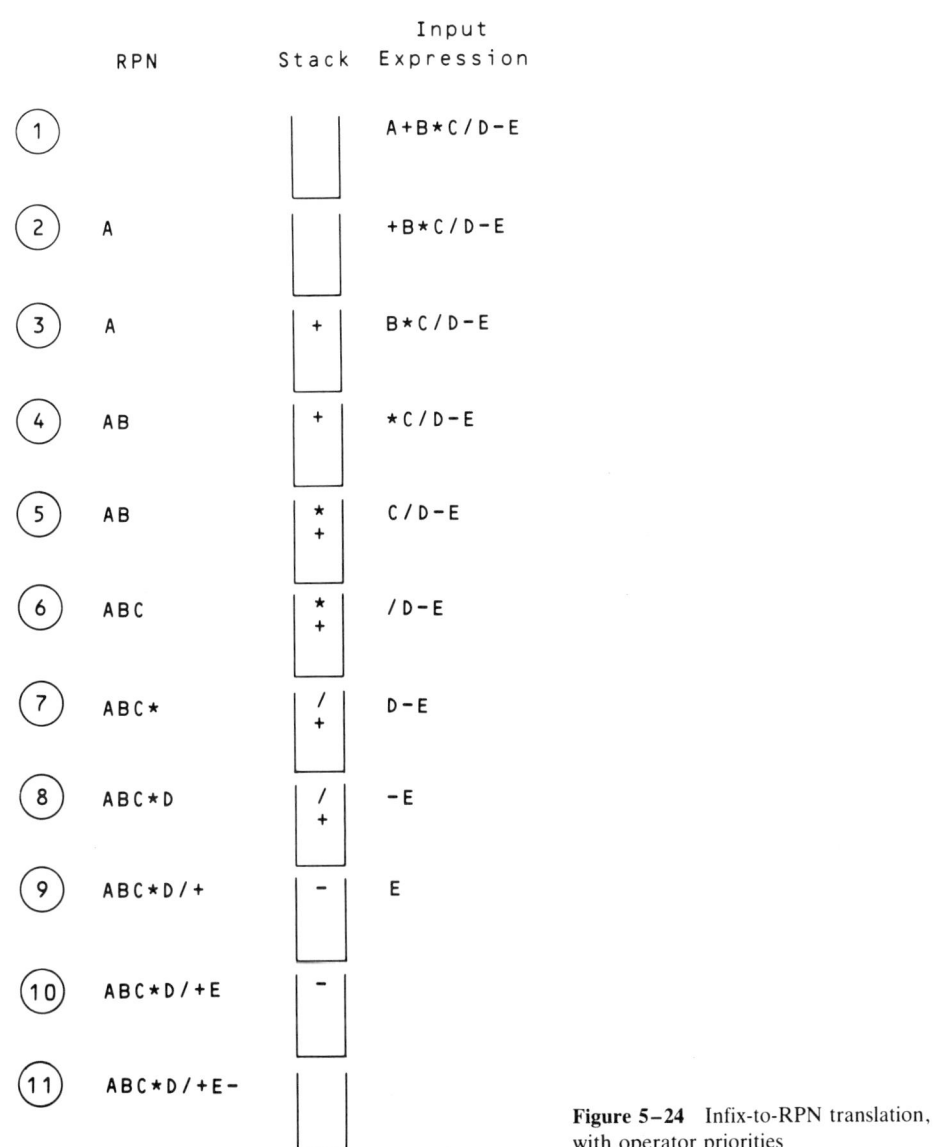

Figure 5–24 Infix-to-RPN translation, with operator priorities

```
PROCEDURE Priority(Operator: CHAR): INTEGER;

BEGIN
    CASE Operator OF
        '+' , '-'  : RETURN 1  |
        '*' , '/'  : RETURN 2
        ELSE         RETURN 0
    END;
END Priority;
```

Figure 5-25 Function to determine operator priority

```
PROCEDURE RPN(X: Text): Text;

    VAR
        C:         CHAR;
        D:         CHAR;
        T:         Text;
        Result:    Text;
        WeirdChar: BOOLEAN;
        ST:        Stack;

    BEGIN
        T := NullText();
        StackInit(ST);
        TextCopy(T,X);
        Result := NullText();
        IF TextEmpty(T) THEN
            RETURN NullText();
        END;

        LOOP
            C := TextHead(T);
            CASE C OF
                'A'..'Z':
                    Result := AppendChar(Result,C)
              | 'a'..'z' :
                    Result := AppendChar(Result,C)
              | '0'..'9' :
                    Result := AppendChar(Result,C)
              | '+' , '-' , '*' , '/' :
                    LOOP    (* CLEAR HIGHER-PRIORITY OPERATORS *)
                        IF StackIsEmpty(ST)      OR
                           (Priority(StackTop(ST)) < Priority(C)) THEN
                            EXIT
                        ELSE
                            Result := AppendChar(Result,StackTop(ST));
                            StackPop(ST);
                        END;
                    END;
                    StackPush(ST,C)
                ELSE
                    WeirdChar := TRUE;
            END;
            T := TextTail(T);
            IF TextEmpty(T) THEN EXIT END;
        END;

        WHILE NOT StackIsEmpty(ST) DO
            Result := AppendChar(Result, StackTop(ST));
            StackPop(ST);
        END;

        RETURN Result;

    END RPN;
```

Figure 5-26 Infix-to-RPN translator which supports priorities

Sec. 5.7   Design Two: An Event-Driven Simulation   155

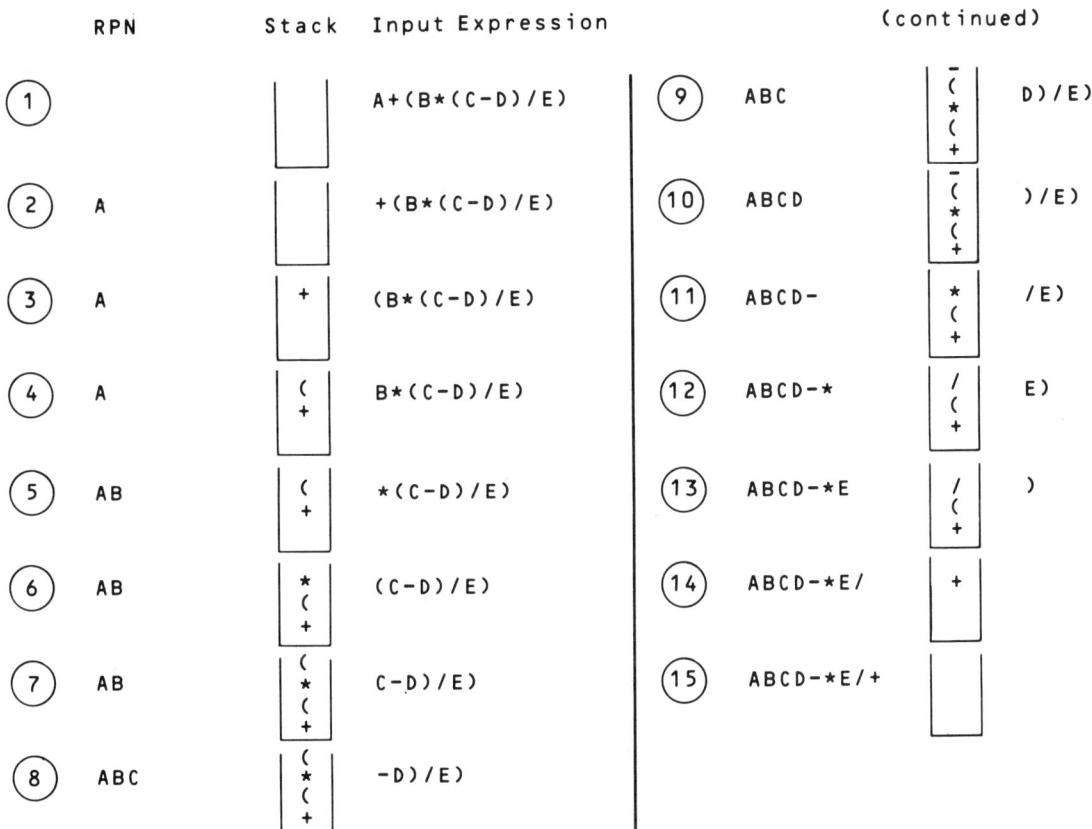

Figure 5-27   Infix-to-RPN translation, with priorities and parentheses

5.7 DESIGN TWO: AN EVENT-DRIVEN SIMULATION

As an example of the application of queues, we consider the simulation of a real-life situation in which people must wait in line for some service. It might be a bank, post office, or supermarket checkout. For definiteness, we choose the last.

A supermarket manager must think carefully about the number of checkout lines which will be open at a given time. Clearly enough lines must be open to permit a customer to check out in a reasonable amount of time, otherwise the shopper will find another store with shorter lines. On the other hand, the cashiers must be paid a wage, so the manager doesn't want unnecessary lanes to be open. A computer simulation of the store at different levels of shopping traffic can aid the manager in finding "just the right number." In a simulation of this type, we try to model the real-world situation as closely as possible with our program objects and algorithms.

Here is the scenario: a shopper arrives at the checkout area of the store at a certain time of day with a certain number of items in the shopping cart. The shopper finds the shortest line and joins it. For simplicity, we shall assume that the shopper cannot see into other shoppers' carts, and therefore the choice of line is not influenced by how full or empty they are. Another simplifying assumption is that the path to the checkout area is narrow and therefore two shoppers cannot enter it at the same instant. We also assume that no shopper gets tired of waiting and abandons a cart, leaving the store without checking out.

We shall represent the time of day as an integer representing the number of time units since the store opened that day, and shall assume that each item requires an average of one time unit to ring up and put in a bag. We define *average checkout time* as the sum of the length of time a shopper waits in line and the length of time to check out all his or her items. The goal of the simulation is to find, for a given store opening period, and a given group of shoppers and cart loads, the average checkout time as a function of the number of open lines.

To set up the simulation we provide a set of NumQueues FIFO queues, each representing one checkout line in the supermarket. We define *departure time* as the time when a customer reaches the front of his or her queue, departs from that queue, and begins being checked out by the cashier. Thus, the first customer in a queue is *waiting* to be served; the customer actually *being* served is thought of as having already left the queue. This assumption might seem unrealistic, but it makes sense if you consider a kind of queueing scheme currently in fashion in banks and post offices, where there is only one queue but many tellers. In such a system, a customer leaves the queue to be served by the next available teller. The assumption that a customer leaves the queue just before being served allows the simulation model to be changed easily to accommodate the single-queue scheme just described.

How shall our simulation program operate? In a real supermarket all the people are independent processes needing no external control; in a program we need a control mechanism. This kind of simulation, in which there are a number of queues all moving at different rates, can be controlled by means of an *event list*, and is called an *event-driven simulation*.

There is no direct supermarket analogy to the event list; it is a special queue containing scheduled arrival and departure events. The event list is not a FIFO queue: its nodes must be sorted by time. The enqueue operation must insert a node into its proper position instead of just adding it to the end of the queue; the dequeue operation remains unchanged. A queue operated in this fashion is an example of a *priority queue*; in this case the item with earliest time is processed with highest priority.

The event list contains, at any given time, no more than one arrival event, and at most NumQueues departure events, one for each of the checkout queues. The event list is initialized with the first arrival record, and the simulation proceeds, processing arrival and departure events, until the event list is empty.

When an arriving shopper record is read from a file, an *arrival event* is placed on the event list (sorted by time, remember, because there may be departures

Sec. 5.8   Summary

already scheduled). When the arrival event reaches the front of the event list, it is removed and joins the shortest checkout queue. If it is the only customer in the queue, it can be served immediately: its arrival and departure times are the same and a *departure event*, indicating the scheduled departure time and queue number, is placed on the event list. At this point another arrival record is read from the file to replace the one just removed from the event list.

When a departure event reaches the front of the event list, we remove the first node from the corresponding queue, say queue k. We know its arrival time, its time of departure from the queue, and its time to process all its items, so we can calculate its checkout time and add it to a grand total from which we will later compute the average. We can also compute the scheduled departure time for the next customer in queue k: since the next customer begins being served just as the previous customer finishes, the next customer's departure time is the sum of the current customer's departure time and the current customer's processing time. Having computed the scheduled departure time for the customer at the front of queue k (the customer waiting to be served), we place the associated departure event on the event list.

Figure 5-28 shows a sketch of the simulation; Figure 5-29 sketches out the necessary types. Note that two different kinds of queues are used here: FIFO's to represent the checkout lines and a priority queue to implement the event list. In Figure 5-30 and Figure 5-31 we show more details of the procedures to process arrival and departure events respectively. Completing the simulation is left as an exercise.

```
MODULE Simulation; (* sketch of event-driven simulation *)

   (* import lists here *)

   (* declarations here *)

   PROCEDURE ProcessArrival(ArrivalTime: INTEGER;
                            NumItems:    INTEGER);
   (* procedure body here *)

   PROCEDURE ProcessDeparture(DepartureTime: INTEGER;
                              k: (*queue*): INTEGER);
   (* procedure body here *)


   BEGIN (* main simulation *)

      (* initialize queues, event list, other variables *)

      (* read ArrivalTime and NumItems from customer file *)
      (* place arrival event on event list *)

      WHILE (* there are still events on the event list *) DO

         (* remove first event from event list *)

         IF (* this event is an arrival event *) THEN
            ProcessArrival( (* pass time, number of items *) );

         ELSE (* it is a departure event *)
            ProcessDeparture( (* pass time, queue number *) );

         END;

      END;

      (* compute and print average checkout time *)

END Simulation.
```

Figure 5-28 Main loop of event-driven simulation

Sec. 5.8   Summary

```
CONST NumQueues = ...; (* number of open checkout lines *)

TYPE CustomerRecord =  (* put these records on the queues *)

   RECORD
      ArrivalTime: INTEGER;
      NumItems:    INTEGER;
   END;

TYPE Queue = ...;      (* choose a suitable implementation for
                          the FIFO queues; remember that the
                          program needs the queue length at
                          frequent intervals *)


VAR  Queues: ARRAY[1..NumQueues] OF Queue;

TYPE EventType = (Arrival, Departure);

TYPE Event =           (* these records go on the event list *)

   RECORD              (* remember that an arrival event holds
     ...                  an arrival time and number of items
   END;                   and a departure event holds a departure
                          time and a queue number *)


TYPE EventList = ...   (* choose a suitable implementation;
                          remember that the event list is a
                          priority queue, so the enqueue
                          operation must insert a node
                          in order by increasing time *)
```

Figure 5-29   Type declarations for simulation

```
PROCEDURE ProcessArrival(ArrivalTime: INTEGER;
                         NumItems:    INTEGER);
(*  sketch of procedure to process arrival event *)

   BEGIN

      (* store ArrivalTime and NumItems in a queue node p *)

      (* find k, index of shortest queue *)
      (* enqueue node p on Queues[k] *)

      IF (* p is the only node on Queues[k] *) THEN

         (* this customer can be served immediately, so
            departure time = arrival time; therefore
            place departure event (from Queues[k])
            on event list (ordered by time) *)

      END;

      IF (* customer file is not empty *) THEN

         (* read ArrivalTime and NumItems from customer file
         (* place arrival event on event list (ordered by time) *)

      END;

   END ProcessArrival;
```

Figure 5-30   Procedure to process an arrival

```
PROCEDURE ProcessDeparture(DepartureTime, k: INTEGER);
(* sketch of procedure to process departure from queue k *)

    BEGIN

        (* dequeue node from Queues[k]; store its info in
           ArrivalTime and NumItems *)

        (* calculate elapsed checkout time *)
        CheckoutTime := DepartureTime + NumItems - ArrivalTime;

        (* update values which contribute to the average *)
        TotalCheckoutTime := TotalCheckoutTime + CheckoutTime;
        NumCustomers := NumCustomers + 1;

        IF (* Queues[k] is not empty *) THEN

            (* compute departure time for next node *)
            NextDepartureTime := DepartureTime + NumItems;

            (* place departure event (from queue k) on event list
               using NextDepartureTime, ordered by time *)

        END;

    END ProcessDeparture;
```

Figure 5-31 Procedure to process a departure

5.8 SUMMARY

Stacks and queues are two important data structures with restricted access. The stack is a last in, first out (LIFO) device; the queue is a first in, first out (FIFO) device. The FIFO and LIFO access methods have many uses in computing applications. For example, stacks are used in implementing general procedure calling and returning, and in language translation involving reverse Polish notation. Queues turn up in operating systems, and in any number of simulation problems where the physical system being simulated involves waiting in line.

A stack or queue can be implemented using either an array, if its maximum size is known at compilation time, or a linear linked list, if its size is not so known.

The next two chapters present two more important structures: the graph and the tree, the latter being a special kind of graph. The stack and queue modules presented here will be re-used in the following chapters, as service modules for the applications to be discussed there.

CHAPTER 5 EXERCISES

1. Explain why, in a circular queue with QueueMaximum positions, no more than QueueMaximum-1 of these positions can actually be occupied.
2. Write a module which implements stacks by one-way linked lists.
3. Some authors define Pop(VAR S: Stack; VAR I: Stack Item) as a combined selector and constructor which adjusts the stack *and* returns a value. Show that this can be implemented in terms of our StackTop and StackPop, and therefore that ours are more "primitive."
4. Some authors recommend that an operation to remove an item from a queue also return the item. Show that this can be implemented in terms of our QueueFirst and Dequeue operations.
5. In some applications where two stacks are necessary, a bit of space can be saved by using an array representation but allowing the two stacks to share the same array. This is done by having one stack fill from the low-subscript end of the array forward, and the other stack fill from the high-subscript end backward. An exception must be raised if the two stacks "collide" in the middle somewhere. Design and implement a module to handle such a "double stack." NOTE: strangely, this structure is often called a *deque*, for "double-ended queue."
6. In most programming languages which have a built-in exponentiation (**) operator, this operator associates right-to-left. That is, A ** B ** C is treated as though it were written A ** (B ** C). Explain why this is a sensible convention.
7. A common parenthesis-free notation is forward Polish or prefix Polish notation. In this scheme an operator precedes its operands, so that for example A + B becomes + AB. For the infix expressions of sections 5.5 and 5.6, find the forward Polish forms.
8. Modify the hand-held calculator simulation of section 5.5.1 so that the numbers in the input can have more than one digit. (You'll have to write a scanner that reads the digits in an integer literal and converts it to an integer number.)
9. Modify the Infix-to-RPN translator of section 5.6 so that parenthesized expressions are handled correctly.
10. Modify the Infix-to-RPN translator of section 5.6 so that the exponentiation operator is allowed and is handled correctly.

11. Write a translator that converts an infix expression to its forward Polish notation (FPN) form.
12. Write a translator that converts an RPN expression to its forward Polish notation (FPN) form.
13. Complete the event-driven simulation of section 5.7.
14. Modify the simulation so that the standard deviation of the average checkout time is computed along with the mean.
15. Many banks, post offices, and airline ticket counters have adopted a scheme in which there is only one waiting line and a customer reaching the front of the line goes to whichever server is available. Change the simulation to support this scheme. For the same set of customer records, are the mean and standard deviation the same as in the multi-queue scheme?
16. Change the simulation so that instead of arrivals coming from a file created in advance, they are generated by a random number generator. Let the transaction time be uniformly distributed over some reasonable interval; also let an arrival time occur a random (but reasonable) number of time units after the previous one.

Chapter 6
DIRECTED GRAPHS

6.1 GOAL OF THE CHAPTER

The graph is an important mathematical structure, and one applied widely in computing problems. A directed graph consists of a set of *points* or *vertices*, and a set of *edges* or *arcs*, which represent connections between the points. In this chapter, we shall consider a number of important mathematical properties of directed graphs and then examine the *adjacency matrix, adjacency list, weighted adjacency matrix*, and *state table* implementation techniques.

A *traversal* is a "walk" around a graph in a systematic fashion, in such a way that each vertex is officially "touched" or *visited* exactly once. We shall consider two important traversal algorithms: the *depth-first search* and the *breadth-first search*. These algorithms use the module for queues developed in earlier chapters.

The chapter concludes with the design of a very simple *lexical scanner*, represented as a state table.

One of the most important characteristics of the directed graph is that the *tree* is just a special case. Armed with an understanding of directed graphs, you will be able more readily to see in chapter 7 just what makes a tree a tree.

6.2 INTRODUCTION

A *graph* G is an ordered pair of sets $\langle V,E \rangle$, where V is a set of *vertices* (which may be thought of as points) and E is a set of *edges* (which may be thought of as lines

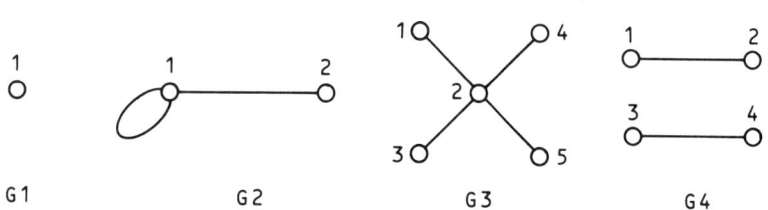

Figure 6-1 Some undirected graphs

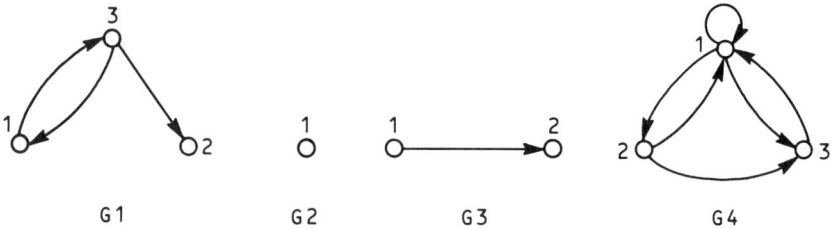

Figure 6-2 Some directed graphs

connecting the points). Some authors refer to vertices as *points* or *nodes*, and to edges as *arcs*. An edge is given as a pair {*m,n*}, where *m* and *n* are in the vertex set *V*. Notice that no direction is given to the edge, so that {*m,n*} and {*n,m*} represent the same edge. Figure 6-1 shows some (undirected) graphs.

A *directed graph G*, or *digraph*, is a graph $G = \langle V, E \rangle$ whose edges are *ordered pairs* $\langle s, d \rangle \in E$, where *s* and *d* are in the vertex set *V*. The fact that the pairs of edges are ordered imposes a directionality to each edge, which is why *G* is called a *directed* graph. The vertex *s* is called the *source* vertex, and the vertex *d* is called the *destination* vertex.

Figure 6-2 shows some directed graphs. Note particularly that in graph G1 the edge $\langle 1,3 \rangle$, for example, is *not* the same as the edge $\langle 3,1 \rangle$, because even though they connect the same pair of vertices their directions are different.

For convenience, we shall write "*sGd*" to mean "the edge $\langle s, d \rangle$ is in the edge set of *G*." If *sGd*, we say that *d is adjacent to s*. The set of all vertices adjacent to *s* is called the *adjacency set of s*.

6.3 PROPERTIES OF DIGRAPHS

There are a number of properties of digraphs which have important applications to computer programming. In defining these properties, we shall use *G* to refer to an arbitrary digraph, and lowercase letters to refer to vertices in *G*'s vertex set. Also, the abbreviation *iff* will be used to mean "if and only if," as is common in mathematics.

Sec. 6.3   Properties of Digraphs

Reflexivity. *G* is *reflexive* iff *xGx* for all vertices *x* in *V*. Intuitively, if we refer to ⟨*x,x*⟩ as a *self-loop*, the *G* is reflexive iff *every* vertex in *G*'s vertex set has a self-loop.

Irreflexivity. *G* is *irreflexive* iff *none* of its vertices has a self-loop. Note that it is quite possible for *G* to be neither reflexive nor irreflexive—namely, if *some*, but not *all*, of its vertices have self-loops. Be careful not to confuse the assertion "*G* is *not* reflexive" with the assertion "*G* is *ir*reflexive."

Figure 6–3 shows some digraphs which are reflexive, some which are irreflexive, and some which are neither.

Symmetry. *G* is *symmetric* iff, whenever *xGy*, then *yGx*. Note carefully that this statement does *not* assert that every pair of vertices must be connected

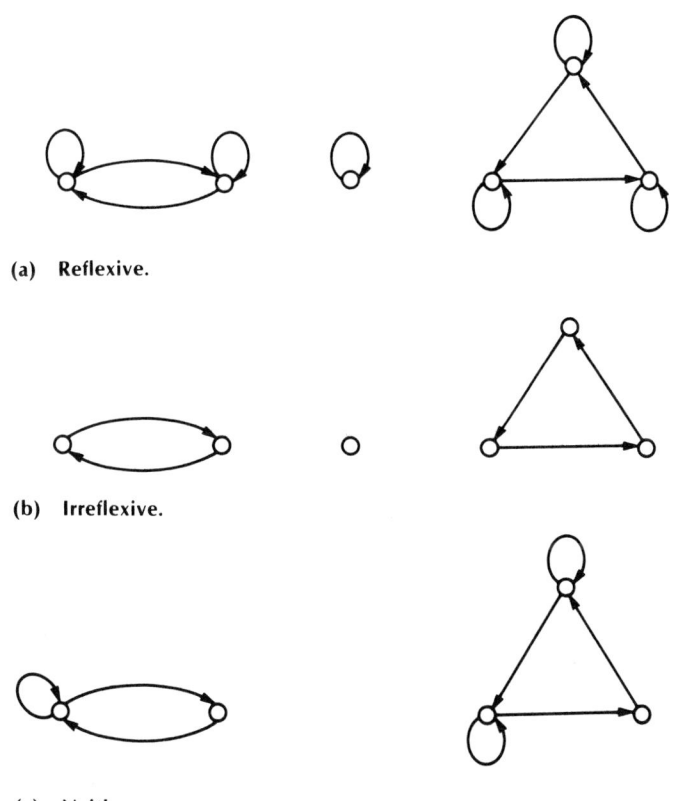

(a) Reflexive.

(b) Irreflexive.

(c) Neither.

Figure 6–3 Reflexivity and irreflexivity

by an edge; it says only that *if* there is an edge $\langle x,y \rangle$, then there must be an edge $\langle y,x \rangle$ for G to be symmetric. A consequence of this definition is that a digraph consisting of a single vertex with no edges (this is possible because nothing in the definition requires E to be nonempty) is symmetric. Such a case might be "pathological," but it does make the point.

Antisymmetry. G is *antisymmetric* iff, whenever xGy and yGx, then $x = y$. This is a way of saying that no two distinct vertices have edges in both directions, but that self-loops are permitted. As in the case of irreflexivity, be careful with language: saying "G is *not* symmetric" is not the same as saying "G is *anti*symmetric," since G may have *some* pairs of vertices with edges directed both ways and *some* pairs with edges directed only one way. Thus, G is neither symmetric nor antisymmetric. Speaking pathologically again, the digraph with one vertex and no edges is *both* symmetric *and* antisymmetric.

Some authors do not permit antisymmetric graphs to have vertices with self-loops. This assumption would make the definition of antisymmetry simpler: we could just say that if we have xGy, then we cannot have yGx. On the other hand, then a reflexive graph could never be antisymmetric—indeed, an antisymmetric graph would be *irreflexive*—and this would mix up two properties that we prefer to keep independent of each other.

In Figure 6-4, there are some symmetric digraphs, some antisymmetric ones, and some which are neither.

Transitivity. G is *transitive* iff, for each triple of vertices, whenever xGy and yGz, then xGz. In other words, if we can get from x to z by way of y, then we can get there directly if G is transitive. Note again that this does *not* mean that there must ever be edges $\langle x,y \rangle$ and $\langle y,z \rangle$: it says only that *if* there are such edges, then if G is to be transitive, there must be an edge $\langle x,z \rangle$. Also, since there is no requirement that x, y, and z be distinct, self-loops must be considered in determining transitivity. Is it possible for a digraph to be symmetric and transitive without being reflexive?

Figure 6-5 shows some transitive digraphs, and some others where it is explained why they are not transitive.

Paths. A *path* is a sequence of edges $\langle v_1,v_2 \rangle, \langle v_2,v_3 \rangle, \ldots, \langle v_{k-1},v_k \rangle$, i.e., a sequence of edges such that the destination of one edge is the source of the next. The path is *simple* iff all its vertices, except possibly the first and the last, are distinct. The *length* of the path is the number of *edges* (not vertices) in it. Thus, a single edge $\langle x,y \rangle$ is a path of length 1, and hence, so is a self-loop $\langle x,x \rangle$. If there is a path *from x to y*, we say that *y is reachable from x*.

Cycles. A *cycle* is a path such that the destination of its last edge is the source of its first edge. (It gets back to where it started.) Thus, a self-loop is a

Sec. 6.3   Properties of Digraphs

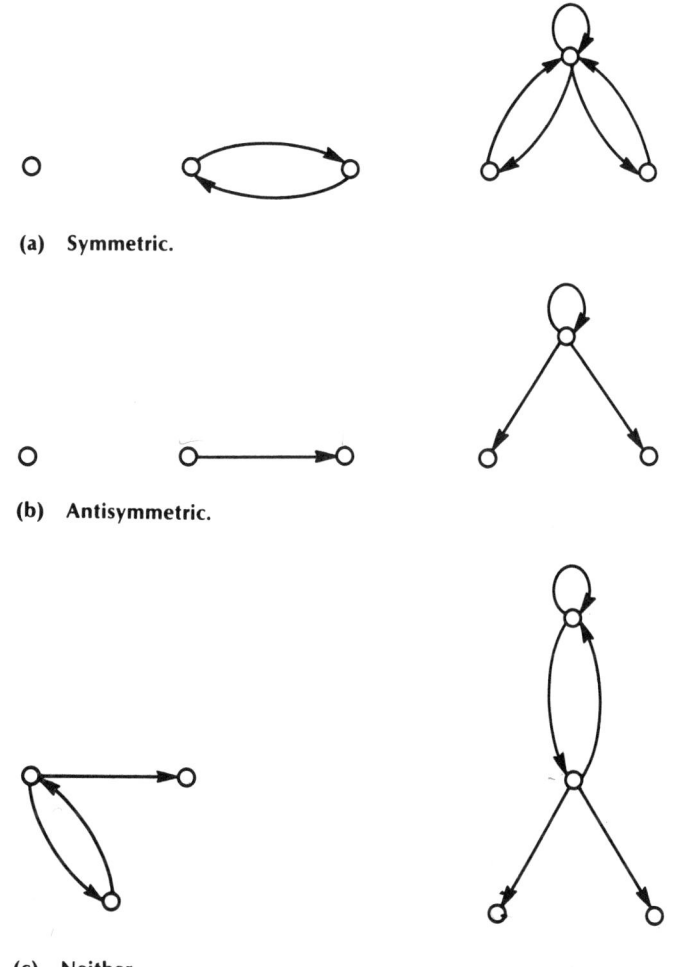

(a) Symmetric.

(b) Antisymmetric.

(c) Neither.

Figure 6-4 Symmetry and antisymmetry

cycle of length 1. A digraph is *acyclic* iff it has no cycles in it. A *simple cycle* is a simple path which is a cycle.

Connectivity. Intuitively, a graph, (directed or not), is *connected* iff it is "all one piece." In other words, treating all edges as though they were two-way, or, equivalently, adding edges to make the graph symmetric, a digraph is connected

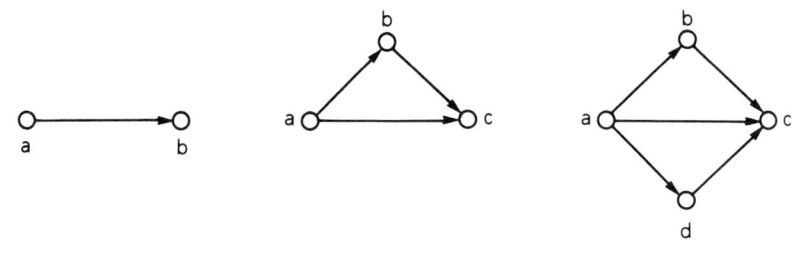

(a) Transitive.

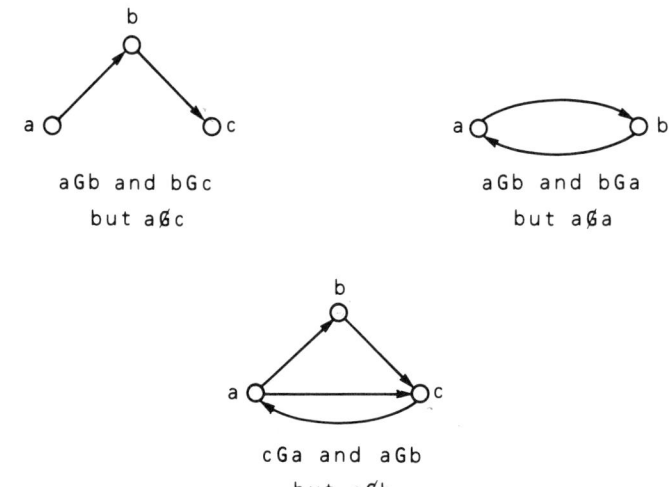

(b) Not transitive.

Figure 6-5 Transitivity

iff we can find a path from any vertex to any other vertex. The "pieces" of a graph which is in several pieces are called *connected components*.

Strong connectivity. A digraph G is *strongly connected* iff, from each vertex, there is at least one path (not necessarily of length 1) to each of the other vertices. Can a strongly connected digraph ever be acyclic?

Figure 6-6 illustrates both connectivity and strong connectivity.

In-degree and out-degree. The *in-degree* of a vertex z in a digraph G is the number of edges which have z as their destination (visually, the number of

Sec. 6.4 Implementations of Directed Graphs

(a) Not connected.

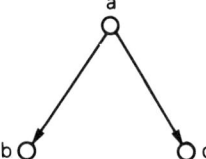

(b) **Connected but not strongly connected** (a cannot be reached from b or c).

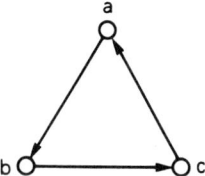

(c) **Strongly connected.**

Figure 6–6 Connectivity and strong connectivity

arrowheads arriving at z). The *out-degree* of a vertex z is the number of edges with z as their source (the number of arrow tails leaving z).

6.4 IMPLEMENTATIONS OF DIRECTED GRAPHS

6.4.1 Adjacency Matrix

The most straightforward way to represent a digraph G with K vertices is by a $K \times K$ Boolean matrix G' called the *adjacency matrix*, where $G'(x,y)$ = TRUE iff xGy, and FALSE otherwise. Thus, row x of the matrix indicates the adjacency set of vertex x.

For this matrix, it is easy to determine whether y is adjacent to x, and this is done in $O(1)$ time, since only a subscript calculation is involved. But a disad-

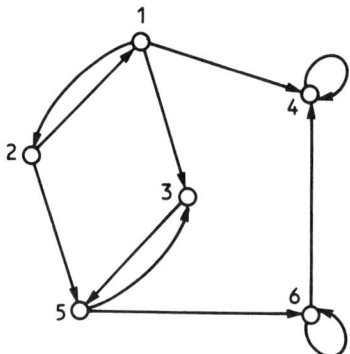

(a) A digraph.

	1	2	3	4	5	6
1	F	T	T	T	F	F
2	T	F	F	F	T	F
3	F	F	F	F	T	F
4	F	F	F	T	F	F
5	F	F	T	F	F	T
6	F	F	F	T	F	T

(b) Adjacency matrix for this digraph.

Figure 6-7 Adjacency matrix for a digraph

vantage of using the matrix is that, even if the graph has few edges, K^2 cells are needed to store it, and any algorithm to examine or even merely read or print the whole graph must have a performance $O(N^2)$. Figure 6-7 shows a digraph and its adjacency matrix.

6.4.2 Adjacency List

In most graphs, the vertices have relatively small adjacency sets, so the adjacency matrix is sparse and most of its elements are FALSE. For this reason, a variant of the sparse-matrix technique called an *adjacency list* is often used to implement

Sec. 6.4 Implementations of Directed Graphs

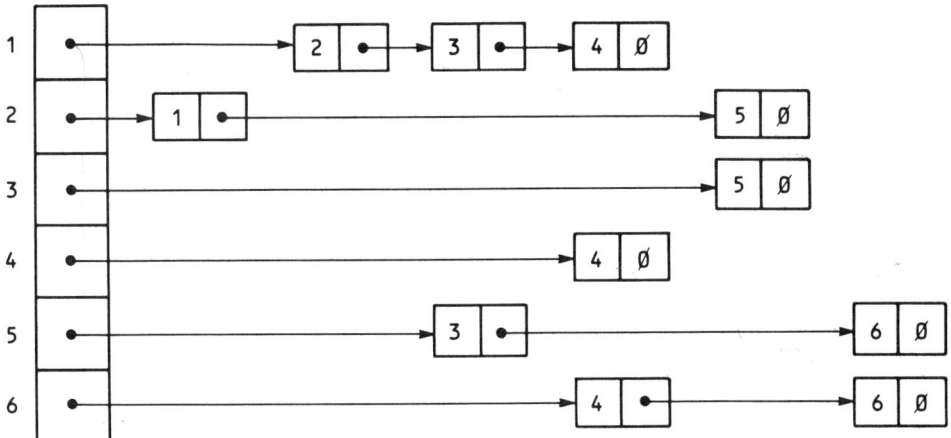

Figure 6-8 Adjacency list structure for the digraph of Figure 6-7

a digraph. In an adjacency list, each vertex x is a header for a linear list, each node of which represents a destination vertex for edges leaving x. The headers can be stored in an array. The structure of an adjacency list is shown in Figure 6-8.

Assuming that Booleans, pointers, and integers identifying vertices all occupy the same number of bytes of storage, when is the adjacency list more economical than the adjacency matrix? To answer this question, let L be the average number of cells in a single vertex list. Then the pointer array requires K cells, each of one storage unit, each list cell requires two storage units, and there are $K \times L$ such cells. So altogether, the structure requires $K + 2 \times K \times L = K \times (1 + 2 \times L)$ storage units. To find the crossover point, we merely set $K^2 = K \times (1 + 2 \times L)$, or $L = (K - 1)/2$.

Often we cannot in fact assume that Booleans, pointers, and vertex identifiers are all the same size in storage: many programming languages give the programmer a way of implementing an array of Booleans such that each array entry is represented by a single bit. In such a situation, the "dense" matrix (two-dimensional array) can be considerably more economical in terms of space required than the "sparse" matrix (list).

There is still the question of performance, however. To print the entire adjacency list takes $O(K \times L)$ operations, which is usually much less than $K \times K$. On the other hand, just determining whether xGy takes $O(L)$ operations (on average), whereas it was $O(1)$ in the adjacency matrix representation. We have here another clear example of the tradeoffs inherent in selecting implementations of abstract objects.

6.4.3 Weighted Adjacency Matrix

The implementations just described give the *structure* of a digraph, but provide no information about its *content*. In many graphs, applications of vertices or edges are associated with data values of one kind or another, often called *weights*. For example, Figure 6-9 shows a digraph with numbers attached to the edges. One interpretation of such numbers might be that they represent the distances between points on a graph that in turn represents a road map. Another might be that they represent the time required to perform a certain task in a complex project. Implementing such a weighted graph is a straightforward extension of the adjacency matrix: each entry of the matrix contains the weight instead of a Boolean, and entries which correspond to missing edges contain some indication to that effect, for example zero or NIL.

Weights can also be used in an adjacency list implementation: weights for edges emanating from a given vertex are simply stored in the vertices of the corresponding adjacency list.

6.4.4 State Table

A special kind of weighted digraph, the *state graph*, comes from the field of abstract machine theory and is useful in hardware design and in building language trans-

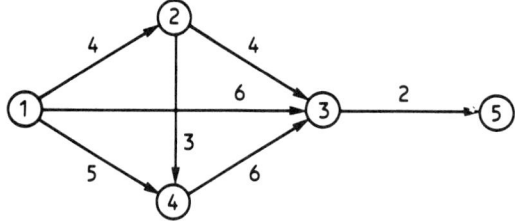

(a) A weighted digraph.

	1	2	3	4	5
1	0	4	6	5	0
2	0	0	4	3	0
3	0	0	0	0	2
4	0	0	6	0	0
5	0	0	0	0	0

Figure 6-9 Weighted digraph and adjacency matrix

Sec. 6.5   Graph Traversals

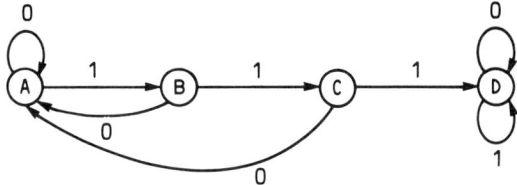

Figure 6-10   A state graph

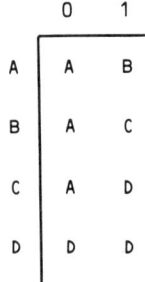

Figure 6-11   A state table

lators. We shall discuss an application of state graphs in building scanner programs shortly, but for the moment we limit ourselves to a description of the structure.

In most cases, the weights in a weighted digraph can be arbitrary values. In a state graph, however, it is required that the weights be a (usually small) discrete set of values—for example, the letters of the alphabet or the numeric digits. The graph is implemented as a two-dimensional array, with a row for each vertex and a column for each *weight*. Each row represents a source vertex, and entries in the matrix represent *destination vertices*, not weights as before. Figure 6–10 shows a state graph whose vertices are *A, B, C,* and *D* and whose weights are just the digits 0 and 1. The corresponding state table is shown in Figure 6–11.

6.5 GRAPH TRAVERSALS

Some applications of graphs require the graph to be *traversed*. This means that, starting from some designated vertex, the graph is "walked around" in a systematic way such that every vertex reachable from that starting vertex is officially "touched" or *visited* exactly once. Two often-used traversal algorithms are called the *depth-first search* and the *breadth-first search*.

The depth-first search algorithm finds all graph vertices reachable from a particular starting vertex in a way that explores a given path from the starting vertex before starting another path. The search strategy, then, is to probe deeper and deeper along a path, hence the designation "depth-first."

The breadth-first search algorithm visits all vertices adjacent to the starting vertex, then visits all vertices adjacent to those vertices, and so on. Since all adjacent vertices are visited before probing further away, the search is broad rather than deep, hence the name "breadth-first."

6.5.1 Depth-First Search

A traversal algorithm requires that each vertex be "officially" visited exactly once. Since a vertex can be adjacent to many other vertices, and since graphs can have cycles, we need a way of keeping track of the vertices that have already been visited. Accordingly, the depth-first search algorithm uses an auxiliary set, called Visited, that is initially empty. A vertex of G is added to the set when it is visited. The algorithm is recursive, and operates as follows:

DEPTH-FIRST SEARCH:

1. Place the designated starting vertex x in the set Visited.
2. Do whatever application-dependent things need to be done upon visiting a vertex.
3. For each vertex y adjacent to x, if y has not been visited, call the depth-first search algorithm recursively with y as the starting vertex.

Thus, the algorithm pursues a given path until a previously visited vertex is reached, and then it returns to the original vertex and pursues another path. If it terminates with the entire vertex set in Visited, then all vertices are reachable from the given starting vertex. Figure 6–12 shows an example of the depth-first search in action.

If we want a program to implement this algorithm, we number the vertices of our graph and then use the adjacency matrix. The line of the algorithm beginning "for each vertex y adjacent to x" becomes a loop which walks across x's row of the matrix. For the set Visited, which we let be a VAR parameter, so that we can keep adding members to it, we call upon the Modula-2 SET operations.

The depth-first search is really just a framework for a program whose real work is application dependent. Therefore, we implement step two of the algorithm as a simple call to a procedure Visit(x), which has the job of carrying out this application-dependent task, whatever it is. A Modula-2 procedure for a depth-first search appears in Figure 6–13.

6.5.2 Breadth-First Search

In a breadth-first search, we start from a vertex x and visit all vertices adjacent to x. Then we visit all vertices adjacent to *those* vertices, and so on. We use a *queue* to keep track of vertices we have visited, but whose adjacent vertices we haven't yet visited. The same set Visited is used to keep a record of visited vertices. The algorithm is as follows:

Sec. 6.5   Graph Traversals

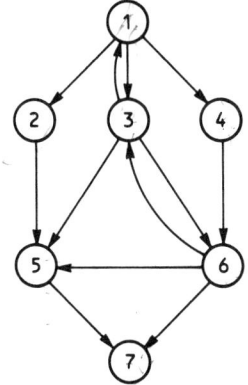

(a)   A directed graph.

```
starting at 1 : 1-2-5-7-3-6-4
starting at 2 : 2-5-7
(graph isn't strongly connected, so only
    three nodes are visited!)
starting at 3 : 3-1-2-5-7-4-6
```

(b)   Some depth-first searches of the digraph.

Figure 6–12 Depth-first search on a graph

BREADTH-FIRST SEARCH:

1. Clear the queue Q.
2. Place the designated starting vertex x in the set Visited.
3. Enqueue x in Q.

```
PROCEDURE DepthFirstSearch(G:          Graph;
                          start:       VertexRange;
                          VAR Visited: VertexSet) ;
    VAR
        dest: VertexRange;

    BEGIN
        Visit(start);
        INCL(Visited,start);

        FOR dest := 1 TO MaxVertices DO
            IF (IsAdjacent(G,start,dest))
                AND NOT (dest IN Visited) THEN
                    DepthFirstSearch(G, dest, Visited);
            END;
        END;

    END DepthFirstSearch;
```

Figure 6–13 Procedure for depth-first search

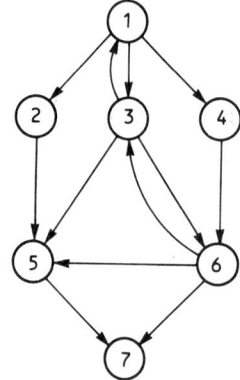

(a) A directed graph.

```
starting at 1 : 1-2-3-4-5-6-7
starting at 3 : 3-1-5-6-2-4-7
```

(b) Some breadth-first searches of the digraph.

Figure 6–14 Breadth-first search on a digraph

```
PROCEDURE BreadthFirstSearch(G:          Graph;
                             start:      VertexRange;
                             VAR Visited: VertexSet) ;
    VAR
        source, dest: VertexRange;
        Q:            Queue;

    BEGIN

        QueueInit(Q);
        Visit(start);
        INCL(Visited,start);
        Enqueue(Q,start);

        WHILE NOT QueueIsEmpty(Q) DO
           source := QueueFirst(Q);
           Dequeue(Q);

           FOR dest := 1 TO MaxVertices DO
              IF (IsAdjacent(G,source,dest))
                 AND NOT (dest IN Visited) THEN
                 Visit(dest);
                 INCL(Visited,dest);
                 Enqueue(Q,dest);
              END;
           END;

        END;

    END BreadthFirstSearch;
```

Figure 6–15 Procedure for breadth-first search

4. Do whatever application-dependent things need to be done upon visiting a vertex.
5. Repeat the remaining steps as long as Q is not empty.
6. Dequeue a value y from Q.
7. For each vertex z adjacent to y, if z has not been visited, place z in Visited. Then do the application-dependent task for z and enqueue z in Q.

An example of a breadth-first search appears in Figure 6–14; a procedure, which imports a Queues module as developed in Chapter 5, is given in Figure 6–15.

6.6 DESIGN: A SIMPLE LEXICAL SCANNER

A Modula-2 identifier consists of a letter followed by zero or more letters and digits. In this section we describe a program or algorithm capable of deciding whether an arbitrary string of characters is a valid Modula-2 identifier. The program is a simple *lexical scanner*; lexical scanners are used for the initial phase of language translation, for checking the validity of commands in an interactive system, and for other similar applications.

The scanner is implemented in the form of a state graph in which one vertex is designated as the *start state* and two other vertices are designated as the *accepting* and *rejecting* states. A vertex which is a source is called a *current state*, while a vertex which is a destination is called a *next state*. A weight is used to represent each possible character in the character set.

The state graph operates as a little machine: it is started in its start state, "reads" the first character of the string, and then moves to the next state corresponding to the character just seen. The next state thus becomes a current state. The machine then reads another character, moves to a new state, and so on. If the machine is in its accepting state when the input string is empty, the string was a valid identifier; if the machine is in its rejecting state, the string had an invalid character in it.

To keep the example simple, we use a very small alphabet for our identifiers: the only *letter* allowed is 'L', the only *digit* '5'. All illegal characters are represented by @. Figure 6–16 gives a number of legal and illegal identifiers in this limited alphabet.

Figure 6–17 shows the state graph for this machine, while Figure 6–18 gives a diagram of the state table. In Figure 6–19 are shown some type declarations for the state table implementation. Note the use of *enumeration types* to list the states (the vertex set V of the graph) and the input alphabet (the discrete set of weights).

To make this data structure work as does a machine, we need a program to

```
L        valid
L5       valid
5        invalid (starts with 5)
LL       valid
L5       valid
5L       invalid (starts with 5)
LL55L    valid
L5@L     invalid (contains @)
```

Figure 6–16 Valid and invalid words over a limited alphabet

"run" it. We write this program as a Modula-2 function ValidIdentifier which accepts a Text object, starts the state graph in its start state, and then reads characters, returning a Boolean which indicates whether the input string was a valid identifier. Formally, such a machine is called a *finite-state machine*. In this example the finite-state machine keeps running until its input string is empty; if it ever gets to the rejecting state, it keeps reading characters and cycling in that state until, again, the input is empty. The program is shown in Figure 6–20.

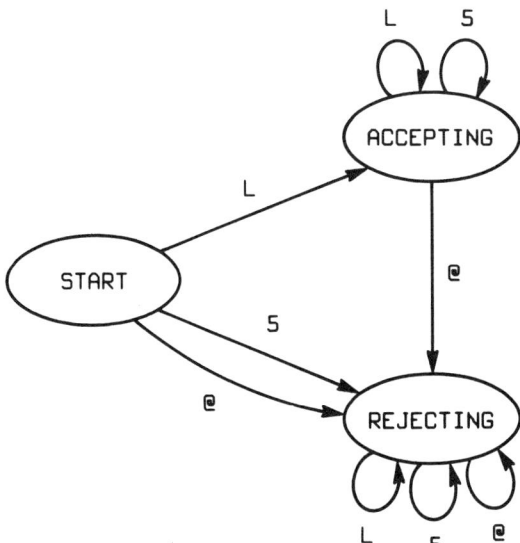

Figure 6–17 State graph for simple scanner

	L	5	@
START	ACCEPTING	REJECTING	REJECTING
ACCEPTING	ACCEPTING	ACCEPTING	REJECTING
REJECTING	REJECTING	REJECTING	REJECTING

Figure 6-18 State table for simple scanner

We shall return to the lexical scanner idea in Chapter 7, where we introduce a finite-state machine for scanning English text, in order to build a cross-reference generator.

6.7 SUMMARY

Graphs have many uses, among which are to show relationships between elements in a set, to establish sequencing of activities in a project, and to scan sequences of characters in a string. Although we cannot treat graphs in this text in a completely general way, we have presented a number of important concepts pertaining to directed graphs, including reflexivity, symmetry, transitivity, connectedness, and depth-first and breadth-first searches. A lexical scanner that was a finite-state machine served as an application.

```
TYPE
    State     = (Start, Accepting, Rejecting);
    InputClass = (Letter, Digit, Underscore, Illegal);

VAR
    StateTable: ARRAY State,InputClass OF State;
    StateTable[Start]     [Letter]     := Accepting;
    StateTable[Start]     [Digit]      := Rejecting;
    StateTable[Start]     [Underscore] := Rejecting;
    StateTable[Start]     [Illegal]    := Rejecting;

    StateTable[Accepting][Letter]     := Accepting;
    StateTable[Accepting][Digit]      := Accepting;
    StateTable[Accepting][Underscore] := Accepting;
    StateTable[Accepting][Illegal]    := Rejecting;

    StateTable[Rejecting][Letter]     := Rejecting;
    StateTable[Rejecting][Digit]      := Rejecting;
    StateTable[Rejecting][Underscore] := Rejecting;
    StateTable[Rejecting][Illegal]    := Rejecting;
```

Figure 6-19 Type declarations for state table

```
PROCEDURE ValidIdent(T: Text): BOOLEAN;

   (* FSM TYPE DEFINITIONS GO HERE *)

   VAR
      S:             Text;
      C:             CHAR;
      Class:         InputClass;
      CurrentState:  State;

   BEGIN

      (* ARRAY INITIALIZATION GOES HERE *)

      TextInit(S);
      TextCopy(S,T);
      CurrentState := Start;

      IF TextEmpty(S) THEN
         RETURN FALSE;
      END;

      REPEAT
         C := TextHead(S);

         IF    C = 'L' THEN
            Class := Letter;
         ELSIF C = '5' THEN
            Class := Digit;
         ELSIF C = '_' THEN
            Class := Underscore;
         ELSE
            Class := Illegal;
         END;

         CurrentState := StateTable[CurrentState][Class];

         S := TextTail(S);

      UNTIL TextEmpty(S);

      IF CurrentState = Accepting THEN
         RETURN TRUE;
      ELSE
         RETURN FALSE;
      END;

   END ValidIdent;
```

Figure 6–20 Program simulating a simple finite-state machine

CHAPTER 6 EXERCISES

1. One interpretation of a digraph is a relation on a set. The vertices in the graph represent elements of the set; an edge from vertex x to vertex y means "x is related to y." A relation is called an *equivalence relation* if it is reflexive, transitive, and symmetric. Clearly a relation has these properties iff its digraph representation does. Write a function to determine if a graph G, implemented as an adjacency matrix, represents an equivalence relation.

2. A relation is called a *partial ordering* if it is reflexive, transitive, and antisymmetric. Using the graph interpretation from the preceding problem, write a function to determine whether a graph G represents a partial ordering.

3. Given a digraph with vertex set {A,B,C,D} and edge set {⟨A,A⟩, ⟨A,B⟩, ⟨A,D⟩, ⟨B,B⟩, ⟨C,B⟩, ⟨C,D⟩, ⟨D,C⟩}, draw the graph and its adjacency matrix and adjacency list forms.

4. For the digraph specified in the preceding problem, indicate whether or not the graph has each of the following properties: reflexive, irreflexive, symmetric, antisymmetric, transitive, connected, strongly connected, acyclic. For each property the graph *doesn't* have, make a list of the *minimum* number of changes necessary to give the graph that property.

5. For the digraph specified above, find the depth-first and breadth-first searches starting with each of the four vertices.

6. Repeat the preceding three problems for the digraph with vertex set {A,B,C,D} and edge set {⟨A,B⟩, ⟨A,C⟩, ⟨B,B⟩, ⟨B,C⟩, ⟨C,C⟩, ⟨C,A⟩, ⟨C,C⟩, ⟨C,D⟩}.

7. Show that it is possible to renumber the vertices of a digraph G so that its adjacency matrix is lower-triangular iff G is acyclic.

8. Given a graph G represented by its unweighted adjacency matrix M. Consider the matrix product of M with itself (the square of M) gotten by using "or" and "and" as the addition and multiplication operators in the matrix product. Calling this matrix MM, show that M(r,c) = true iff there is a path of length two or less from vertex r to vertex o.

9. Starting from the previous problem, show that in the matrix representing the p-th power of M, a true entry in the r-th row and c-th column indicates that there is a path of length p or less from vertex r to vertex c, of the matrix M.

Chapter 7
TREE STRUCTURES

7.1 GOAL OF THE CHAPTER

A tree is a special case of a directed graph, with many applications in computing. More formally, a tree is just a connected digraph such that exactly one vertex (called the *root*) has an in-degree of zero and all other vertices have an in-degree of one. The consequence of this definition is that, starting from the root, there is exactly one path to each of the other vertices. This makes a tree useful for representing hierarchical relationships.

This chapter focuses mainly on the special case of the *binary tree*, in which no vertex has more than two outgoing edges. Two important applications of binary trees are the *expression tree*, which is used in translating or interpreting programming language statements, and the *binary search tree*, or BST, which is yet another implementation of a mapping or dynamic table.

An important concept in the study of trees is the *traversal*. As in the case of general digraphs, a traversal is an algorithm for "walking around" the tree so that all its vertices are visited exactly once in some systematic sequence. Out of the many possible traversals of a tree, we shall study three, all of which are written as recursive algorithms.

Even though we concentrate on binary trees, section 7.6 briefly introduces two important applications of more general trees: *digital search trees*, used in applications like spelling checkers, and *B-trees*, often used as a basis for organizing large structured files on secondary storage devices.

There are two design sections in this chapter. Design One shows how to construct a parser for simple arithmetic expressions; design Two shows how an indexing or cross-reference program can be constructed using a binary search tree. The style guide in this chapter focuses on some alternative implementations of binary trees to reduce algorithm run times and avoid recursive traversal programs.

7.2 TREES

A tree is a special case of a directed graph whose main application is expressing purely hierarchical relationships of some kind or other. For example, Figure 7-1 shows the basic structure of a hypothetical company, with a single president, a few vice-presidents, some managers, and some workers; Figure 7-2 shows a family tree representing three generations of descendants of a person; and Figure 7-3 illustrates the operator-operand relationship in a programming language assignment statement. The common characteristic of all these trees is that there is a single vertex—the *root*—that can be identified as the "top" of the tree, and that from the root to any other vertex in the tree there is *exactly one path*. Formally, a *tree* is a connected digraph such that:

1. There is exactly one vertex, called the *root*, with in-degree = 0.
2. All other vertices have in-degree = 1.

Notice that the definition says nothing about out-degree. In a general tree there is no restriction on the out-degree of a vertex, nor indeed on whether the vertex set must even be finite. In most important applications, however, the tree has a finite number of vertices, and so there is necessarily a subset of the vertex set with out-degree = 0. These vertices are at the "bottom" of the tree and are termed *leaves*, or sometimes, *terminal vertices*. A vertex which is neither a root nor a leaf is called an *interior vertex* (sometimes *nonterminal vertex*) of the tree. We shall assume throughout that all trees are finite, i.e., that all their paths, or *branches*, terminate in leaves.

Figure 7-4 shows some structures that are trees and some that are not.

Since a tree is a digraph, it makes sense to consider which properties of graphs pertain to trees. From the definition of a tree, since a vertex has at most one edge leading to it, there are certain properties of graphs that all trees must have, and others that no tree can have. For instance, because there is *at most* one path from any vertex to any other vertex, and because there is *exactly* one path from the root to any vertex, a tree is necessarily antisymmetric and irreflexive. We leave consideration of other graph properties of trees to an exercise.

From the way we have defined trees, every vertex at the destination end of an edge for which the root is the source is itself the root of another, smaller tree. We shall call this structure a *subtree*. Note that a single vertex, by itself, is a tree.

The *depth* of a tree is defined to be the length of the longest path from the

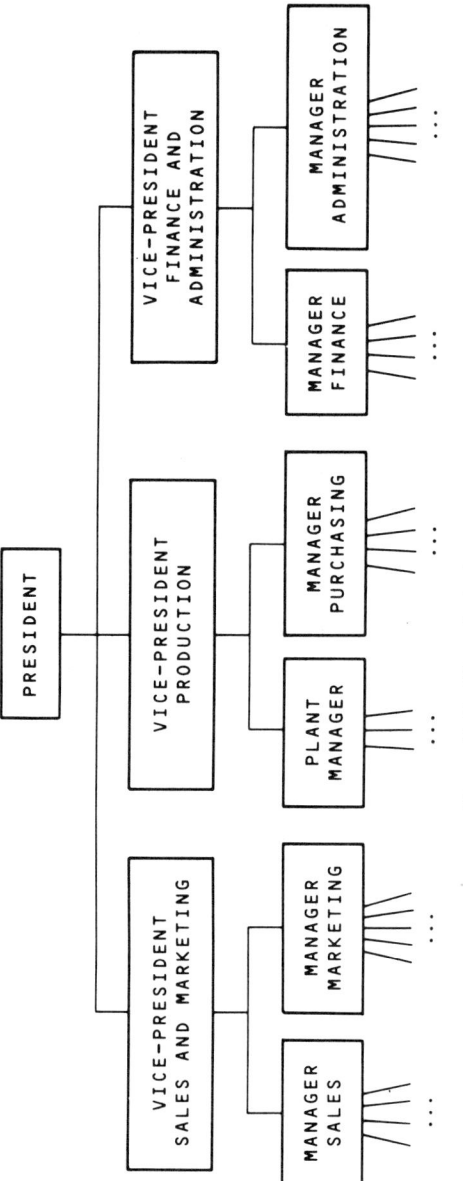

Figure 7–1 Hypothetical corporate structure

Sec. 7.2   Trees   185

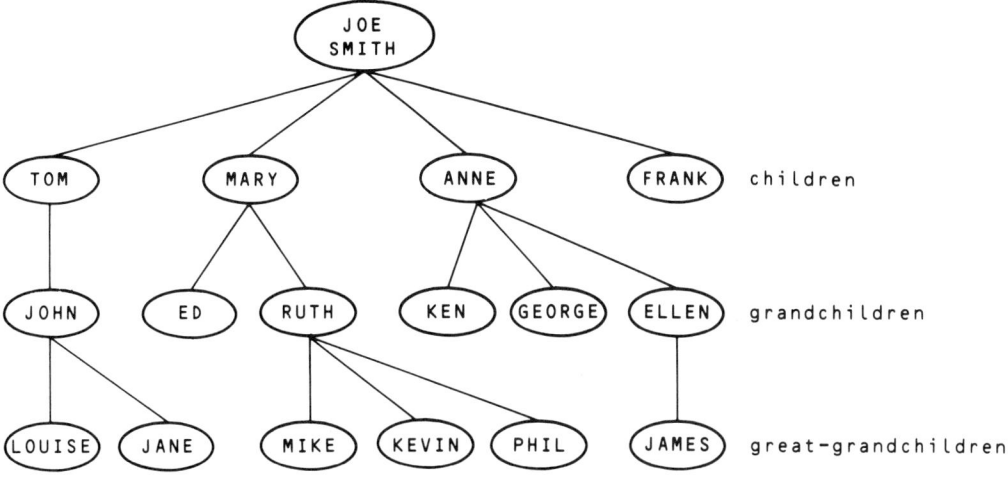

Figure 7-2 Descendants of Joe Smith

root to a leaf. The *level* of a vertex is the length of the path (remember, there is only one path!) from the root to that vertex. The level of the root itself is thus 0. Figure 7-5 shows some trees and indicates their depths.

X := Y+Z-(A*B/W)+G

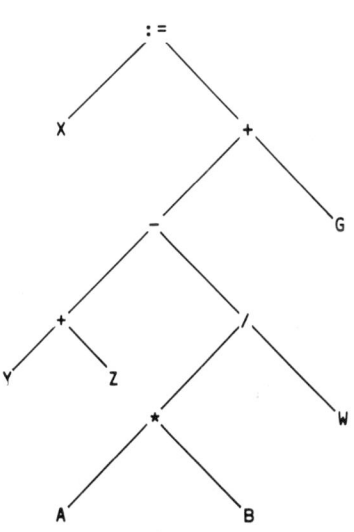

Figure 7-3 Operator-operand relationships in an arithmetic assignment statement

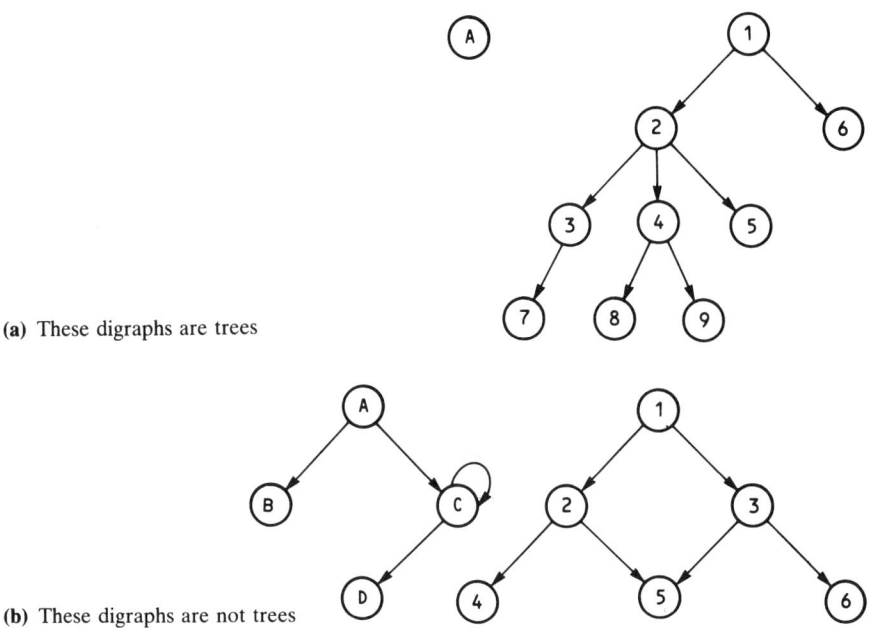

(a) These digraphs are trees

(b) These digraphs are not trees

Figure 7-4 Some trees and some digraphs that are not trees

Drawing some terminology from genealogical (family) trees, we shall refer to the destination vertices of a vertex as the *children* of that vertex, and a vertex from which one or more children grow as the *parent* of those children. Children of the same parent are referred to as *siblings*, and all vertices reachable from a given vertex are called that vertex's *descendants*. Note that a child of any vertex is the root of a *subtree* of that (parent) vertex.

Despite the genealogical terminology, the analogy with family trees is imperfect because, while humans and most animals have precisely two parents, a vertex in the type of trees we have been discussing has precisely *one* parent.

In applications of trees to computing, there often is information associated with each of the vertices of a tree. Obviously, the nature and interpretation of this information depend on the application; we shall refer to it generically by a number of names, for example, *label*, *data*, *value*, or *key*.

In general, we do not bother to draw arrowheads on the edges of a tree, but instead write the root at the top and "grow" the tree in a downward direction on the page. Also, it is sometimes convenient to omit the circle indicating a vertex and write the data there instead, as in Figure 7-6.

We shall return to the subject of general trees later in the chapter; for now, let us limit our attention to the special and useful case of *binary trees*.

Sec. 7.3   Binary Trees

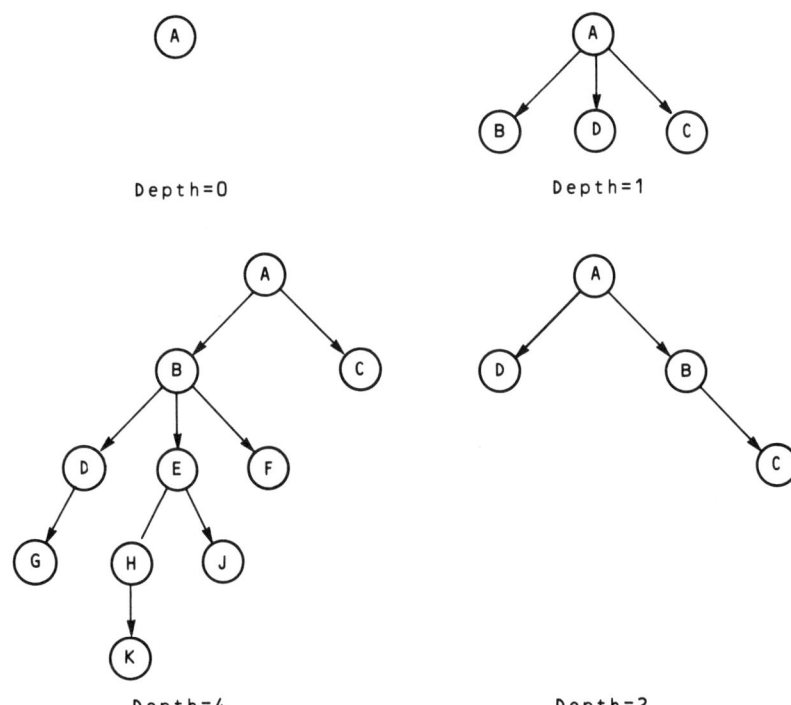

Figure 7–5 Some trees and their depths

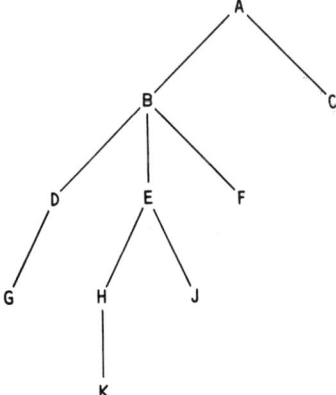

Figure 7–6 Simplified tree notation

7.3 BINARY TREES

A *binary tree* is a tree all of whose vertices have out-degree ≤ 2. Furthermore, the subtrees of a binary tree are *ordered* in the sense that there is a *left* child and a *right* child. Thus, if a vertex has only one child, it must be clearly identified as left or right. The two trees given in Figure 7–7(a) are different binary trees, as are the two in Figure 7–7(b).

7.3.1 Properties of Binary Trees

Strictly binary. *T* is a *strictly binary tree* iff each of its vertices has out-degree = 0 or out-degree = 2. (Strictly binary trees have no vertices with out-degree = 1.)

Complete. *T* is a *complete binary tree of depth K* iff each vertex of level K is a leaf and each vertex of level less than K has nonempty left and right children. This means that a complete binary tree has all its leaves at the same level and every vertex that is not a leaf has both children present. Notice that a complete binary tree of depth K always has exactly $2^{(K+1)} - 1$ vertices. Thus, a complete binary tree of depth = 0 has a single vertex, one of depth = 1 has three vertices, one of depth = 2 has seven vertices, and so on.

Conversely, a complete binary tree of N vertices has depth equal to $\log_2 (N + 1) - 1$.

(a) These are *different* binary trees

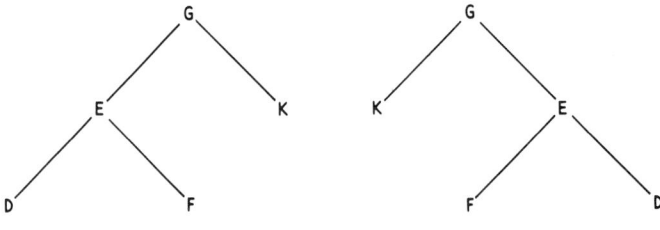

(b) So are these

Figure 7–7 Binary trees have ordered children

Sec. 7.3 Binary Trees

Almost complete. *T* is an *almost complete binary tree* (*ACBT*) of depth *K* iff it is either complete, or fails to be complete only because some of its leaves are at the right-hand end of level *K* − 1. In the latter case, all of the level *K* leaves are concentrated at the left end of level *K*, while all of the level *K* − 1 leaves are concentrated at the right end of level *K* − 1. Figure 7–8 shows complete and almost - complete binary trees, and some trees with neither property.

Almost - complete binary trees are useful in certain sorting applications (e.g., *heap sort*, which we shall examine in chapter 9) because an ordinary array may be

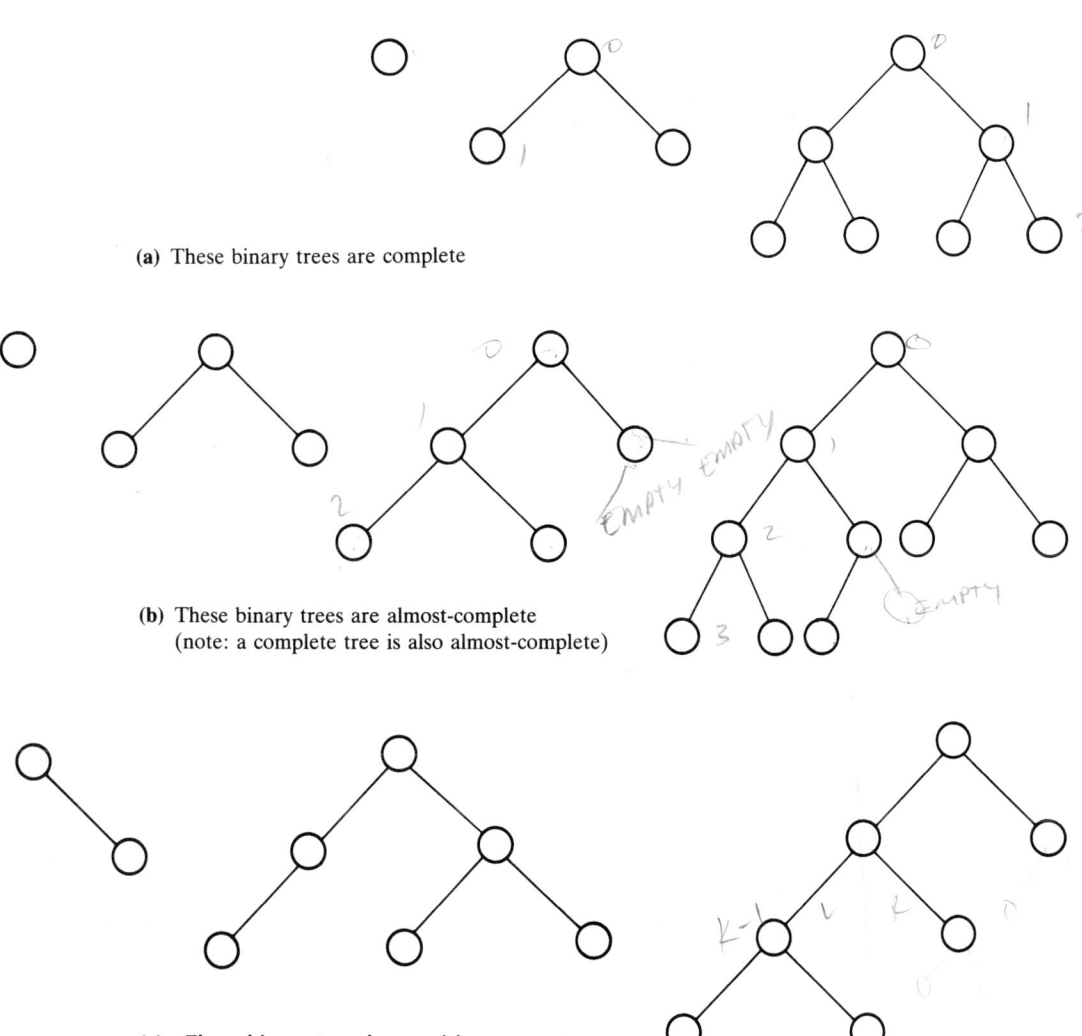

Figure 7–8 Complete and almost-complete binary trees

190 Tree Structures Chap. 7

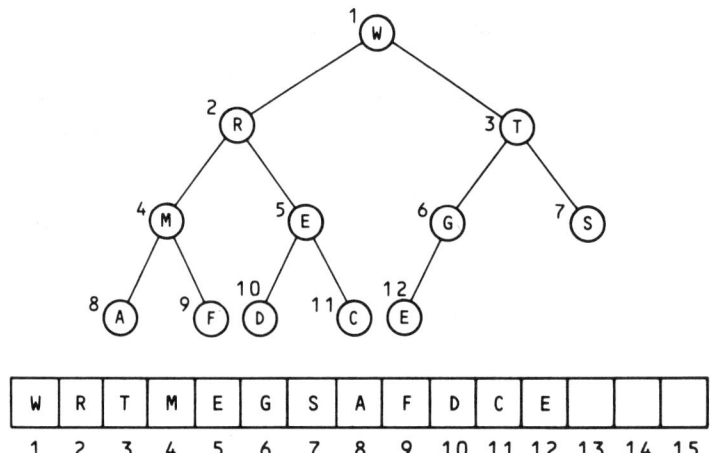

Figure 7–9 An array, viewed as an almost-complete binary tree (it would be complete if elements 13, 14, and 15 were present)

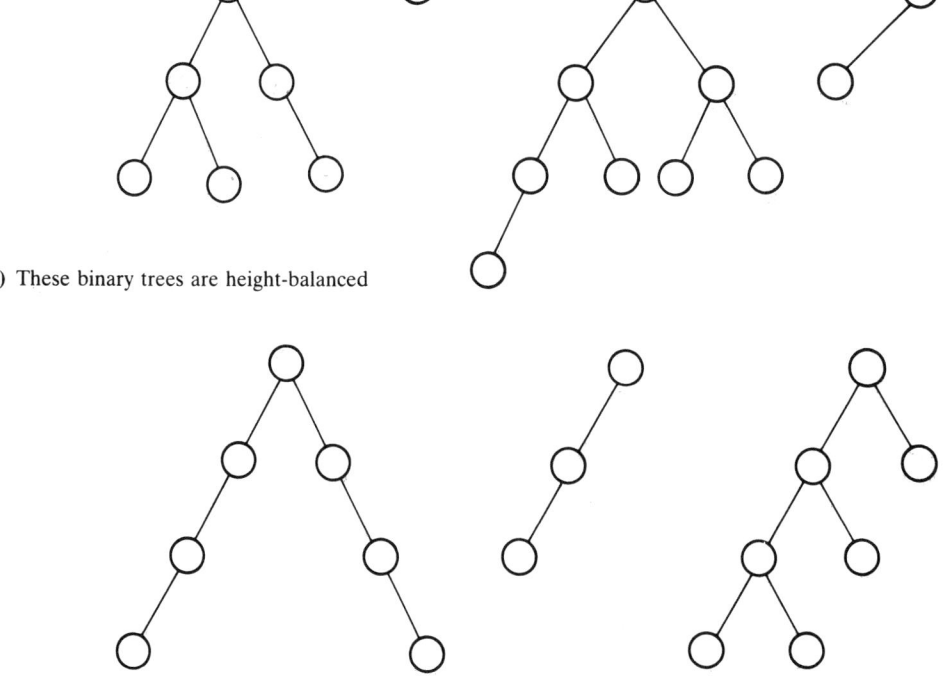

(a) These binary trees are height-balanced

(b) These are not

Figure 7–10 Balanced (or height-balanced) binary trees

Sec. 7.3 Binary Trees

viewed as an implementation of an ACBT. According to this scheme, the first element of the array is considered the root of the tree, the second and third elements are the children of the root, and so on. If we number the vertices of an ACBT, starting at the root and proceeding level by level and left to right within a level, the numbers correspond to the subscripts of the array, as indicated in Figure 7-9.

Note that some authors use the term "full binary tree" to mean what we have called a complete binary tree, and the term "complete" to mean what we have called almost complete.

Balanced. T is a *balanced* (sometimes called *height-balanced*) binary tree iff, for each vertex t in T, the depths of t's right and left subtrees differ by at most one. (If one subtree is null, the other subtree must either be null or a leaf.)

It is important to understand that for a tree to be balanced, the property must hold for every vertex in the tree, not just its root. Study the trees in Figure 7-10, and state why each is either balanced or not balanced. The notion of balance in a binary tree is important in the study of binary search trees.

The definition of balance can also be stated recursively:

1. A binary tree consisting of a single vertex is balanced.
2. A vertex with a single subtree is balanced iff that subtree is a leaf.
3. A binary tree is balanced iff its left and right subtrees are balanced and their depths differ by at most one.

7.3.2 Implementing Binary Trees

Since a binary tree is really just a digraph, we could implement it using one of the graph representations. However, it is usually more practical to make use of our knowledge that a binary tree has right and left subtrees and create a more specialized structure. Accordingly, we shall implement each vertex as a linked node, i.e., a record with an information field, a pointer to the left subtree, and a pointer to the right subtree. Thus, a tree can be built as a linked structure using either dynamic storage allocation if such a feature is available in the coding language, or cursor allocation otherwise. Henceforth, we shall refer to a tree's *nodes*, meaning either the vertices of the mathematical structure or the linked nodes of the data structure used to implement the tree, depending on the context.

In Figure 7-11 we show some Modula-2 type definitions for the nodes and

```
TYPE KeyType = ...

TYPE Tree = POINTER TO BinaryTreeNode;

TYPE BinaryTreeNode =
    RECORD
        key:    KeyType;
        left:   Tree;
        right:  Tree;
    END;
```

Figure 7-11 Type declarations for binary tree node

Statement   Resulting Structure

```
VAR T1, t : Tree;
```
Must put nils in by initialization
```
Allocate(T1,SIZE(BinaryTreeNode));
T1^.key := 'D';
```

```
Allocate(T1^.left,SIZE(BinaryTreeNode));
T1^.left^.key := 'B';
Allocate(T1^.left^.right,SIZE(BinaryTreeNode));
T1^.left^.right^.key := 'C';
```

```
Allocate(t,SIZE(BinaryTreeNode));
t^.key := 'F';
```

```
T1^.right := t;
```

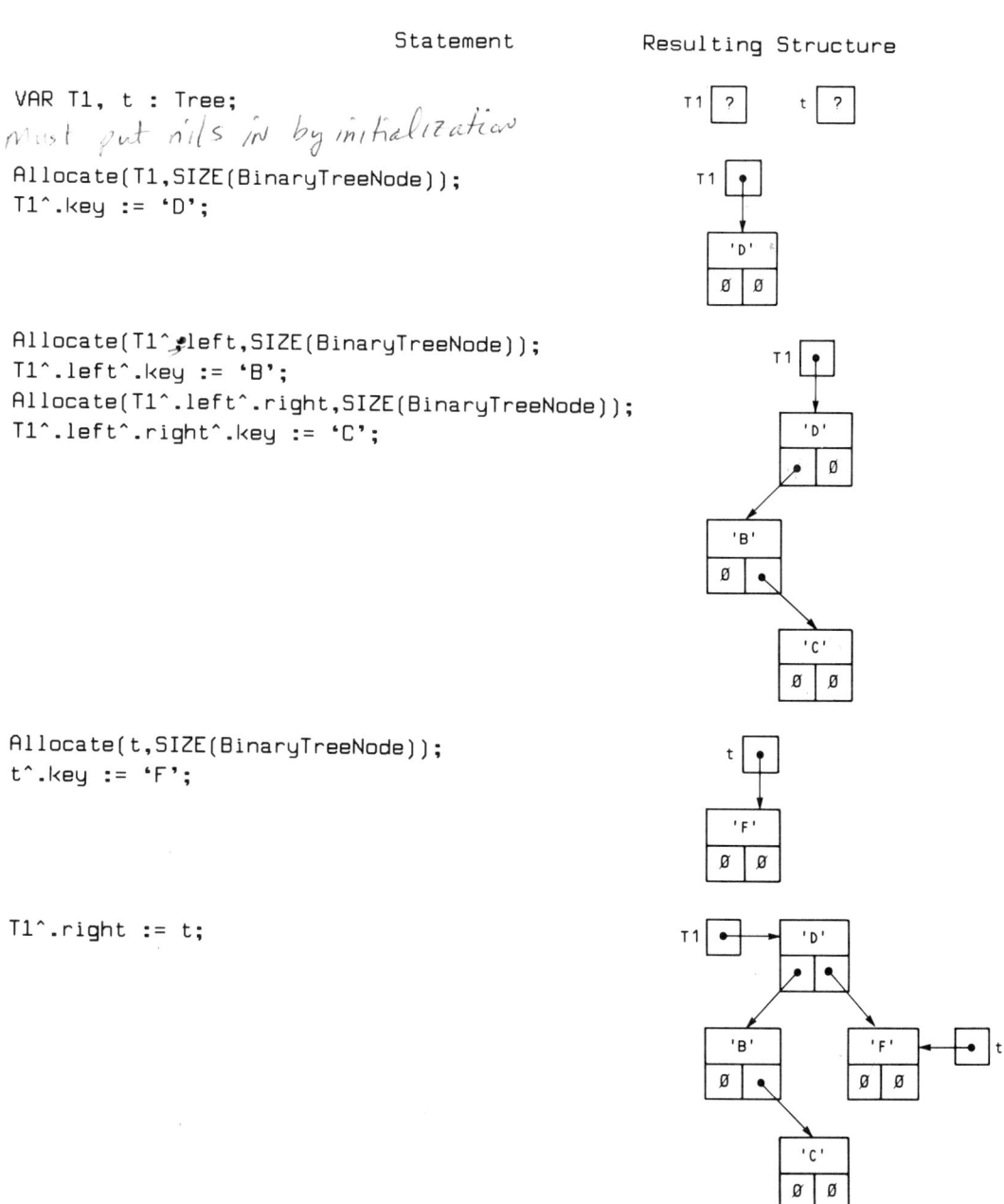

Figure 7–12 Statements manipulating binary tree nodes

Sec. 7.4   Expression Trees

pointers of a binary tree. These simple definitions assume that the information field in each node is just a key; for many applications, this field will of course be a composite record of some kind. A sequence of statements declaring and manipulating nodes is given in Figure 7-12, along with diagrams showing the results of each operation.

Generally, we shall avoid drawing boxes to represent the nodes and just use the more abstract diagrams as in previous examples.

7.3.3 Traversals of Binary Trees

Many applications require *traversing* a tree in a particular way so that all the nodes are *visited* in a certain order. Three traversal algorithms are particularly useful in dealing with binary trees. These are often called *preorder*, *inorder*, and *postorder*,

```
PROCEDURE TraverseNLR(T: Tree);

   BEGIN
      IF T = NIL THEN
         RETURN;
      ELSE
         VisitTree(T);
         TraverseNLR(T^.left);
         TraverseNLR(T^.right);
      END;
   END TraverseNLR;



PROCEDURE TraverseLNR(T: Tree);

   BEGIN
      IF T = NIL THEN
         RETURN;
      ELSE
         TraverseLNR(T^.left);
         VisitTree(T);
         TraverseLNR(T^.right);
      END;
   END TraverseLNR;



PROCEDURE TraverseLRN(T: Tree);

   BEGIN
      IF T = NIL THEN
         RETURN;
      ELSE
         TraverseLRN(T^.left);
         TraverseLRN(T^.right);
         VisitTree(T);
      END;
   END TraverseLRN;



PROCEDURE VisitTree(T: Tree);

   BEGIN
      WriteKey(T^.key);
   END VisitTree;
```

Figure 7-13 Recursive tree traversals

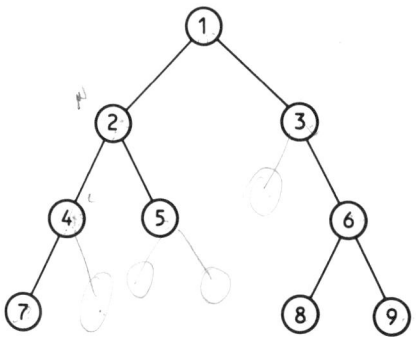

(a) A binary tree

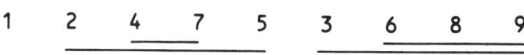

(b) NLR (node-left-right or preorder) traversal

$$\underline{\underline{7 \quad 4}} \quad 2 \quad 5 \quad 1 \quad 3 \quad \underline{\underline{8 \quad 6}} \quad 9$$

(c) LNR (left-node-right or inorder) traversal

$$\underline{\underline{7 \quad 4}} \quad 5 \quad 2 \quad \underline{\underline{8 \quad 9}} \quad 6 \quad 3 \quad 1$$

(d) LRN (left-right-node or postorder) traversal

Figure 7–14 Three traversals of a binary tree

but different authors occasionally disagree on what the terms mean. To avoid confusion, we shall call the algorithms TraverseNLR, TraverseLNR, and TraverseLRN, respectively, where L stands for "left subtree," R stands for "right subtree," and N stands for "node." The order of the letters indicates the traversal order. For example, in TraverseNLR a node is visited, then its left subtree is traversed; in TraverseLNR, the left subtree is traversed before the node is visited, and then the right subtree is traversed.

How do these algorithms work? Since a binary tree is recursively defined (every subtree of a binary tree is a binary tree), a traversal defined for a tree is also defined for any subtree. We can thus write the three traversal algorithms recursively, as shown in Figure 7–13.

Figure 7–14 shows the steps required to perform the three traversals for a given tree. In each traversal, the details of the visit operation are deferred, since precisely what the operation should accomplish is application dependent. For completeness, the figure includes a visit operation that simply prints the key. The applicability of each traversal will become apparent shortly.

7.4 EXPRESSION TREES

One common application of binary trees is in interpreters or compilers for programming languages, where the statements of a source program are converted into trees to make their structure apparent. As a simple case of this, we shall consider *expression trees*, which are transformations of arithmetic expressions into binary trees. For simplicity, we use the same restricted expressions that we used in chapter 5, in the discussion of stacks and RPN. Recall that an expression consists of a combination of single-letter identifiers or variable names, one-digit integer constants, the four arithmetic operators $+$, $-$, $\times$, and $/$, and parentheses.

7.4.1 Constructing Expression Trees

Later on, we shall consider how to build a scanner or parser program that can construct an expression tree for the simple expressions just described. For now, let us see how to construct an expression tree manually. The general technique is similar to the way we constructed an RPN expression from an infix one.

We consider first only fully parenthesized expressions. An expression tree always has an operator at its root and identifiers or constants at its leaves. (The exception is the tree for an expression consisting only of a single identifier or constant: there is just one node which is both a root and a leaf.) The root operator is the "main" operator of the expression, that is, the operator which is performed *last* as the expression is evaluated. Interior nodes are the operators of subexpressions.

To give a few examples, Figure 7–15 shows the expression trees for A, $A - B$, $(A - B) + C$, $A - (B + C)$, and $(A + B) \times (C - D)$. Notice carefully how these trees are constructed, and how $(A - B) + C$ and $A - (B + C)$ give rise to *different* trees: in $(A - B) + C$, $+$ is the main operation, since it is performed *last*; in $A - (B + C)$, $-$ is the main operation. As an exercise, try building expression trees from $(A \times B) - (C + (D/E))$ and $((A - B) + (C/D)) \times E$.

Let us now relax the condition that expressions be fully parenthesized. We use the association and priority rules developed in chapter 5: $+$ and $-$ are priority 2 operators; $\times$ and $/$ are priority 1 operators, and adjacent operators of equal priority associate left to right. Thus, the expression $A + B \times C$ will be treated as though it were parenthesized $A + (B \times C)$, and $A/B - C$ will be treated as though it were parenthesized $(A/B) - C$. Consequently, in the first expression $+$ is the main operator, while in the second it is $-$. The expression trees for the two expressions are as shown in Figure 7–16.

As an example of what happens when equal-priority operators are involved, $A - B - C$ is treated as though it were written $(A - B) - C$, and $A/B \times C$ is treated as though it were written $(A/B) \times C$.

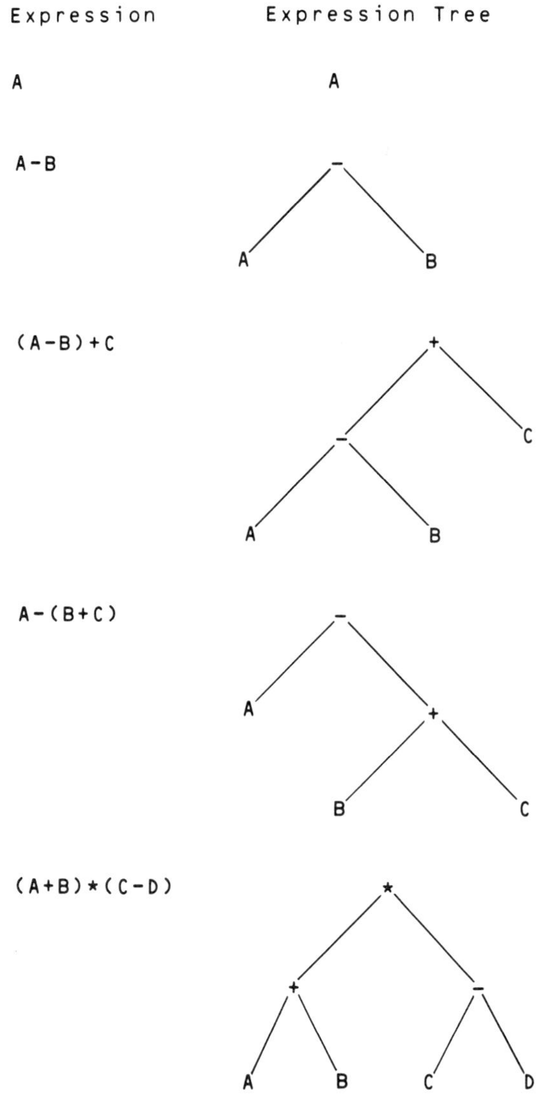

Expression	Expression Tree
A	A
A−B	(tree: − with children A, B)
(A−B)+C	(tree: + with left child − (A,B) and right child C)
A−(B+C)	(tree: − with left child A and right child + (B,C))
(A+B)∗(C−D)	(tree: ∗ with left child + (A,B) and right child − (C,D))

Lower precedence

Figure 7–15 Some expression trees

Consider next the expression $A + B - C + D$. Since adjacent operators of equal priority are handled in a left-to-right fashion, we treat it as though it were $((A + B) - C) + D$. Now look at $A - (B + C) \times D$. Here, the two operators of interest are $-$ and $\times$ (the $+$ doesn't count because it's inside a subexpression), and the $\times$ is done first because its priority is one. So this expression is handled as though it were $A - ((B + C) \times D)$. The corresponding expression trees are shown in Figure 7–17. Try diagramming $A - B \times C/(D - E)$ and $A \times B - (C + D) + E$ as an exercise.

Sec. 7.4 Expression Trees

Expression   Expression Tree

A + B * C

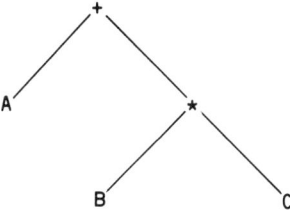

A / B - C

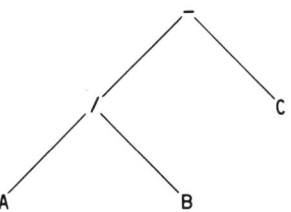

Figure 7-16 More expression trees

Expression   Expression Tree

A + B - C + D

A - (B + C) * D

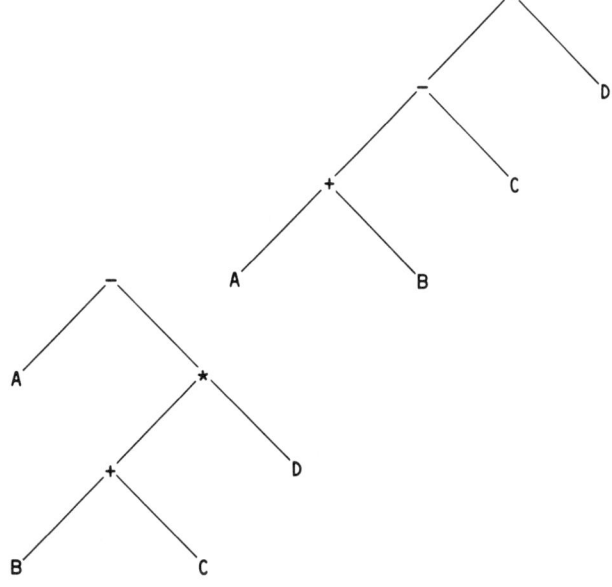

Figure 7-17 Still more expression trees

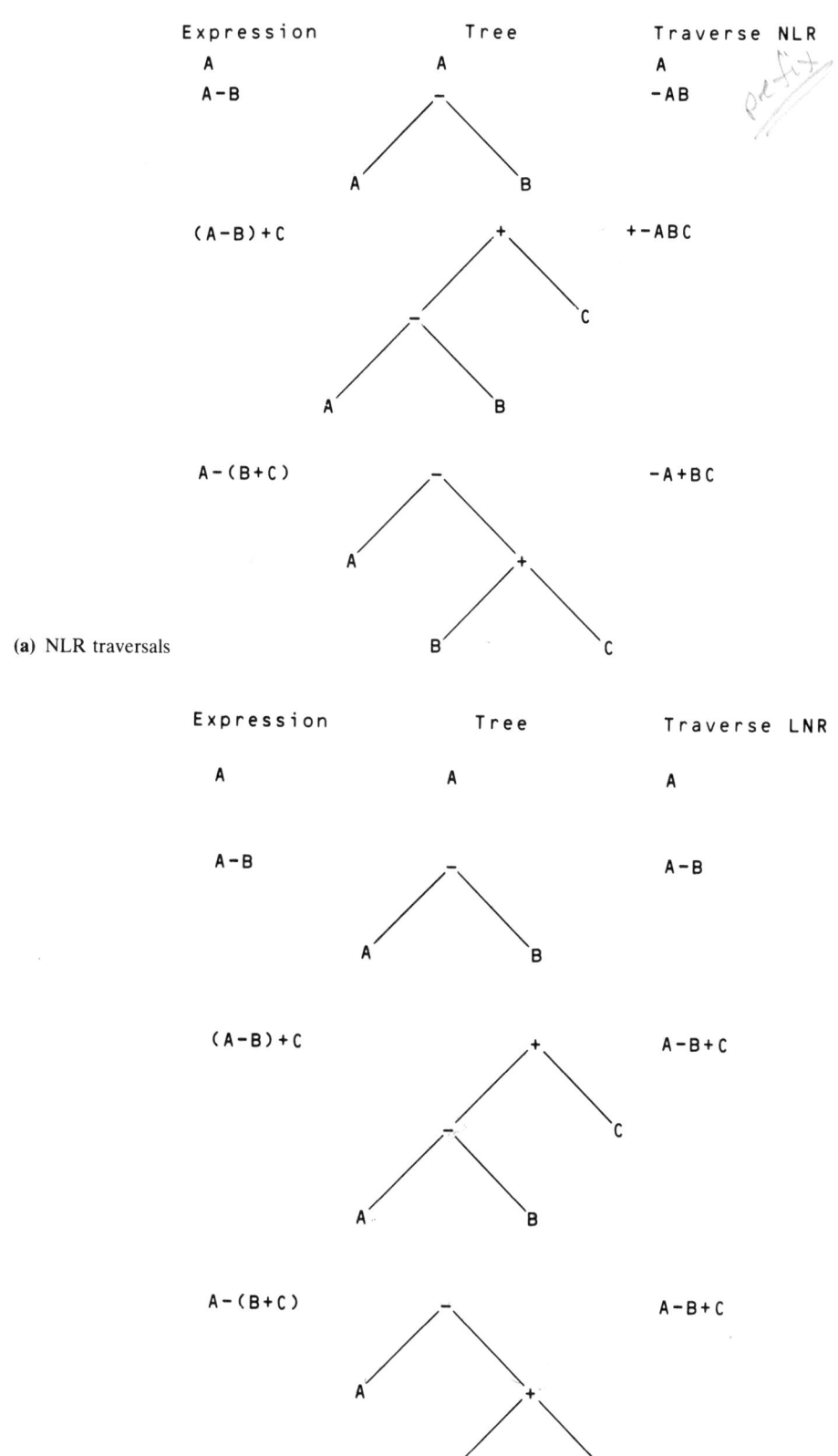

(a) NLR traversals

(b) LNR traversals (note ambiguity!)

Sec. 7.4   Expression Trees

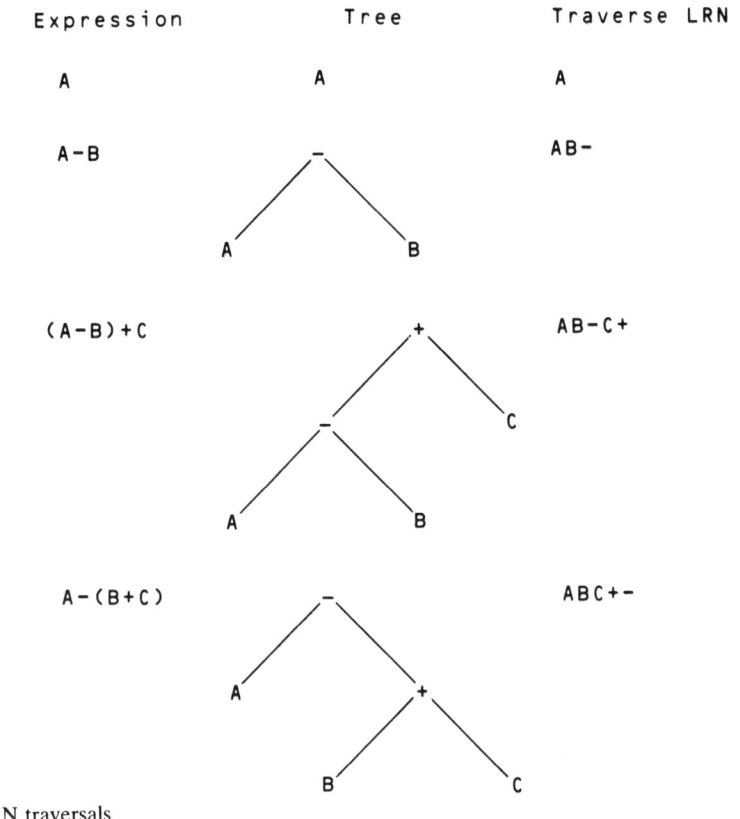

(c) LRN traversals

Figure 7–18 Traversals of expression trees

7.4.2 Traversing Expression Trees

Figure 7–18 shows the three traversals, TraverseNLR, TraverseLNR, and TraverseLRN, performed on an expression tree for the expression $A - (B + C)$. It is interesting that TraverseNLR produces the forward Polish or prefix form of the expression, whereas TraverseLRN produces the RPN form.

What about TraverseLNR? This traversal turns out not to be terribly useful for expression trees, since it produces an infix form of the expression *with the parentheses removed*. Thus, it can lead to ambiguities, since, for example, the expressions $(A - (B - C))$ and $((A - B) - C)$, which clearly have different expression trees, have the same TraverseLNR infix form. Indeed, if numerical values were substituted for *A*, *B*, and *C*, an evaluation of the two expressions would yield different results, only one of which would correspond to evaluating the infix form. Note that such ambiguities do not arise in the prefix and postfix cases. Even though TraverseLNR is not very useful for expression trees, it does have a useful application, as we shall subsequently see.

You have seen that there is an intimate relationship between an infix expression, its tree, and its forward and reverse Polish forms. In compiler applications, some form of the expression tree is often used as a convenient intermediate internal representation of a program. An expression tree is a structure that can easily be manipulated by a program, and even restructured to optimize the object-program instructions that are generated.

7.5 BINARY SEARCH TREES (BSTs)

Another useful application of binary trees is in the implementation of efficient insertions and deletions in tables with dynamically varying entries. We define a *binary search tree* (*BST*) as a binary tree with the property that the value of the key at any node is greater than all values in that node's left subtree, and less than or equal to all values in that node's right subtree. Recursively speaking, we have:

1. A leaf node is a BST.
2. A node is the root of a BST if its key value is greater than that of its left child and less than or equal to that of its right child, and if both of its children are either null or the roots of BSTs.

Figure 7–19 shows some trees that are BSTs and some that are not. Note carefully how to distinguish the former from the latter.

Figure 7–20 gives the specification for a dynamic table handler shown earlier in chapter 2. It turns out that the BST is often an effective implementation of such a table; the remainder of this section explains how each of the table operations can be implemented as operations on a BST. We have kept the examples uncluttered by using records consisting only of a single-letter key; the generalization to records with value fields is immediate.

7.5.1 The Report Operation: Traversing a BST

Earlier it was mentioned that the LNR traversal of a binary tree, although not very useful for expression trees, has an interesting use. Here, we examine that use: to implement the "report" or "print in order" operation of a dynamic table.

A BST has the interesting property that *an LNR traversal of it will visit the nodes in the order of their key values*. This can be readily understood by realizing that every key in the root's left subtree is necessarily less than the root key (otherwise it wouldn't be a BST), and so visiting all the nodes in the left subtree prior to visiting the root will visit smaller keys. Similarly, visiting nodes in the right subtree after visiting the root will visit the root before visiting any keys greater than or equal to the root.

Now, since LNR traversal is recursive, and since the left and right subtrees of the root are themselves BSTs, it turns out the nodes are visited in key-sorted order. The situation is illustrated in Figure 7–21.

Sec. 7.5 Binary Search Trees (BSTs)

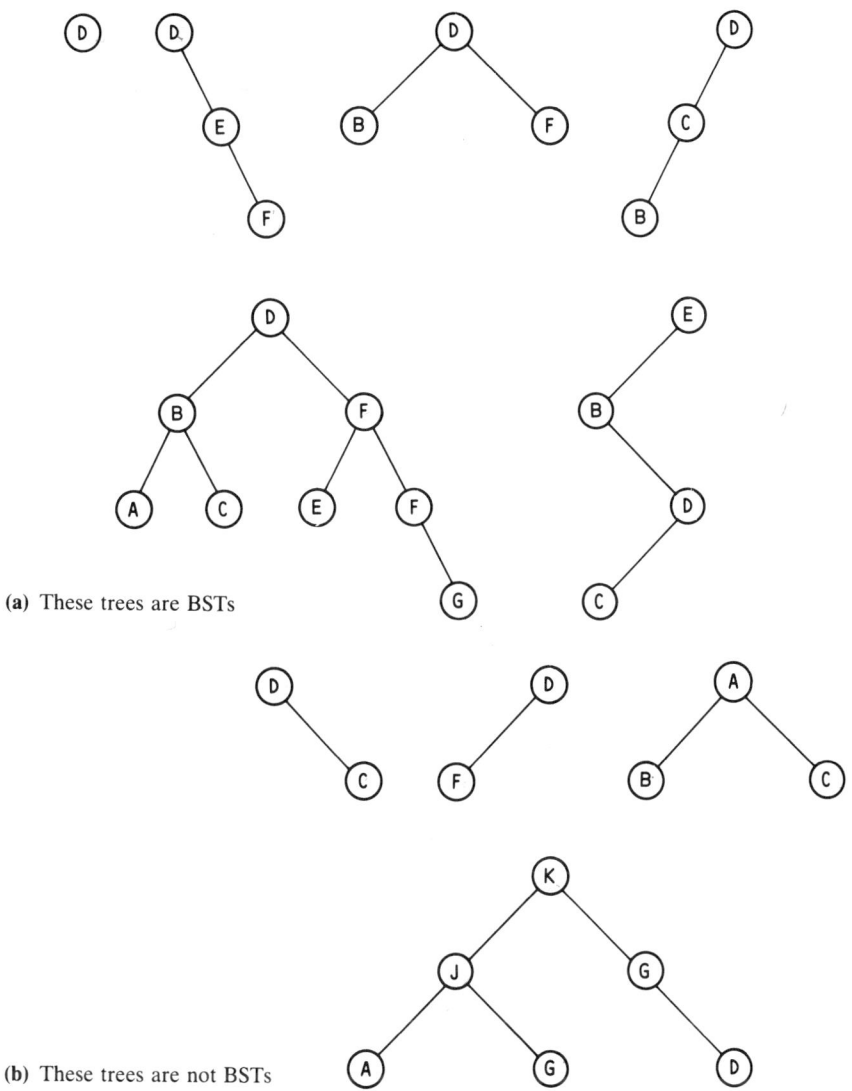

Figure 7-19 Binary search trees

7.5.2 The Update Operation: Inserting a Record in a BST

Since BSTs are recursively defined, we can discover a very natural recursive algorithm for inserting a new key in the tree. Assuming that the tree is not empty, we first test the key against the root key. If it is less than the root key, we insert it in the left subtree; if it is equal to or greater than the root key, we insert it in the right subtree. Eventually, after several recursive calls, we will reach a point

```
DEFINITION MODULE TableHandler;

    TYPE KeyType   = ... ; (* to be determined *)
    TYPE ValueType = ... ; (* same here         *)
    TYPE TableType = ... ; (* this one too      *)

    PROCEDURE Create(VAR T: TableType);
    PROCEDURE Update(VAR T: TableType;
                         K: KeyType;
                         V: ValueType);
    PROCEDURE Search(T:  TableType;
                     K:  KeyType) : ValueType;
    PROCEDURE Delete(VAR T: TableType;
                         K: KeyType);
    PROCEDURE Report(T: TableType);

END TableHandler;
```

Figure 7–20 Sketch of table handler (repeated from chapter 2)

where the subtree into which the new key is to be inserted is empty. At this point, we just create a new node for the key and link it to the appropriate pointer in the parent node. Figure 7–22 gives a procedure for inserting a key in a BST.

In Figure 7–23, we show four "auxiliary routines" called by the insertion procedure. Two of these procedures, ConnectLeft and ConnectRight, are responsible for connecting a leaf node, created by the function procedure MakeNode, to its parent node as the left or right child of the parent, respectively. The fourth procedure is called ProcessDuplicate and handles the case where a "duplicate key" is encountered, that is, where a given key is seen for the second time. The action

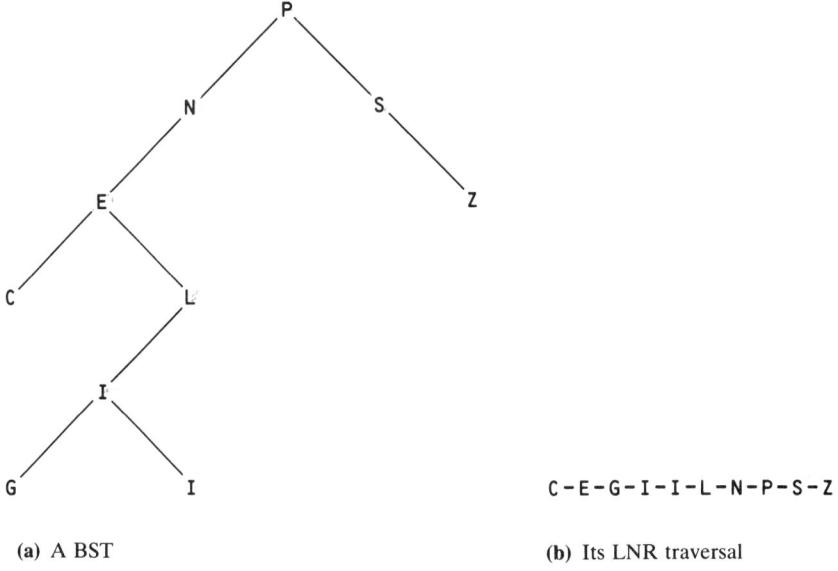

(a) A BST

(b) Its LNR traversal

C-E-G-I-I-L-N-P-S-Z

Figure 7–21 LNR traversal of a binary search tree

Sec. 7.5 Binary Search Trees (BSTs)

```
PROCEDURE UpdateBST(VAR T:   Tree; K: KeyType);

   VAR
      R: Relation;

   BEGIN

      IF T = NIL THEN
         T := MakeNode(K);

      ELSE
         R := KeyCompare(K, T^.key);
         CASE R OF

            Less:
               IF T^.left = NIL THEN
                  ConnectLeft(T,K);
               ELSE
                  UpdateBST(T^.left,K);
               END

          | Greater:
               IF T^.right = NIL THEN
                  ConnectRight(T,K);
               ELSE
                  UpdateBST(T^.right,K);
               END

          | Equal:
                ProcessDuplicate(T,K);
         END;
      END;

   END UpdateBST;
```

Figure 7-22 Binary search tree update procedure

to be taken upon encountering a duplicate key is application dependent. In some applications, duplicate keys are not allowed, and ProcessDuplicate would signal an exception. In the present application, we treat the second occurrence of a key as though it were *greater* than the original, thereby forcing it into the right subtree. This strategy gives what is known as a *stable sort*, in which equal keys appear in the LNR traversal in precisely the order in which they arrived.

What is the performance of the insertion algorithm? Suppose the BST is balanced. Then if there are K nodes in the tree, the number of levels will be approximately log K, and finding the right place for a new arrival will take approximately log K comparisons. But will the BST be balanced? In fact, we have no guarantee that it will be: balance depends on the order of arrival of the new keys, which is independent of the algorithm that built the BST.

How bad can it get? Suppose that the keys arrive *in sequential order*, say, sorted in an ascending fashion. Then each new arrival will be greater than the previous one, and will thus go into the right subtree. No arrival will ever go into a left subtree! Thus, the tree will be badly deformed, rather resembling a linear list. Consequently, adding a new arrival will be a *linear* function of the number of keys already there, instead of a *logarithmic* one. Figure 7-24 shows this worst-case situation.

```
PROCEDURE MakeNode(K: KeyType) : Tree;

    VAR
       Result: Tree;

    BEGIN

       Allocate(Result,SIZE(BinaryTreeNode));

       WITH Result^ DO
           key   := K;
           left  := NIL;
           right := NIL;
       END;

       RETURN Result;

    END MakeNode;


PROCEDURE ConnectLeft(VAR T:  Tree; K: KeyType);

    BEGIN
       T^.left := MakeNode(K);
    END ConnectLeft;


PROCEDURE ConnectRight(VAR T:  Tree; K: KeyType);

    BEGIN
       T^.right := MakeNode(K);
    END ConnectRight;


PROCEDURE ProcessDuplicate(VAR T:  Tree; K: KeyType);

    BEGIN
       UpdateBST(T^.right,K);
    END ProcessDuplicate;
```

Figure 7-23 Auxiliary routines for BST update

The best possible performance of Update, then, is logarithmic, the worst linear. The average will be somewhere in between. In practice, BSTs are not useful for applications in which there is a high probability that the incoming data items are already sorted. In situations where the data is reasonably "mixed up," the average performance is acceptable.

As it happens, algorithms exist for balancing BSTs. A balanced BST is sometimes called an *AVL* tree, after Adel'son-Vel'skii and Landis, the two researchers who discovered it. The algorithm they discovered is rather complicated and unintuitive, and in any event beyond the scope of this book.

7.5.3 The Search Operation: Finding a Record in a BST

The algorithm for the procedure Search is similar to the first part of Update. Given a key to search for, we simply start at the root of the tree, comparing the key with the one we find at that node. If the two are equal, we have found the record we are looking for and return its location. If our key is less, we search in the left

Sec. 7.5 Binary Search Trees (BSTs)

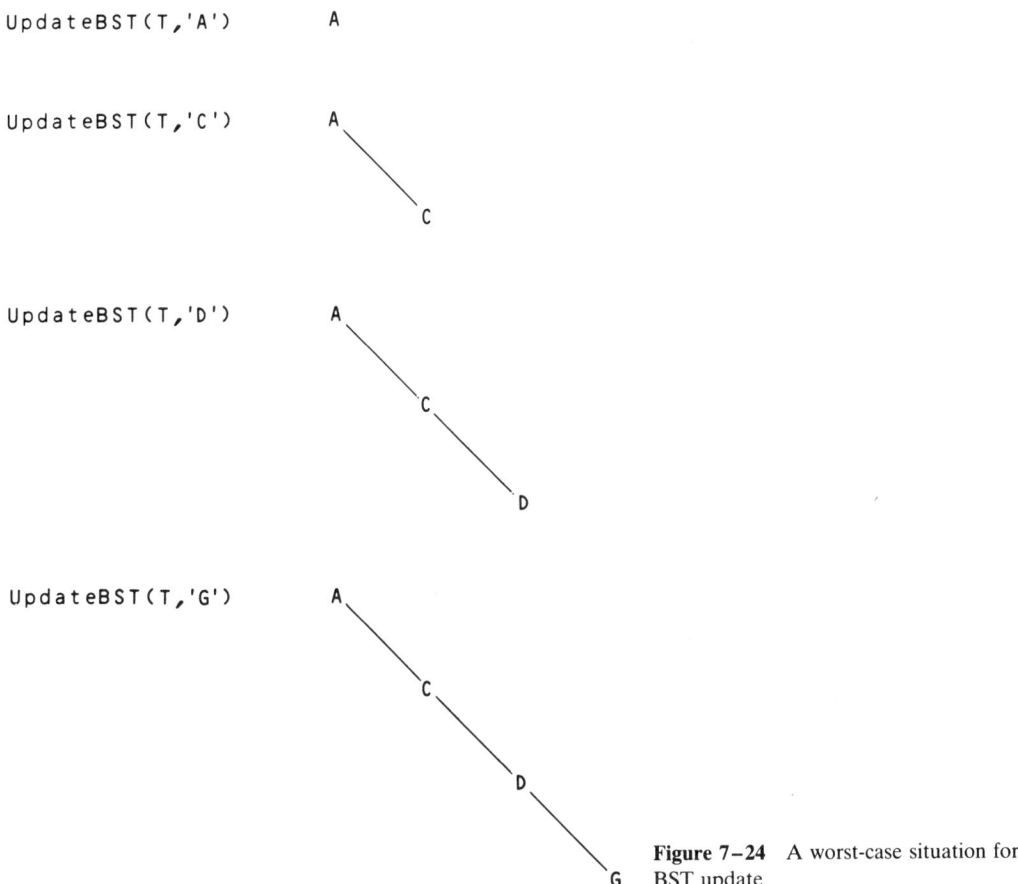

Figure 7-24 A worst-case situation for BST update

subtree; if it is greater, we search in the right subtree. If we reach a null subtree, we know the key we are looking for is not in the tree. The algorithm is shown as a Modula-2 function in Figure 7-25. Clearly, its performance depends upon the structure of the particular tree, varying from logarithmic in the best case to linear in the worst.

Update and Search show clearly where a binary search tree got its name from: they are the tree equivalents of their counterparts for ordered arrays and are very similar to a binary search. The binary search tree can thus be seen as a binary tree used for searching (*binary* search tree) or as a tree which implements a binary search (*binary search* tree).

7.5.4 The Delete Operation: Deleting a Record from a BST

In storing dynamically varying sets of records, deletions need to be performed in such a manner that the BST property of the remaining tree is preserved. Assuming,

```
PROCEDURE SearchBST(T: Tree; K: KeyType): Tree;
   VAR
      R: Relation;
   BEGIN
      IF T = NIL THEN
         RETURN NIL;
      ELSE
         R := KeyCompare(K, T^.key);
         CASE R OF
            Less:
               RETURN SearchBST(T^.left,K)
          | Greater:
               RETURN SearchBST(T^.right,K)
          | Equal:
               RETURN T
         END;

      END;

   END SearchBST;
```

Figure 7-25 BST search operation

then, that deletion of a node from a BST always takes the form "delete the record containing a given key," what is the algorithm? If the desired node is a leaf, we have an easy problem: just cut it off the tree. Otherwise it has subtrees, which we need to rearrange so that the BST property is not disturbed. If only one subtree is present, we can just delete the node by making its parent point to whichever child is there. If both subtrees are present, we replace the node by its *LNR* or *Inorder Successor*, which is so called because it is the next node to be traversed in an LNR or inorder traversal. Formally, we have:

TO DELETE A NODE FROM A BST:

1. Locate the desired node by a search; call it *t*.
2. If *t* is a leaf, disconnect it from its parent. (Set the pointer in the parent's node to NIL.)
3. If *t* has a left child but no right child, remove *t* from the tree by making *t*'s *parent* point to *t*'s *left child*.
4. If *t* has a right child but no left child, remove *t* from the tree by making *t*'s *parent* point to *t*'s *right child*.
5. Otherwise, find *t*'s LNR successor, i.e., the node in the *t*'s *right* subtree with the *smallest* key. Copy this node's information into *t*, and then delete the node.

Figure 7-26 shows some deletions from a BST which illustrate each case of the preceding algorithm.

To arrive at a deletion procedure, consider the last case of the algorithm. To handle that case, we write an auxiliary procedure DeleteSmallest(T,K) which finds the node in a tree T with the smallest key, deletes that node, and returns the

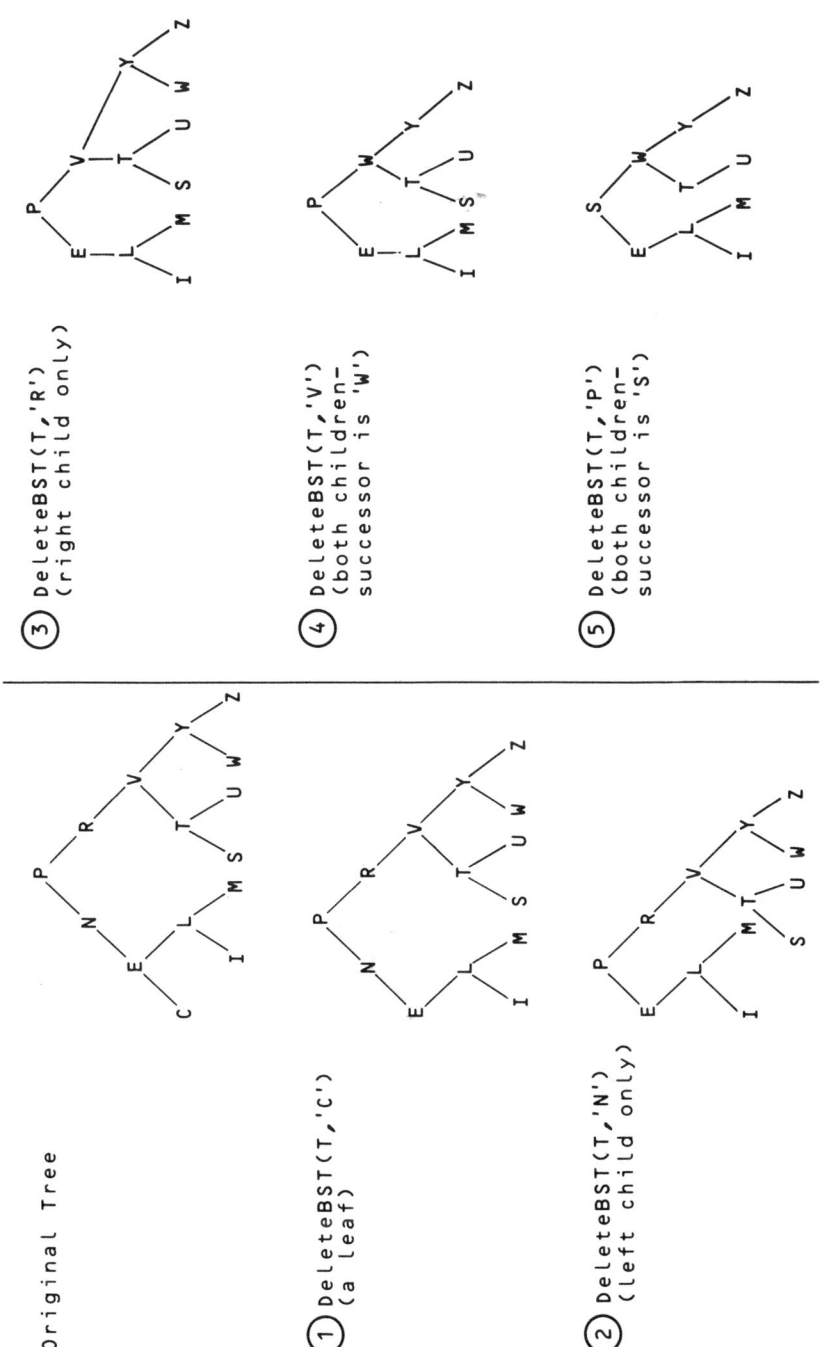

Figure 7-26 Deletions from a BST

```
PROCEDURE DeleteSmallest(VAR T: Tree; VAR K: KeyType);

    VAR
       TempT: Tree;

    BEGIN
       IF T^.left = NIL THEN

          (* T ALREADY POINTS TO THE SMALLEST KEY *)
          TempT := T;
          K :=    T^.key;
          T :=    T^.right;
          Deallocate(TempT);

       ELSE

          (* T HAS LEFT CHILD, SO LOOK THERE *)
          DeleteSmallest(T^.left,K);

       END;

    END DeleteSmallest;
```

Figure 7-27 Procedure to delete and return the smallest item in a BST

key in K. The procedure is shown in Figure 7-27; the full DeleteBST procedure is given in Figure 7-28.

There are really two possible algorithms for Delete: we could just as well have used *t*'s LNR *predecessor*, the node in its *left* subtree with the *largest* key. Finding this other algorithm is left as an exercise.

Notice that a deletion clearly affects a tree's balance. Experimental results have shown that in a "real-world" BST, with many insertions and deletions, all coming in random order, the tree's balance is best maintained by alternating successor and predecessor deletions.

7.6 GENERAL TREES

In this section, we present two examples of the use of more general tree structures than binary trees. The first example is the *digital search tree*, which is an application of a tree in which a node has a potentially large and highly variable number of children. The second example is the B-tree, a structure used frequently to organize large files on secondary storage devices. In a B-tree each node has a variable number of children, but with a fixed, usually relatively small, maximum. The balanced BST turns out to be a special case of the B-tree.

7.6.1 Digital Search Trees

Consider the problem of designing a program to check whether the words in a report are spelled correctly. This is usually solved by creating a dictionary of all those words likely to be used in the report, and then writing a program that scans

Sec. 7.6 General Trees

```
PROCEDURE DeleteBST(VAR T: Tree; K: KeyType);

VAR
   TempK:  KeyType;
   TempT:  Tree;
   R:      Relation;
BEGIN
   R := KeyCompare(K, T^.key);
   CASE R OF
      Less:
         DeleteBST(T^.left,K)

    | Greater:
         DeleteBST(T^.right,K)

    | Equal:
         IF   (T^.left = NIL) AND (T^.right = NIL) THEN
         (* T IS A LEAF; DELETE IT *)
            T := NIL;
            Deallocate(T);

         ELSIF T^.right = NIL THEN
         (* REPLACE T BY ITS PREDECESSOR *)
            TempT := T;
            T     := T^.left;
            Deallocate(TempT);

         ELSIF T^.left  = NIL THEN
         (* REPLACE T BY ITS SUCCESSOR *)
            TempT := T;
            T     := T^.right;
            Deallocate(TempT);

         ELSE
         (* BOTH CHILDREN THERE *)
            DeleteSmallest(T^.right,TempK);
            T^.key := TempK;

         END;

   END;

END DeleteBST;
```

Figure 7-28 BST delete operation

the report, word by word, and reports all words not appearing in the dictionary to the user as possible spelling errors. Thus, a word will be reported if it is misspelled, but also if it is a valid word that just isn't in the dictionary.

Theoretically, any kind of table can be used to represent the dictionary—an ordered array or a balanced BST, for example. The difficulty is that for real-world dictionaries, the amount of space required would be enormous, since in the usual tables complete words have to be stored. The *digital search tree* provides a solution: only a single character is stored in each node. There are as many separate trees as there are possible first letters. (Such a collection of trees is usually called a *forest*.) Thus, each tree has a different first letter at its root, and the children of the root contain the second letters of all the words with the given first letter, the children of a given second-letter node contain the third letters of words with

210   Tree Structures   Chap. 7

CAN, CANE, CON, CONE, COP, COPE, CURE, CURT, CUT, CUTE, CUTS

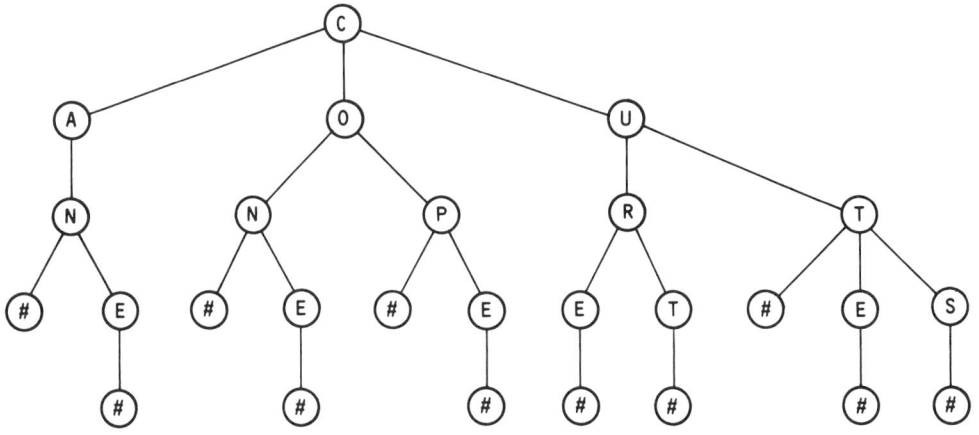

DEBT, DEBTOR, DEEP, DEEPLY, DO, DON, DONATE, DONE

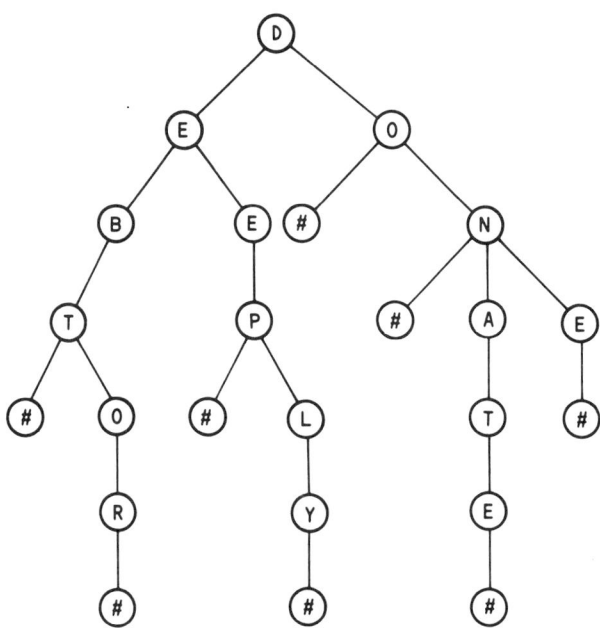

Figure 7–29 Two digital search trees

the given second letter, and so on. A search for a word in such a forest then involves starting with the first letter of the word and trying, letter by letter, to find a path through the appropriate tree. If one is found, the word is valid; otherwise it is reported.

Figure 7–29 shows a diagram of a pair of digital search trees for some words beginning with C and D. Notice that we have added a special character '#' to indicate the end of a word, so that, for example, the word "DEC" (not a valid English word) would not be erroneously reported as correct by going part way down the path for "DECIDE." If we can find an appropriate implementation of this tree, great storage savings can be achieved, making it feasible to build a dictionary which can be loaded into primary memory in its entirety, thus avoiding time-consuming disk accesses.

One possible implementation is to represent a node by an array of 27 pointers, one for each letter and the end-of-word character, so a parent can have up to 27 children. This has the advantage of letting us determine in constant time whether, say, the letter s in a given node has the letter q for a child: we just check to see whether the pointer for q in the s node is NIL or not. On the other hand, such an implementation uses space very inefficiently, since its arrays will normally be sparse: whatever the language of the dictionary, many letter combinations will simply not appear. A given letter, at a given "level" of the words being indexed, will have only a few successors.

A better approach is to treat the node analogously to a sparse vector: represent the children of a given parent as an ordered linear list, as shown in Figure 7–30. Now each node has only two pointers, one to its leftmost child, the other to its own immediate right sibling. The trees in the forest are all connected at the top level to an artificial "super-root," representing the beginning of the word. (We can use the same artificial end-of-word sign '#' here.) A part of the dictionary used in the Figure 7–30 is shown in this implementation in Figure 7–31.

As in other sparse-vector techniques, we have traded time for space: determining whether a certain letter has another given letter as a child requires a linear search through the child list, but the child lists are likely to be short. Writing an appropriate module for the digital search tree, as well as examining the space and time requirements therefor, are left as an exercise.

The use of a linked list to represent siblings is not limited to the spelling-checker application: it is a common implementation structure for all kinds of general trees. Sometimes each node carries a pointer back up to its parent as well.

7.6.2 B-Trees

The *B-tree* is a generalization of the balanced BST (or AVL tree), frequently used as a basis for structuring large files on external devices like disks. The balanced BST developed in section 7.5 can be generalized to allow the nodes of the tree to be stored on disk instead of memory merely by using a disk input/output package that permits addressing individual records on disk and then letting node pointers

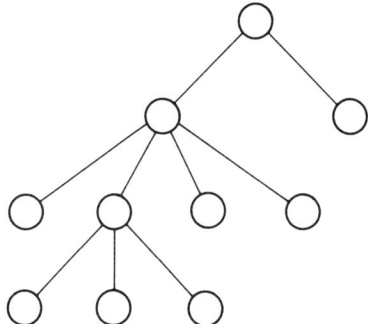

(a) Abstraction

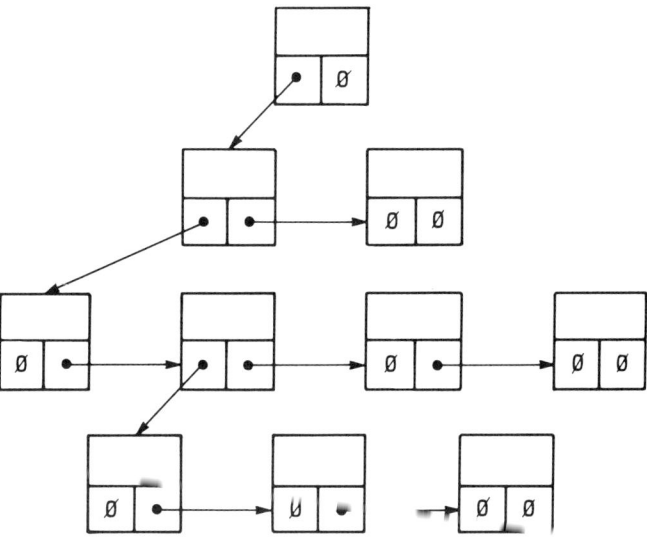

(b) Implementation

Figure 7-30 Left child/right sibling implementation of a general tree

represent disk-record addresses rather than main-memory locations. However, for a BST large enough to arouse consideration of storing it externally, this scheme could use too many disk accesses, which are slow because of the time required to search for a given record on the device. However stored, a balanced BST with N nodes requires $O(\log(N))$ record accesses in the worst case, and for really large files, 10, 20, or 30 disk operations to carry out a search is just too many.

On the other hand, even when a tree is large enough to justify disk storage, we do not usually need more than a few records at a time in main memory.

Sec. 7.6 General Trees

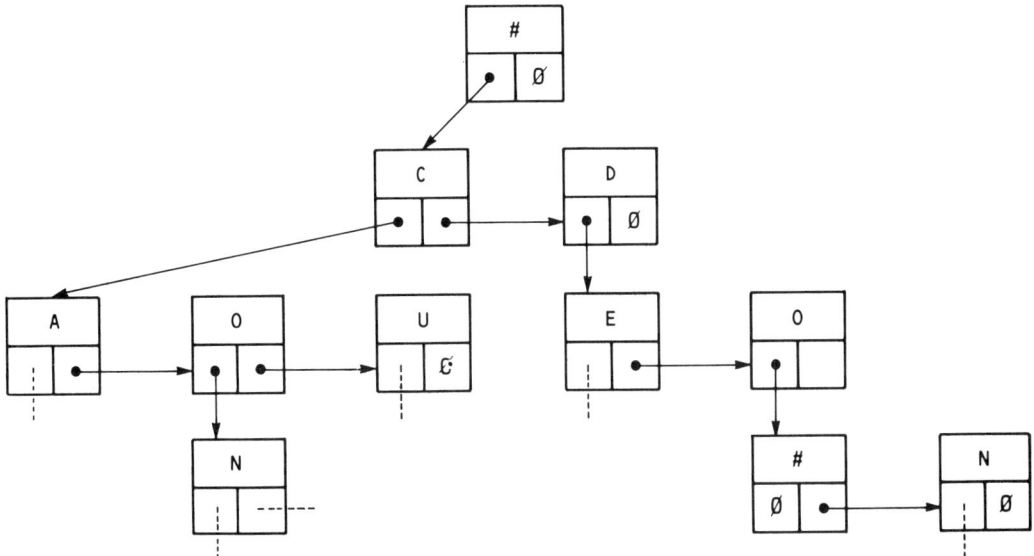

Figure 7-31 Digital search tree implementation

Consequently, these records can be rather large. Moreover, disk storage is relatively inexpensive, and the time for retrieving a large record from disk is about the same as the time for retrieving a small one, since most of the time is used to *find* the record, not to transfer it to main memory.

All this gives rise to the idea of a *B-tree of order K*, in which each node is of *fixed size*, capable of holding K keys and $K + 1$ pointers to children, as shown in Figure 7-32. Under this scheme, a balanced BST is simply a B-tree of order

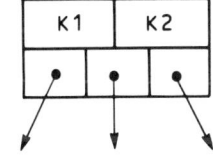

(a) Order 2 node

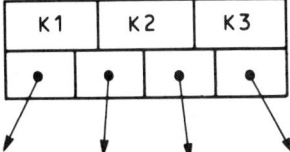

(b) Order 3 node

Figure 7-32 B-tree nodes

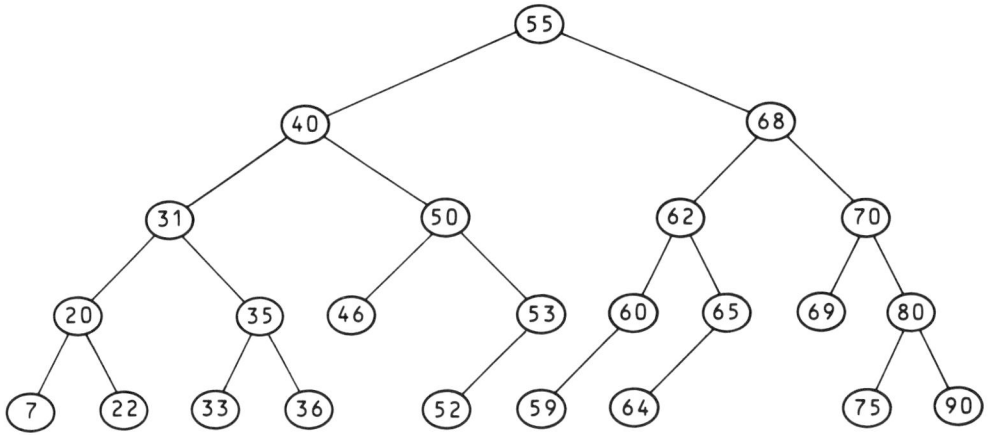

(a) A balanced BST (AVL tree or B-tree of order 1)

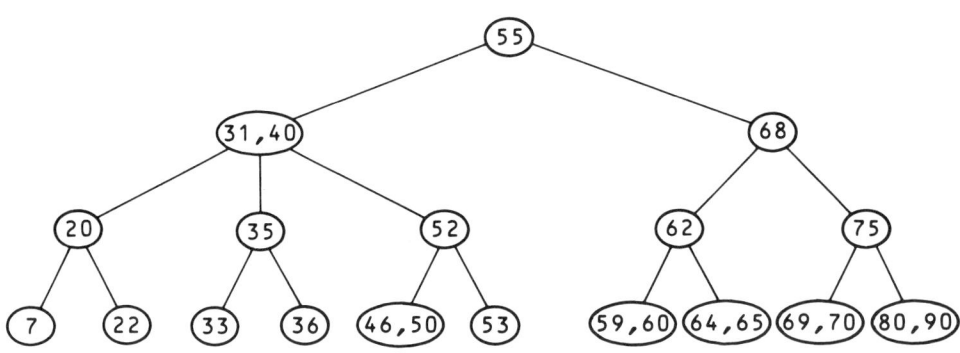

(b) A 2-3 tree (or B-tree of order 2) for the same data

Figure 7-33 Balanced BST compared with 2-3 tree

one. Another special case, the B-tree of order two, often goes by the name "2-3 tree," because each node has *two* keys and *three* pointers.

The keys in a given node of a B-tree are *ordered*. Looking at Figure 7-32, we construct the tree so that the two pointers surrounding a given key point to subtrees in such a way that the BST property is preserved: all the values in a given key's left subtree are less than that key, and all the values in its right subtree are greater than it *but less than the adjacent key*. A 2-3 tree (B-tree of order two) is shown in Figure 7-33(b); its corresponding balanced BST is shown in Figure 7-33(a) for comparison. Note the difference in the depths of the two trees. In this particular case, the depths differ by only one, but there is also a good bit of "extra capacity" for keys in the 2-3 tree, which can be used before more levels are required. Generally speaking, we maintain the balance in a B-tree by requiring that a node always be at least half full. Placing several keys in each node leads to a "flatter" tree, and thus to fewer disk accesses.

For completeness, we should add that B-tree nodes do not usually carry the entire record around, since that would require more space per node, much of it unused. Rather, the tree is used as a directory structure: along with the K keys are often stored the disk addresses of the corresponding records. Addresses take a lot less space than full records!

Detailed implementation of the B-tree structure is left to the exercises.

7.7 DESIGN ONE: BUILDING AN EXPRESSION TREE

In Chapter 5, we developed a function to translate an arithmetic expression into RPN. It turns out that the algorithm to produce the expression tree for that expression is very similar to that function, and the decision process for pushing and popping operators on and off a stack is exactly the same.

There is a difference, though: in the case of the function, when an operand (letter or number) was scanned, it was immediately output (concatenated to the RPN string); similarly, an operator popped from the stack was immediately output. In the case of an expression tree, however, we need to retain those operands and operators, connecting them together in the tree. We do this by maintaining a separate stack for intermediate *tree* results, letting items in the stack be pointers to subtrees instead of just characters. Likewise, our operator stack is converted to hold pointers to nodes; an operator is placed in such a node before being pushed. At the end of the algorithm, a pointer to the root of the resultant tree is left on top of the node stack. Figure 7–34 shows the conversion of an expression to a tree, with all the details of the nodes illustrated.

The main loop of a Modula-2 function for the expression translator is shown in Figure 7–35; Figure 7–36 gives the detailed code. The translator uses a local procedure, PopConnectPush, which pops an operator node from the operator stack, pops the two top nodes from the node stack, connects the operator node as the root of the new tree, and then pushes this node back onto the node stack. This procedure is the core of the difference between the expression-to-RPN translator and the current expression-to-tree translator.

The similarity of the two algorithms illustrates once again the intimacy of the relationship between infix expressions, trees, and Polish notation.

7.8 DESIGN TWO: A CROSS-REFERENCE GENERATOR

A cross-reference generator is an example of an indexing program. Two applications come from the fields of programming and text analysis.

A programmer uses a cross-reference listing of a program to help debug that program. The cross-reference listing indicates, for each identifier in the program,

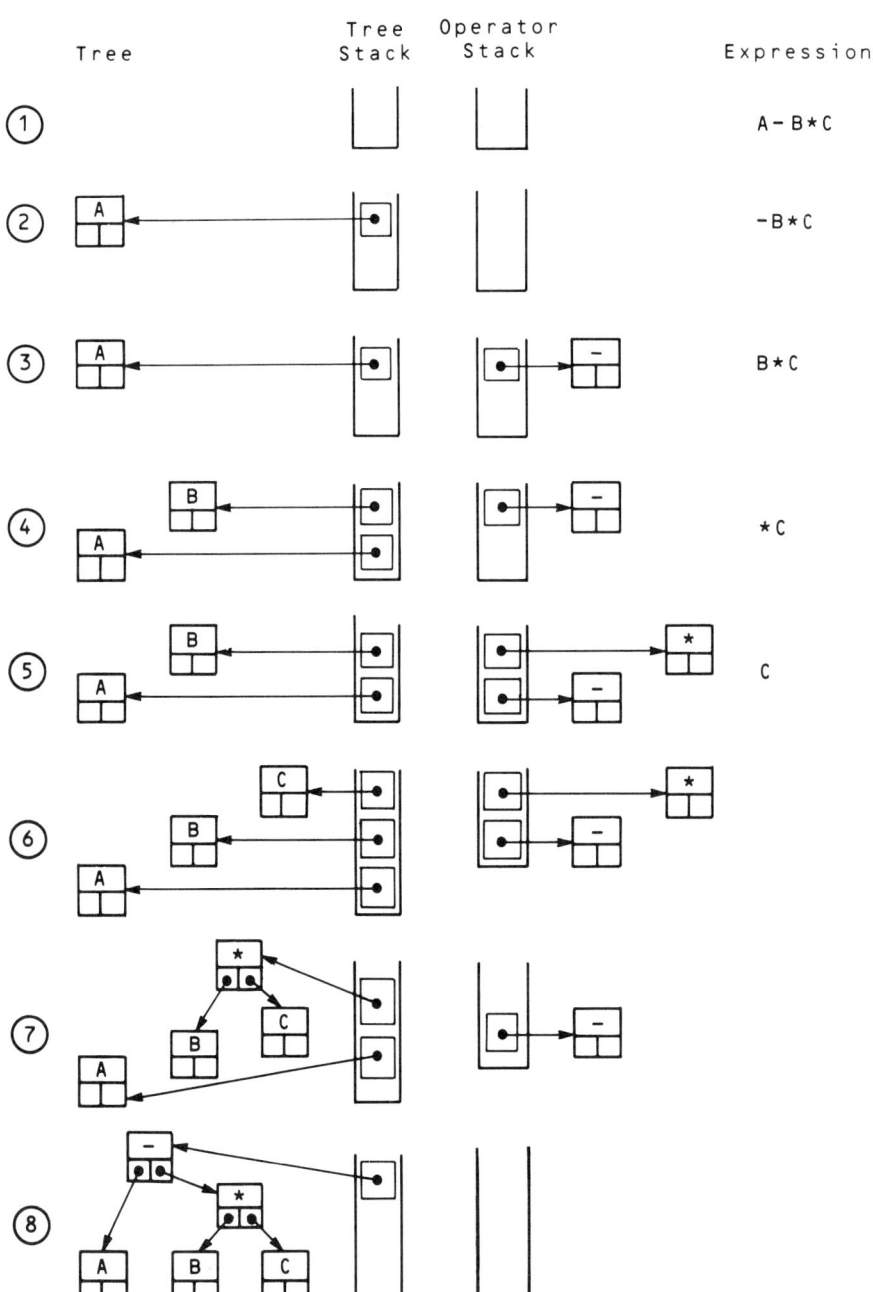

Figure 7-34 Translation of infix expression to a tree

Sec. 7.8 Design Two: A Cross-Reference Generator

```
PROCEDURE ExpTree(X: Text) : Tree;

VAR
    C:           CHAR;
    D:           CHAR;
    T:           Text;
    Temp:        Tree;
    Ops, Nodes:  Stack;
    WeirdChar:   BOOLEAN;

PROCEDURE PopConnectPush;
    BEGIN
        Temp := StackTop(Ops); StackPop(Ops);
        ConnectRight(Temp,StackTop(Nodes)); StackPop(Nodes);
        ConnectLeft (Temp,StackTop(Nodes)); StackPop(Nodes);
        StackPush(Nodes,Temp);
    END PopConnectPush;

BEGIN
    InitTree(Temp);
    T := NullText();
    StackInit(Ops);
    StackInit(Nodes);
    TextCopy(T,X);
    IF TextEmpty(T) THEN RETURN Temp; END;

    LOOP
        C := TextHead(T);
        CASE C OF

            (* CASE STATEMENT BODY GOES HERE *)

        END;
        T := TextTail(T);
        IF TextEmpty(T) THEN EXIT END;
    END;

    WHILE NOT StackEmpty(Ops) DO
        PopConnectPush;
    END;

    RETURN StackTop(Nodes);

END ExpTree;
```

Figure 7-35 Expression-to-tree translator

which statements that identifier appears in. A person analyzing text in a natural language uses a cross-reference listing of that text (sometimes called a *concordance*) to indicate how frequently and in which lines each important word occurs. For example, one or two hundred years ago—long before computers existed, in any case—a number of monks in England produced a concordance of the entire Bible, all by hand, of course!

Whatever its application, a cross-reference generator consists of two parts: a *dynamic table handler* to hold the words read from the text, together with all their references, in some efficient way; and a *scanner* or *parser*, which knows the specifics of the language being analyzed and, therefore, how to distinguish a meaningful word from gibberish in the language.

```
CASE C OF

   'A'..'Z',
   'a'..'z',
   '0'..'9':
      StackPush(Nodes,MakeNode(C))

|  '+' , '-' ,
   '*' , '/' :

      LOOP (* CLEAR HIGHER-PRIORITY OPERATORS *)
         IF StackEmpty(Ops)                       OR
            (KeyPart(StackTop(Ops)) = '(') OR
            (Priority(KeyPart(StackTop(Ops))) < Priority(C))
         THEN
            EXIT;
         END;
         PopConnectPush;
      END;

      StackPush(Ops,MakeNode(C))

|  '(' :
      StackPush(Ops,MakeNode(C))

|  ')' :
      (* CLEAR STACK BACK TO THE LEFT PAREN *)
      WHILE KeyPart(StackTop(Ops)) <> '(' DO
         PopConnectPush;
      END;
      StackPop(Ops);

   ELSE
      WeirdChar := TRUE;

END;
```

Figure 7-36 Body of translator CASE statement

The main loop of the cross-reference generator is shown in Figure 7-37. The scanner procedure GetWord is called to read text from the input file and return the next word it finds; GetWord also keeps the line counter up to date and reports the end of the file at the right time. A word and its line number are put in the table by a call to Update. When the input file is exhausted, Report is called to print out the cross-reference listing. Notice how all the details of the scanning and table handling are hidden.

7.8.1 The Table Handler

Let us first look at the requirements for the table handler. We assume that the distinct words in the text or program are few enough in number that the table can be constructed in main memory. The input text will be scanned, the cross-reference built, and then the results reported *once*. We suppose that we do not know either

Sec. 7.8 Design Two: A Cross-Reference Generator

```
MODULE EnglishXref;

    FROM FileSystem IMPORT File, Open;
    FROM InOut IMPORT ReadString, WriteString, WriteLn;
    FROM Indexer IMPORT GetWord;
    FROM Texts IMPORT Text, TextEmpty;
    FROM XrefTrees IMPORT Tree, UpdateBST, InitTree, TraverseLNR;

    VAR
        FileName:   ARRAY[0..31] OF CHAR;
        F: File;

        T:          Tree;
        LineNumber: INTEGER;
        ThisWord:   Text;

        EOF: BOOLEAN;
        EOL: BOOLEAN;

    BEGIN
        InitTree(T);
        EOF := FALSE;
        EOL := FALSE;
        LineNumber := 1;

        WriteString("Please enter name of data file");
        ReadString(FileName);
        Open(F,FileName,FALSE);

        REPEAT

            GetWord(F,ThisWord,EOL,EOF);

            IF NOT TextEmpty(ThisWord) THEN
                UpdateBST(T,ThisWord,LineNumber);
            END;

            IF EOL THEN
                INC(LineNumber);
            END;

        UNTIL EOF;

        WriteString("Cross Reference Listing for ");
        WriteString(FileName); WriteLn;
        TraverseLNR(T);

    END EnglishXref.
```

Figure 7-37 Main program for cross-reference generator

precisely how many different words will arrive or how many references each will have. Furthermore, words and references will only be added, never deleted. All these facts argue for a table structure whose update and search operations are efficient. The report operation is not worrisome, since it is done only once per run, and no deletion is ever performed.

Unless the total number of words is very large, the BST structure will satisfy the requirements. Also, since people don't often write programs or essays with

```
1 We wish to point out the difference
2 between the terms "data type," "abstract
3 data type," and "data structure."
```

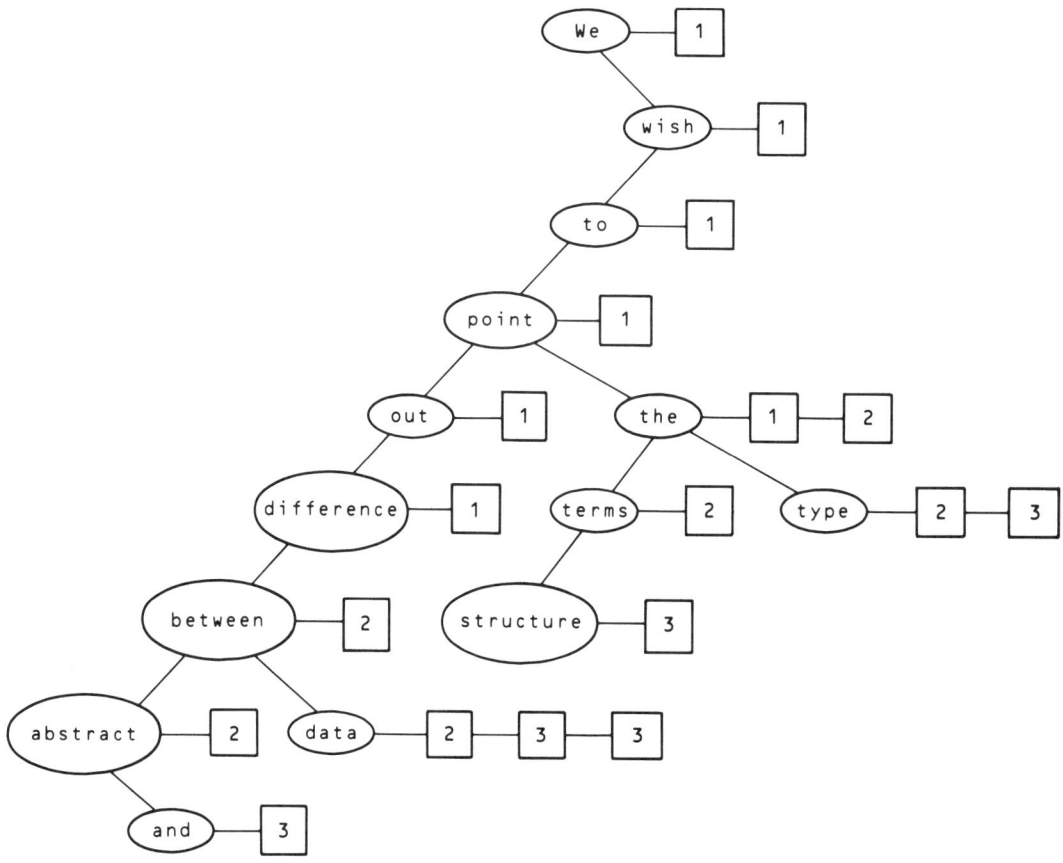

Figure 7–38 A cross-reference tree for English text

the words in alphabetical order, the chances of getting a badly unbalanced tree are slim, so that updating and reporting will perform in roughly $O(\log(N))$ time.

A first attempt to build a BST package would carry a node for each word and reference (line number): the word would be the key, the reference the value. A moment's thought, however, reveals that this is wasteful of space, since we really need only one copy of each word. Accordingly, let us put all the references to a given word in a one-way list and then use the value part of the tree node as the

Sec. 7.8 Design Two: A Cross-Reference Generator

list header. A diagram for this scheme, for a small number of words of English text, is shown in Figure 7-38.

How should update work? Recall that the Update procedure for BSTs, given in section 7.5.2, calls two lower level procedures, ProcessFirst and ProcessDuplicate. For a cross-reference generator, ProcessFirst should create a node for the newly arrived word—which is being seen for the first time—and then set up the reference list with the current line number in the first node of the list. ProcessDuplicate should just create a list node for a new reference, adding it to the tail of the list. Since additions are made only in this manner, it helps if the list header keeps a pointer to the tail.

Type definitions for the tree and list nodes are shown in Figure 7-39. Figure 7-40 gives the modified Modula-2 code for Update, and Figure 7-41 shows the necessary auxiliary routines. The other important operation is Report, which, for a BST, is just a TraverseLNR operation. The Visit procedure in TraverseLNR prints the word in the tree node and then traverses the reference list, printing out line numbers as it goes. The modification of TraverseLNR is left to an exercise.

7.8.2 The Scanner

Developing scanners for languages is a science in itself, and a general treatment is well beyond the scope of this book. Rather, we shall simplify the scanner by relying on some key assumptions about the text to be scanned. We assume that the text is English, that upper and lowercase letters are treated separately, and that numeric digits are treated just like letters, so that dates, phone numbers, etc., will be indexed along with words. Punctuation is not to be indexed; nor is there embedded punctuation like an apostrophe or a hyphen. Finally, a word is never broken across two lines. We shall relax some of these assumptions in the exercises.

We shall implement our scanner using a structure, namely the *finite-state machine*, which generalizes nicely to many other scanning applications. Figure 7-42 shows a simple diagram for this structure, which we introduced in chapter 6 as a *state graph*, or *transition graph*. The circles represent *states* of the machine; the arrows represent *transitions* from one state to another. An arrow is labeled with two things: the left part is the *class* of input character just scanned, and the right part is an *action* to be taken just before the machine moves to a new state.

Figure 7-43 gives the actual state graph for the scanner. It begins in its *Start* state, and continues to cycle in that state until it sees a letter. If a carriage return (CR) is seen, it updates the line counter and returns to the Start state. (We need to account for the possibility of a line containing all blanks or all punctuation.) Once a letter is seen (remember, digits count as letters!), the machine executes an action called *StartWord*, which initializes a string in which to store the word, stores the letter in the string, and transfers to a state called *Build*. While in the Build state, the machine reads characters, adding the letters it finds onto the word string

```
TYPE List = POINTER TO OneWayListNode;

TYPE OneWayListNode =

   RECORD
      val:   INTEGER;
      next:  List;
   END;

TYPE ListHeader =

   RECORD
      head:  List;
      tail:  List;
   END;

TYPE XrefTable = POINTER TO XrefTreeNode;

TYPE XrefTreeNode =

   RECORD
      key:   Text;
      val:   ListHeader;
      left:  Tree;
      right: Tree;
   END;
```

Figure 7-39 Type declarations for cross-reference

```
PROCEDURE UpdateBST(VAR T: Tree; K: Text; V: INTEGER);

   VAR
      R: Relation;

   BEGIN
      IF T = NIL THEN
         T := MakeNode(K,V);

      ELSE
         R := KeyCompare(K, T^.key);
         CASE R OF
            Less:
               IF T^.left = NIL THEN
                  ConnectLeft(T,K,V);
               ELSE
                  UpdateBST(T^.left,K,V);
               END

         |  Greater:
               IF T^.right = NIL THEN
                  ConnectRight(T,K,V);
               ELSE
                  UpdateBST(T^.right,K,V);
               END

         |  Equal:
                  ProcessDuplicate(T,K,V);
         END;
      END;

   END UpdateBST;
```

Figure 7-40 BST update, modified for cross-reference

Sec. 7.8 Design Two: A Cross-Reference Generator

```
PROCEDURE MakeListNode(V: INTEGER) : List;

   VAR
      L: List;

   BEGIN
      Allocate(L,SIZE(OneWayListNode));
      L^.val  := V;
      L^.next := NIL;
      RETURN L;
   END MakeListNode;


PROCEDURE MakeNode(K: Text; V: INTEGER) : Tree;

   VAR
      Result: Tree;
      L:      List;

   BEGIN
      L := MakeListNode(V);
      Allocate(Result,SIZE(XrefTreeNode));

      WITH Result^ DO
         TextInit(key);
         TextCopy(key,K);
         val.head := L;
         val.tail := L;
         left  := NIL;
         right := NIL;
      END;

      RETURN Result;

   END MakeNode;


PROCEDURE ProcessDuplicate(VAR T: XrefTable;
                               K: Word; V: Reference);
   VAR
      L: List;

   BEGIN
      L                  := MakeListNode(V);
      T^.val.tail^.next  := L;
      T^.val.tail        := L;
   END ProcessDuplicate;
```

Figure 7–41 Auxiliary routines for cross-reference

by means of an action called *AddLetter*. When a character that is not a letter is seen, the word is complete and the machine transfers to its *Finish* state. If the latter character was a carriage return, the line counter is incremented.

We implement the finite-state machine by hiding it in a module which exports only the procedure GetWord. The state names, input classes, and actions are written as *enumeration types*, as shown in Figure 7–44. The transition graph is implemented as a two-dimensional array which uses the states as its row subscripts and classes of inputs as its column subscripts. Each entry in the array is itself a

224 Tree Structures   Chap. 7

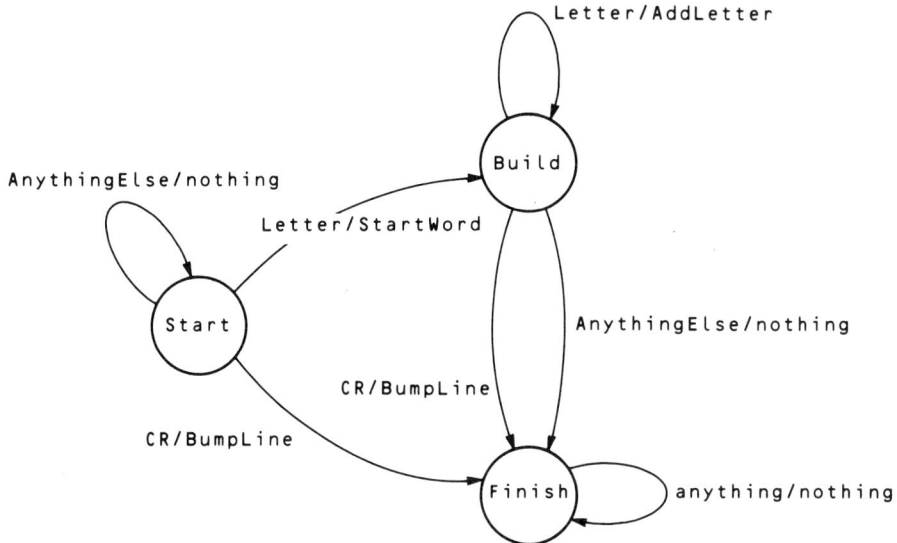

Figure 7-42 Graph and table notations for finite-state machine

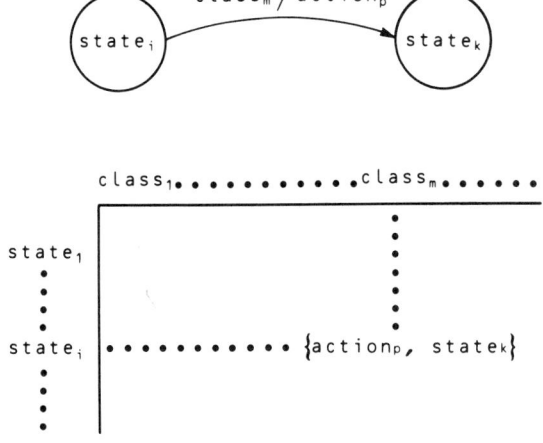

Figure 7-43 State graph for scanner for simple English

Sec. 7.8 Design Two: A Cross-Reference Generator

```
TYPE State         = (Start, Build, Finish);

TYPE InputClass    = (Letter, CR, AnythingElse);

TYPE Action        = (Nothing, StartWord, BumpLine, AddLetter);

TYPE LexicalEntry =

  RECORD
    ThisAction: Action;
    NewState: State;
  END;

TYPE FSM_Table     = ARRAY State, InputClass
                        OF LexicalEntry;

VAR EnglishText: FSM_Table;
```

Figure 7–44 Type declarations for lexical scanner

```
              (* entries for current state = Start *)

EnglishText [Start] [Letter]          .ThisAction := StartWord;
EnglishText [Start] [Letter]          .NewState   := Build;

EnglishText [Start] [CR]              .ThisAction := BumpLine;
EnglishText [Start] [CR]              .NewState   := Finish;

EnglishText [Start] [AnythingElse]    .ThisAction := Nothing;
EnglishText [Start] [AnythingElse]    .NewState   := Start;

              (* entries for current state = Build *)

EnglishText [Build] [Letter]          .ThisAction := AddLetter;
EnglishText [Build] [Letter]          .NewState   := Build;

EnglishText [Build] [CR]              .ThisAction := BumpLine;
EnglishText [Build] [CR]              .NewState   := Finish;

EnglishText [Build] [AnythingElse]    .ThisAction := Nothing;
EnglishText [Build] [AnythingElse]    .NewState   := Finish;

              (* entries for current state = Finish *)

EnglishText [Finish] [Letter]         .ThisAction := Nothing;
EnglishText [Finish] [Letter]         .NewState   := Finish;

EnglishText [Finish] [CR]             .ThisAction := Nothing;
EnglishText [Finish] [CR]             .NewState   := Finish;

EnglishText [Finish] [AnythingElse]   .ThisAction := Nothing;
EnglishText [Finish] [AnythingElse]   .NewState   := Finish;
```

Figure 7–45 FSM state table for lexical scanner

```
PROCEDURE Classify(Ch: CHAR): InputClass;
  BEGIN
    CASE Ch OF
      EOL          : RETURN CR
    | 'A'..'Z'     : RETURN Letter
    | 'a'..'z'     : RETURN Letter
    | '0'..'9'     : RETURN Letter
      ELSE           RETURN AnythingElse;
    END;
  END Classify;
```

Figure 7-46 Character classification function

```
PROCEDURE GetWord(F:           File;
              VAR Word:        Text;
              VAR EndLine:     BOOLEAN;
              VAR EndFile:     BOOLEAN) ;
  VAR
    Ch:           CHAR;
    ThisClass:    InputClass;
    PresentState: State;
    ThisEntry:    LexicalEntry;
    NewAction:    Action;

  BEGIN
    TextInit(Word);
    EndLine := FALSE;
    EndFile := FALSE;
    PresentState := Start;

    LOOP

      (* CODE FOR MAIN LOOP GOES HERE *)

    END;

  END GetWord;
```

Figure 7-47 Framework for GetWord

record, containing an action field and a new-state field. The filled-in array for the scanner is shown in Figure 7-45.

The procedure GetWord is the "machine" that actually moves around the state graph. It reads a character, classifies it (by calling a function shown in Figure 7-46), determines the action to be taken by looking in the array using its current state and input class as subscripts, executes the action, and then goes to its new state. If the new state is Finish, the procedure returns to the caller. The main loop for GetWord is given in Figure 7-47; details appear in Figure 7-48.

Designing the cross-reference generator has shown the advantages of separating the independent functions of scanning and table handling into manageable pieces. It has also illustrated the clarity with which structures like tables can be written using enumeration types. A number of the exercises invite various modifications to this basic design.

```
            LOOP
                IF    PresentState = Finish THEN
                    RETURN;
                ELSIF Eof(F) THEN
                    EndFile := TRUE;
                    RETURN;
                ELSE
                    ReadChar(F,Ch);
                    Write(Ch);
                    ThisClass := Classify(Ch);
                END;

                ThisEntry := EnglishText[PresentState][ThisClass];
                NewAction := ThisEntry.ThisAction;

                CASE NewAction OF
                    Nothing :
                        (* NOTHING! *)
                |   StartWord :
                        Word := CharToText(Ch);
                |   AddLetter :
                        Word := AppendChar(Word,Ch);
                |   BumpLine :
                        EndLine := TRUE;
                END;

                PresentState := ThisEntry.NewState;
            END;
```

Figure 7-48 Details of GetWord

7.9 STYLE GUIDE: THREADING TREES FOR EFFICIENCY

Sometimes it is useful to have a nonrecursive algorithm available for tree traversal: not every language supports recursion directly, and even in those that do, recursion requires extra storage and time for all those subprogram calls. Accordingly, we shall develop a technique called *threading* that traverses a tree nonrecursively. Although we illustrate for the case of a BST, threading is equally applicable to expression trees, and the details are left to an exercise.

Threading is a very simple idea: as we build a BST, we fill empty pointer fields with pointers that help us move *up* the tree as well as down. Moving up the tree helps us to find the successor of a node during a Report or TraverseLNR operation. Figure 7-49 shows several threaded BSTs with the threads shown as dashed lines. Such trees are often called *right in-threaded*, because they contain threads to facilitate their right inorder traversal.

Where are the threads generated in threading stored? If a node has a right child, then its LNR successor is *below* it, somewhere in the right subtree. Otherwise, its LNR successor is *above* it in the tree. Since a node needs a thread only

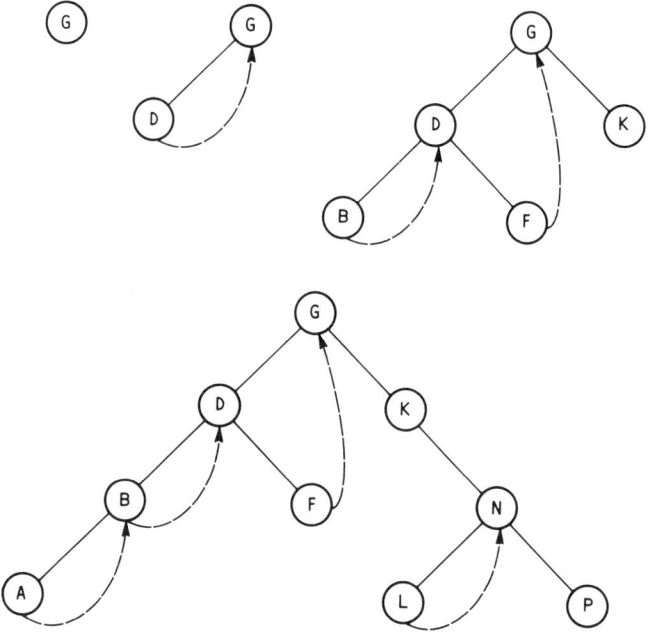

Figure 7-49 Some threaded binary search trees

if it has no right child, common practice is to store the thread in the right-child field of a node with a null right child, using some kind of flag to indicate that it is a thread and not an ordinary pointer. In a cursor implementation of a tree-node storage pool, the pointers are always positive integers, so a thread can be distinguished by making it negative. Since we are employing a dynamic implementation, we shall add to each node a Boolean field called Thread which is true if a thread is stored in the right child field, and false otherwise. Figure 7-50 gives the modified type definition.

Now let us give a modified TraverseLNR procedure using the threads we have just created. Essentially, the procedure just moves all the way down the left side of the tree to find the first node to be visited, follows the threads back up

```
TYPE Tree = POINTER TO ThreadedBinaryTreeNode;

TYPE ThreadedBinaryTreeNode =
    RECORD
        key:    KeyType;
        left:   Tree;
        right:  Tree;
        thread: BOOLEAN;
    END;
```

Figure 7-50 Threaded binary search tree node

```
PROCEDURE TraverseLNR(VAR T: Tree);

VAR
   Current:  Tree;   (* search pointers *)
   Previous: Tree;   (* Previous trails Current *)

BEGIN
   Current := T;
   REPEAT

      Previous := NIL;

      (* DOWN LEFT BRANCH TO BOTTOM *)
      WHILE Current <> NIL DO
         Previous := Current;
         Current  := Current^.left;
      END;

      IF Previous <> NIL THEN
         VisitTree(Previous);
         Current := Previous^.right;

         (* NOW BACK UP FOLLOWING THREADS *)
         WHILE Previous^.thread DO
            VisitTree(Current);
            Previous := Current;
            Current  := Previous^.right;
         END;

      END;

   UNTIL Previous = NIL;
END TraverseLNR;
```

Figure 7-51 Nonrecursive LNR traversal

until a node with a right child is encountered, and then starts back down the left side of the child subtree. This program is shown in Figure 7-51; note that it is nonrecursive.

Finally, we develop a nonrecursive Update procedure which threads the tree as it goes along. When a node is inserted as the *left* child of another node, its parent is its LNR successor. When a node is inserted as the *right* child of another node, *it* becomes its parent's LNR successor, and the LNR successor of the new node is the *parent's* former LNR successor. Figure 7-52 gives several examples of how new nodes are added. The new Update and its auxiliary routines appear in Figure 7-53 and Figure 7-54, respectively. It is left as an exercise to develop nonrecursive Search and Delete routines for right in-threaded BSTs.

7.10 SUMMARY

In this chapter, we presented a number of definitions pertaining to trees. Binary trees were examined in detail, including two of their applications: expression trees and binary search trees (BSTs).

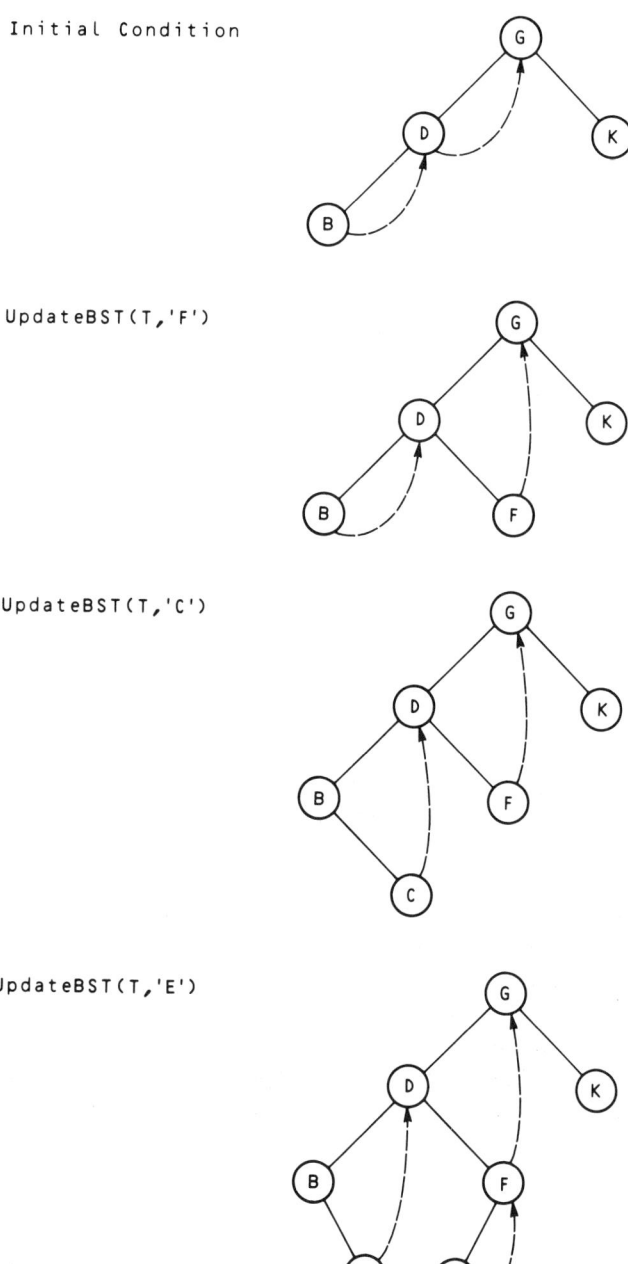

Figure 7-52 Updating a threaded BST

Sec. 7.10 Summary

```
PROCEDURE UpdateBST(VAR T: Tree; K: KeyType);

VAR
   R: Relation;
   Current: Tree;

BEGIN
   IF T = NIL THEN
      T := MakeNode(K);
      RETURN;
   END;

   Current := T;     (* SET UP SEARCH POINTER *)
   LOOP
      R := KeyCompare(K, Current^.key);
      CASE R OF
         Less:
            IF Current^.left = NIL THEN
               ConnectLeft(Current,K);
               EXIT;
            ELSE
               Current := Current^.left;
            END

       | Greater, Equal:    (* TREAT DUPLICATES AS GREATER *)
            IF (Current^.right = NIL) OR Current^.thread THEN
               Current^.thread := FALSE;
               ConnectRight(Current,K);
               EXIT;
            ELSE
               Current := Current^.right;
            END

      END;
   END;
END UpdateBST;
```

Figure 7-53 Nonrecursive BST update

Traversal of a binary tree—that is, visiting each node of the tree in some specified order—was presented, and the usefulness of three traversal schemes, the NLR, LNR, and LRN algorithms, was considered. The close connection between trees and expressions in infix or Polish form was examined, as well.

To illustrate some of the uses of general trees, the digital search tree and the B-tree were introduced. Then, the designs of an expression parser and a cross-reference generator were discussed, and finally, some nonrecursive tree-handling programs were shown, and their associated data structures developed.

This chapter is the last one in which data structures *per se* are presented. The remaining chapters take up two important applications, namely sorting files and searching tables. In these chapters much use is made of all the structures we have used until now; there is also important emphasis on performance issues.

```
PROCEDURE MakeNode(K: KeyType) : Tree;

    VAR
       Result: Tree;

    BEGIN
       Allocate(Result,SIZE(ThreadedBinaryTreeNode));
       WITH Result^ DO
          key    := K;
          left   := NIL;
          right  := NIL;
          thread := FALSE;
       END;
       RETURN Result;
    END MakeNode;


PROCEDURE ConnectLeft(VAR T: Tree; K: KeyType) ;

    BEGIN
       WITH T^ DO
          left           := MakeNode(K);
          left^.thread   := TRUE;
          left^.right    := T;
       END;
    END ConnectLeft;


PROCEDURE ConnectRight(VAR T: Tree; K: KeyType) ;

    VAR
       Temp: Tree;

    BEGIN
       Temp := MakeNode(K);
       IF T^.right <> NIL THEN
          Temp^.thread := TRUE;
          Temp^.right  := T^.right;
       END;
       T^.right := Temp;
    END ConnectRight;
```

Figure 7-54 Auxiliary routines for nonrecursive BST update

CHAPTER 7 EXERCISES

1. Given a connected digraph represented by its adjacency matrix G, write a Boolean function IsTree(G) which returns true iff G represents a tree. Hint: review the definition of a tree!

2. Given a connected digraph represented by its adjacency matrix G, write a Boolean function IsBinaryTree(G) which returns true iff G represents a binary tree.

3. Given a connected digraph represented by its adjacency matrix G, write a Boolean function IsStrictlyBinaryTree(G) which returns true iff G represents a strictly binary tree.

4. Which properties of digraphs must all trees have? Which properties does no tree have?

5. Given a binary tree T, write a Boolean function Balanced(T) which returns true iff T is height-balanced, false otherwise. Hint: Think recursively.
6. Given a binary tree T, write a function Depth(T) which returns the depth of the tree. Hint: think recursively.
7. In a binary tree T, each leaf vertex can be reached by only one path from the root. Write a function MinPathLength(T) which returns the length of the shortest of all such paths. Hint: think recursively.
8. Write a procedure implementing a Delete operation for a binary search tree in which, if the element to be deleted possesses both children, it is replaced by its LNR predecessor instead of its successor.
9. Write a Delete operation for a BST in which successive deletions are done alternately by the successor and predecessor methods.
10. Develop a procedure implementing the Report operation for a digital search tree.
11. Develop a procedure implementing the Update operation for a digital search tree.
12. Develop a procedure implementing the Delete operation for a digital search tree.
13. An interesting application of the digital search tree is the implementation of a multidimensional array. One of the difficulties with row- and column-major implementations is that the storage mapping functions contain multiplications, which may be rather slow to execute. Instead, use a digital search tree which has as many levels as the array has dimensions. For example, an array dimensioned [1 . . 10], [1 . . 5], [1 . . 8] has three levels. The root has ten children; each child points to a node with five children; each of these points to a 1-dimensional, 8-element array. Storing and retrieving values becomes a matter of following pointers instead of doing a subscript calculation. Design a module implementing such a scheme.
14. Write a module implementing the abstract table operations for a 2-3 tree.
15. Write a module implementing the abstract table operations for a B-tree of order K.
16. Develop a procedure implementing the Report operation for a cross-reference tree.
17. Develop a procedure implementing the Delete operation for a cross-reference tree. Be careful: this depends upon whether all references associated with a key are to be deleted, or only one.
18. Develop a nonrecursive Search operation for a threaded BST.
19. Develop a nonrecursive Delete operation for a threaded BST.
20. Develop a threading scheme suitable for NLR traversal of an expression tree and a corresponding nonrecursive traversal procedure.
21. Develop a threading scheme suitable for LRN traversal of an expression tree and a corresponding nonrecursive traversal procedure.

Chapter 8

HASH TABLE METHODS

8.1 GOAL OF THE CHAPTER

In this chapter we reconsider the problem of searching for and updating items in a table, implemented as an array, whose contents vary dynamically. In virtue of performance considerations, we develop the idea of a *hash table*, or *scatter storage* method. This is a table scheme in which updates, searches, and deletions are done, ideally in *constant* time. As we shall see, in actuality the performance of these operations can be made to approximate constant time, but rarely to achieve it exactly.

In a hash table scheme, we identify a record by its key field and assume that there are many more possible key values than there are storage positions in the table. We then seek a mathematical function called a *hash function* or *key-to-address transformation* which produces a table address when supplied with a key. Since there are many more possible key values than addresses, this is a many-to-one function in which many different key values can lead to the same table address. Moreover, since we do not know which keys will actually arrive for placement in the table, it is possible that two keys with the same address actually *will* arrive. Two such keys are called *synonyms* of each other, and the arrival of a second key after this synonym has already been placed in the table is called a *collision*, or sometimes a *hash clash*.

There are many different hash functions, grouped into several different classes, the details of which depend upon the structure and distribution of the keys. Designing a hash table involves two essential parts: finding a hash function that

Sec. 8.3   The Hash Table   235

minimizes the likelihood of collisions, and finding an appropriate scheme for resolving those collisions which do occur.

8.2 SEQUENTIAL AND BINARY SEARCH REVISITED

Let us go back to the table searching strategies we considered in chapter 2. Remember that these are grouped into two main strategies: sequential (or linear) and binary (or logarithmic).

In a sequential search, the items in the array to be searched are assumed to have a key part and a value part, and are maintained in unordered form. Updating the array depends merely upon keeping track of the location of the next empty position in the array and then inserting a new arrival in that position. Searching the array and deleting an entry from it, on the other hand, require looking sequentially through the array, item by item, until either the desired item is found or the end of the array is reached.

In a binary search, we store the table elements in order, sorted by their key. Updating then requires a logarithmic operation (finding the correct position) followed by a linear one (moving the elements to make room for the new one). For tables large enough for us to care about performance, the linear component dominates. Searching the table is purely logarithmic, and deleting an entry from it is similar to updating an entry in terms of performance.

Figure 8–1 gives a summary of the "big Os" of these operations for the two implementations. In the next section we introduce the notion of a hash table, where update, search, and delete operations are carried out in approximately *constant* time.

	UNORDERED	ORDERED
Create	O(1)	O(1)
Update	O(1)	O(N)
Search	O(N)	O(Log(N))
Delete	O(N)	O(N)
Report	O(N x Log(N))	O(N)

Figure 8–1   Comparative performances of table operations for linear and binary search strategies

8.3 THE HASH TABLE

Let us assume, as is the case in most applications, that the set of possible keys K is much larger than the table we wish to maintain. For example, suppose you have around 100 friends whose phone numbers you wish to keep in your list, and you want to retrieve a friend's number according to, say, the first four letters of his or her name. Since you keep making new friends, and you don't know in advance what their names will be, you have to assume a large number of possible four-letter combinations. There are 26^4 or 456,976 four-letter combinations in the English alphabet. Of course, not every combination shows up in people's names—QQQQ would be very unlikely, for instance—but the realistic number is still quite large.

As another example, consider a university with 10,000 students in which each student is assigned, say, a six-digit number upon first arriving at the school. There are one million possible numbers, but only 10,000 students. And a teacher keeping a list of students in a given course may be dealing with only a hundred or so of those. Of course, since the numbers are assigned purely sequentially, the group of numbers in use will tend to drift over time so that at a given moment all *currently registered* students will have numbers with a leftmost digit of, say, 3 or 4. But this still leaves 200,000 possible keys.

Yet a third example is the symbol table used by a compiler or assembler to keep track of the machine addresses it allocates to program variables or identifiers. The keys are the identifiers, the values the assigned addresses. The number of possible identifiers is huge; Fortran, for example, allows a letter followed by up to five letters or digits, and other languages permit even longer names. In practice, of course, a given program will have only a few dozen variables or so, but obviously, the compiler writer cannot predict which ones a programmer will choose.

In the *hash table*, or *scatter storage* technique, the entries are scattered around the table in an approximately uniform fashion, the result of mathematical transformation, called the *hash function* or *key-to-address transformation*, which accepts a key as its input and returns a table address (array subscript) as its output. Such a function is usually designated $h(k)$, where k represents a key; a pictorial representation is shown in Figure 8-2.

In the next section we shall examine a number of hash functions; for the moment all we need to know is that a typical transformation might be simply to take the first few digits or the last few digits of the key, or to multiply the key by some number and select the middle few digits of the result. The point is that these computations generally have constant performance, since arithmetic operations generally do not depend on the values of their arguments and therefore are independent of the number of items in the table and usually of the table size as well. Given a well-chosen transformation, a table address can be delivered in $O(1)$ time.

If, for a given key structure and desired table size, we can invent a good $h(k)$, then updating consists simply of passing the key of an arriving item through this "transformer" to get a table address, usually called the *hash address* or *hash code* and then storing the item at that address (in constant time, of course!). In a similar fashion, a search would pass the key whose value is sought through $h(k)$, and then look in the appropriate table location. And a deletion would just remove the item to be deleted by finding its location and marking that location as available.

All this would work wonderfully—and with guaranteed $O(1)$ performance—

Figure 8-2 A key-to-address transformer

were it not for the fact that there are many more possible keys than there are locations in the table, and we don't know just which keys will arrive. Therefore, the function $h(k)$ *cannot* be one-to-one, and so will deliver the same table address for many different keys. So, potentially, many items will compete for the same table location.

We denote by *synonyms* the elements of the set of keys for which a given $h(k)$ will deliver the same hash address. Mathematically a set of synonyms is an equivalence class. A situation in which a given table location is occupied by one item, and then one of its synonyms arrives, is called a *collision* or *hash clash*. Designing a good hash table depends upon finding good solutions to the following two problems:

1. Find an $h(k)$ that will minimize the number of collisions by spreading arriving records around the table as evenly and uniformly as possible.
2. Since *any* $h(k)$ must be many-to-one, and therefore collisions are inevitable, find a good way of resolving them.

The term "hash function" derives from our desire to "chop up" or "hash together" the characters or digits of the key to get a high degree of randomness in the hash code. The next two sections introduce, respectively, a number of different kinds of $h(k)$ functions, and some methods of resolving collisions. In the meantime, a final word about one operation we haven't mentioned: reporting. The items in a hash table are, by definition, scattered around the table in no particular order. Moreover, in any good hashing scheme they aren't even stored in contiguous locations. So reporting is a rather expensive operation involving a sort. This is, or course, not much worse than reporting an unordered array, but it's worth pointing out.

8.4 CHOOSING HASH FUNCTION

In this section we introduce the four main classes of $h(k)$ functions: *truncation*, *division*, *mid-square*, and *partitioning* or *folding*. No one of these is "best" in general: choice of a particular $h(k)$ depends heavily on the structure of the keys in a table, the degree of unpredictability, and the amount of extra table space the designer is willing to tolerate in the interest of achieving a fast search. Indeed, perhaps the only generalizations that can be made are that certain hash functions can turn out to be disastrous, and that in the end the best way to know whether a hash function is effective is to try it in practice on real data.

8.4.1 Truncation

By *truncation*, we mean just taking the first few or the last few characters or digits of the key as the hash code. However, we cannot do this naively: in some cases the method will work acceptably, while in others it can be disastrous.

Consider, for example, a student ID consisting of six decimal digits, as described earlier. The school assigns these numbers on a first-come, first-served basis, so of the million possible numbers only a fairly dense subset will be in active use at a given time. For example, at the author's university at this writing, almost all active student ID's have a high-order digit of 4 or 5.

Now take the three high-order digits of the ID as a hash code into a 1,000-item table. Since almost all codes begin with 4 or 5, only about 200 of the 1,000 possible codes from 000 to 999 will actually be generated by this scheme. Frequent collisions are guaranteed by the fact that arriving items are really competing for only 20 percent of the available positions. On the other hand, taking the low-order three digits is much better, since any of the 1,000 combinations has an equal likelihood of occurrence.

This example shows one of the criteria of a good $h(k)$: it must at least be capable of generating the full range of table addresses. Taking the first three digits of the student ID is an approach that is *obviously* wrong because it is so extreme; other key sets can have biases that are less obvious, but just as damaging. It is important, then, in designing an $h(k)$, to study the set of keys thoroughly to determine what bias there might be and then design a function that will minimize the effect of the bias.

The truncation method does have one important advantage: it can be coded to run very fast, since—at least in assembly language—selecting digits or characters from a memory word is just a shifting or masking operation, which is usually faster than an arithmetic operation like multiplication or division.

8.4.2 Division

An alternative to truncation which works reasonably well given that the hardware implements fixed-point division is division of the key by the size of the table, which we shall call MaxItems, and then taking the remainder as the hash code. It can be shown that the best policy here is to choose MaxItems to be a prime number. A good exercise is to consider why this is so.

8.4.3 Mid-Square

In the mid-square method the key is multiplied by itself (*squared*), and then the *middle* few digits of the result are selected as the hash code. An example is given in Figure 8–3; note that, unlike truncation, all digits of the key participate in the calculation, not just a few.

The mid-square approach is advantageous because it guarantees equal par-

Sec. 8.4 Choosing Hash Function

```
          K=510324

           510324
         × 510324
         ────────
           2041296
          1020648
         1530972
        000000
       510324
      2551620
      ─────────────
      26043|058|4976
```

for a 1000-element table, h(k)=058 **Figure 8-3** Mid-square hashing function

ticipation of all digits, so that unbiased digits tend to diminish the effect of biased ones. In this way, we hope to be able at least to generate all table addresses. An obvious disadvantage of the method is that squaring a long quantity like a key involves a long multiplication operation, which can be fairly slow on most computers.

K=510324

(a) A key.

Folding Method 1: "slide" left and right sections

```
     51
     03
   + 24
   ────
     78
```
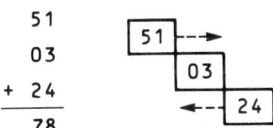

for a 1000-element table, h(k)=078

(b) Folding method 1: "slide" left and right sections.

Folding Method 2: "fold" left and right sections

```
     15
     03
     42
   ────
     60
```
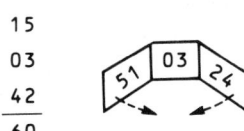

for a 1000-element table, h(k)=060

(c) Folding method 2: "fold" left and right sections.

Figure 8-4 Folding method

8.4.4 Folding or Partitioning

Folding or *Partitioning* is another way of ensuring a good randomizing of the digits of the key. In this method, the key is partitioned or divided into several pieces, then the pieces are operated upon in some way, typically by adding them together, and finally the required number of digits of the result is taken as the hash code. Two different ways of folding are shown in Figure 8-4.

Folding is similar to the mid-square method in regard to advantages and disadvantages: all digits of the key are "hashed" together, but possibly with slow performance because of the amount of arithmetic necessary.

8.5 RESOLVING COLLISIONS IN HASH TABLES

If a collision arises in attempting to place an item in a hash table, we need to search in a systematic and repeatable fashion for an alternative location. The technique is called *probing*, and several different methods exist for doing it. They all depend upon selecting an *increment* function, denoted inc(i), which takes a hash address (not a key) i and produces another hash address. If the new location is occupied, we take *that* hash address and pass it again through the increment function, and so on until we find an open location. With luck and good choices of $h(k)$ and inc(i), we should be able to place the item with only a few probes in most cases, so we still have approximately $O(1)$ performance.

Finding an unoccupied position in the table depends upon our being able to tell that the position is empty. The most common ways of doing this are (1) to have the Create operation initialize all the positions of the table with some value we can use to indicate "unoccupied," and (2) to associate with each table position a flag or code indicating "unoccupied." As a matter of fact, it turns out that we need to make a distinction between "currently unoccupied" and "never occupied," so whatever indicator we use will actually need three states, not two.

The framework for a hash-table module incorporating hash and increment functions (the details of which are not supplied) is given in Figure 8-5. Shown are the Update and Search operations; we also assume that two functions, CurrentlyUnoccupied and NeverOccupied, are available, which hide the details of the occupancy indicator. This kind of hash table scheme is often called *closed hashing*, because all items are stored in the same fixed-size table. Opposed to it is another scheme called *bucket hashing*, to be discussed shortly.

8.5.1 Linear Probing

In linear probing, we let the increment function be

$$\text{inc}(i) = (i + 1) \text{ MOD MaxItems}$$

That is, we just add one to the hash address and "wrap around" if we reach the

Sec. 8.5 Resolving Collisions in Hash Tables

```
IMPLEMENTATION MODULE TableHandler;

(* THESE SIX ROUTINES ARE NOT EXPORTED TO THE USER *)
(* DETAILS ARE OMITTED; STUDENT SHOULD SUPPLY THEM *)

PROCEDURE Hash              (k: KeyType):  IndexType;
PROCEDURE Increment         (i: IndexType): IndexType;
PROCEDURE CurrentlyOccupied(T: Table; i: IndexType): BOOLEAN;
PROCEDURE NeverOccupied     (T: Table; i: IndexType): BOOLEAN;

PROCEDURE Store(VAR T: Table; i: IndexType; K: KeyType);
(* store item in array; turn on occupancy indicator *)

PROCEDURE Remove(VAR T: Table; i: IndexType); ...
(* turn off occupancy indicator *)

(* THE FOLLOWING OPERATIONS ARE EXPORTED *)

PROCEDURE Update( VAR T: Table; K: KeyType);
   VAR
      ProperHome, Probe: IndexType;
   BEGIN
      ProperHome := Hash(K);
      Probe      := ProperHome;
      LOOP
         IF CurrentlyUnoccupied(T,Probe) THEN
            EXIT;
         ELSE
            Probe := Increment(Probe);
            IF Probe = ProperHome THEN
               TableFull := TRUE;
               RETURN;
            END;
         END;
      END;
      Store(T,i,K);
   END Update;

PROCEDURE Search(T: Table; K: KeyType): IndexType;
   VAR
      ProperHome, Probe: IndexType;
   BEGIN
      ProperHome := Hash(K);
      Probe      := ProperHome;
      LOOP
         IF    K = KeyPart(T,Probe)     THEN
            RETURN Probe;
         ELSIF NeverOccupied(T,Probe) THEN
            RETURN Zero;
         ELSE
            Probe := Increment(Probe);
            IF Probe = ProperHome THEN
               RETURN Zero;
            END;
         END;
      END;
   END Search;

PROCEDURE Create(...); (* details omitted *)
PROCEDURE Delete(...); (* details omitted *)
PROCEDURE Report(...); (* details omitted *)

END TableHandler.
```

Figure 8-5 Framework for a hash-table module

end of the table. If the location obtained thereby is occupied, we add one again, continuing to search *linearly* for an open position. As long as there is enough extra space in the table, and it doesn't become too densely filled, we should be able to find a position in a reasonable number of attempts.

Now let us see how searching and deleting work in such a scheme. Intuitively, we should just apply the same sequence consisting of $h(k)$ followed by as many calls to inc(i) as we need, checking the key of every item we find along the way until we arrive at the right one. However, a problem arises when we consider how we know when we've searched long enough: the simple answer—stop when we reach an open location—just isn't enough.

To see why this is so, consider the example in Figure 8-6. Suppose keys K1 and K2 are successfully placed in the table after being transformed by $h(k)$. Now suppose K3, a synonym of K1, arrives. By linear probing, it will be placed adjacent to K1. Next, K4, another synonym of K1, arrives and is of course placed just beyond K2. At this point, a search for any of the items will succeed.

Now suppose we need to delete K3. No problem yet: K1 is in the "official" position for K3, so we try the next position, find K3, and mark the position as "open." Now comes trouble, however: if we search for K4, we'll stop at the position formerly occupied by K3 and think K4 isn't there.

The problem arises because we haven't distinguished between two meanings of "open": "never occupied" and "formerly occupied." We really need *three*, not two, states for the status indicator: one indicating "never occupied," one indicating "formerly occupied," and the third indicating "currently occupied." We use "never occupied" to indicate that we can stop looking in a search or delete operation:

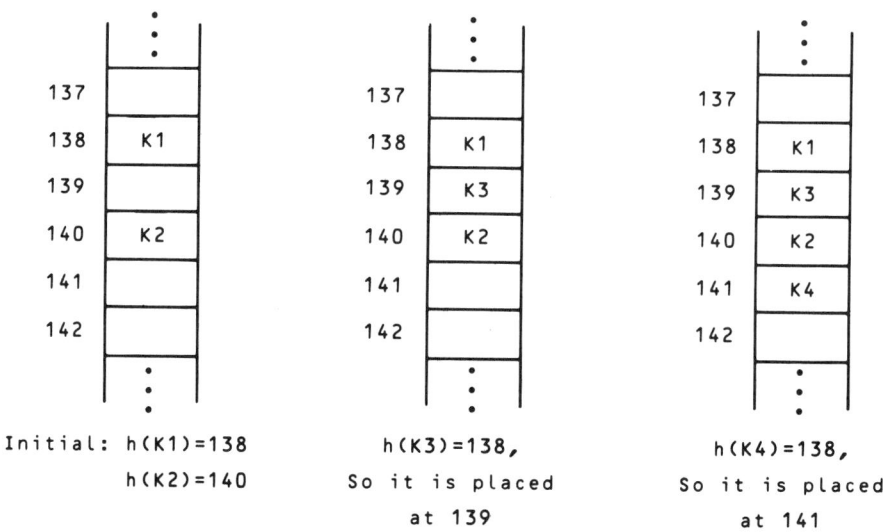

Figure 8-6 Linear probing and the clustering problem

Sec. 8.5   Resolving Collisions in Hash Tables

finding a "never occupied" location indicates that the target item isn't in the table. In an update operation, either the "never occupied" or the "formerly occupied" state can be used to place the arriving item.

Clearly, linear probing will result in a situation called *clustering*, where a number of synonyms will all be placed adjacent to each other and mixed together with some "official" occupants. As the table system runs, these clusters will inevitably grow larger and larger, making the update, search, and delete operations run progressively more slowly.

8.5.2 Nonlinear Probing

Other probing methods have been proposed and analyzed in order to reduce clustering and thus speed up the average search performance. One way is to keep track of the number of probes and then give the increment function *two* arguments: the value of the previous hash address, and the number of probes carried out thus far to place the current item. So instead of

$$\text{inc}(i) = (i + 1) \text{ MOD MaxItems},$$

we get

$$\text{inc}(i,p) = (i + p) \text{ MOD MaxItems}$$

where p is the number of probes thus far. Accordingly, the first increment will be one position away, the next probe will move two positions from the last, the next probe three positions, and so on. This scheme tends to put more space between successive synonyms, thereby reducing clustering.

Another method is the so-called *quadratic* hashing technique, in which

$$\text{inc}(i,p) = (i + ap + bp^2) \text{ MOD MaxItems}$$

where a and b are constants, usually $+1$ or -1. It can be shown that this method spreads items out over the table and will cover exactly half the table before repeating. There is little clustering because, if one search starts at location i_1 and continues over $i_2, i_3, i_4, \ldots, i_N$, then another search which starts at, say, i_2 will not touch any of the locations $i_3, i_4, \ldots, i_N$. (Prove this as an exercise!)

Still another method is to use a pseudorandom number generator as an increment function. This will eliminate clustering, but may be a slower computation than the preceding ones.

8.5.3 Bucket Hashing

All of the methods described thus far assume that successive synonyms are placed in the same closed or fixed-size table. By contrast, the *bucket hashing* method establishes a *bucket* or separate storage area for all members of a given synonym (equivalence) class. Then, $h(k)$ is used just to determine which bucket the new arrival belongs in.

The most common way to implement bucket hashing is to use a linear-list structure for the buckets. In this case, the original table contains, not records, but list headers, and each arriving entry goes into its appropriate list. An illustration appears in Figure 8–7.

Bucket hashing has the obvious disadvantage of requiring extra space for its lists, but this is offset by the fact that the list nodes can be allocated dynamically (given a programming language with that feature), and thus the space not used is shared with other program structures. Furthermore, the amount of space allocated to the original table can be reduced. Note, however, that there is a time/space tradeoff operating: if the number of buckets is B, then the average list length is ActualItems/B (assuming a reasonably random $h(k)$); thus the linear search to find an item in the list is $O(\text{ActualItems}/B)$, so a larger B results in a shorter search.

A minute's thought reveals that bucket hashing is really a miniaturization of the other sequential table-handling strategies. The bucket idea just cuts down the length of the sequential searches by reducing the list length from ActualItems to (an average of) ActualItems/B.

8.5.4 Ordered Hashing

In an application in which a search operation frequently reports an unsuccessful search, and the unsuccessful search is not immediately followed by an update of the item not found, we can cut down the time taken during an unsuccessful search by inserting synonyms into a bucket in an ordered sequence, say, in ascending

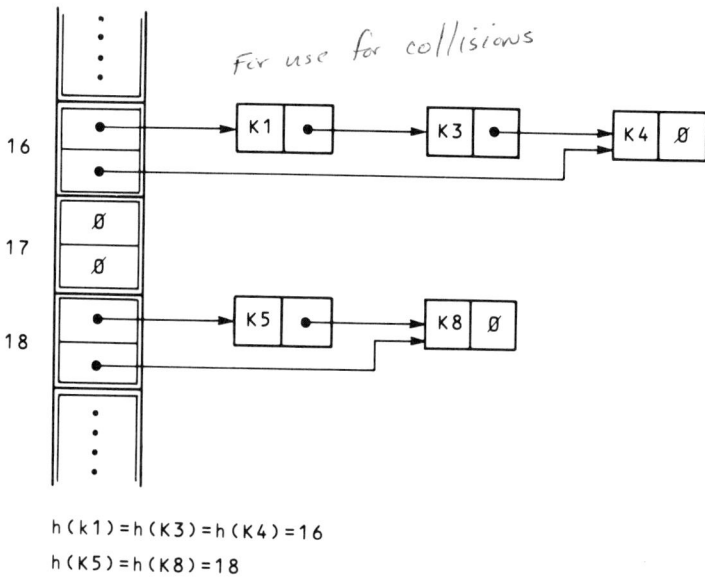

h(k1)=h(K3)=h(K4)=16
h(K5)=h(K8)=18

Figure 8–7 Bucket hashing

order. An unsuccessful search will then take the same time as a successful one, because the search can stop when an item with a key greater than the target is found. Of course, this strategy winds up increasing the time for an update, since a new arrival cannot just be placed at the end of a bucket, but overall, it is likely to be worthwhile. (It is definitely not worthwhile in compiler symbol-table applications, in which an unsuccessful search is almost *always* followed by an update.)

8.6 DESIGN: HYBRID SEARCH STRATEGIES

An application need not use only a single search strategy: often several methods can be combined, perhaps in earlier and later stages of the application's task. A good example of this is a translator (compiler or assembler) symbol table. When the translator makes an early pass over the source program, the main goal is to discover the *first* appearance of an identifier or program variable, so that an object-program address can be assigned to it. A fast update operation is then of interest: unsuccessful searches are always followed by insertions, whereas successful searches (the identifier is already in the table!) are not interesting at all. Later, in the code-generation pass of the translation, each time an identifier is discovered in the source program its address must be looked up in the symbol table. Therefore, a fast search is most desirable: updates do not occur after the table has been built, and deletions never occur at all in this application.

Accordingly, some translator designers use different table structures for these two passes. For example, a binary search tree or bucket hash table is used for the scanning pass, because neither the number of identifiers nor their spelling is known before the source program is scanned, and in most languages an enormous number of possible identifiers exists. The BST update will probably perform reasonably close to $O(\log(N))$ because programmers rarely, if ever, declare or use their identifiers in alphabetical order. Also, both the BST and bucket hash methods handle dynamic space allocation with ease.

On the other hand, once the scanning pass is completed, the table contents are fixed, and only search operations are done from that point on. Therefore, compiler designers sometimes reorganize the symbol table between passes, sorting it and storing it in an ordered array, so that (binary) searching is guaranteed to perform in $O(\log(N))$ time.

Working out the details of this hybrid structure is left to an exercise.

8.7 SUMMARY

By means of an appropriate hash function $h(k)$—a mathematical transformation of a table key—update, search, and delete operations performed on a dynamic table approximate $O(1)$, or constant time. Unfortunately, however, hash functions are inherently many-to-one, with many different keys—synonyms—all yielding the same hash address. Therefore, we have a twofold problem: (1) find an $h(k)$ which minimizes the likelihood of these coincidences or collisions, and (2) then find a good way of resolving those collisions that do occur.

Truncation, division, mid-square, and partitioning are methods that are commonly used to derive hash functions. There is no mechanical way to decide on a best $h(k)$, but two important considerations are the uniformity with which the function spreads items around the table, and the speed of calculation it entails.

The two most common ways of resolving collisions are the closed method and the open, or bucket, method. In the closed method, when the "official" position of an arriving item is already occupied, a search ensues which probes for the first open position by using an increment function $inc(i)$, and then places an arriving item there. Several different increment functions are possible, e.g., linear, quadratic, and random. Each of these methods has its strengths and weaknesses regarding uniform spread and speed; what they all have in common is a sequential search, whose linear performance is directly counter to our desire to place or search for an item in constant time. Hence, in practice, the ideal goal of achieving $O(1)$ performance is only approximated.

In the open collision-resolution method, items in the same synonym class are all placed in a bucket, typically a linear list. While detailed implementations vary and small optimizations are possible, there is usually again a linear search involved in at least one of the operations.

CHAPTER 8 EXERCISES

1. In designing a hash table where h(k) is a division-method function, show why it is best that the divisor and therefore the table size be a prime number.

2. A certain computer does all its arithmetic and array subscripting in binary-coded decimal, not the usual binary integer. A word in this machine consists of eight decimal digits; characters are coded as two decimal digits, according to the following code:

0–9	00 thru 09
A–I	11 thru 19
J–R	21 thru 29
S–Z	32 thru 39

    Now consider this hash-coding problem. There is a table of 100 items, indexed by 2-digit decimal subscripts. The key part of each item is a four-letter sequence. A hashing method is proposed in which h(k) is computed by dividing the key by 10 (integer division!) then taking the rightmost two digits of the result. Find h(k) for each of the keys MARY, JACK, WILL, and MACK. Is this a good hashing method? Why or why not?

3. Suggest strategies for implementing the Report operation for a hash table. Don't forget that Report must print out the table in sorted order by key.

Sec. 8.7 Summary

4. Clearly the midsquare method is easiest to implement if the table size is a power of 2 and h(k) is written in a language (assembler, for example) which allows easy extraction of bits. Design a high-level language implementation of this method, using some programming language that doesn't allow direct bit manipulation.
5. Show that the quadratic method of collision resolution eliminates clustering and covers exactly one half the table before repeating.
6. Starting with the table handler framework given in Section 8.3, fill in the details for specific hash and re-hash functions and implement the handler module.
7. Design a symbol-table scheme for a high-level or assembler language with which you are familiar, using a bucket hash method for the scanning pass of the translator and an ordered array for the code-generation pass. Show how you will reorganize the table between passes.

Chapter 9
INTERNAL SORTING METHODS

9.1 GOAL OF THE CHAPTER

Sorting, or putting a list of records in sequence, is an important part of computing. To take a somewhat extreme example, one data-processing installation with which the author is familiar conducted a survey of its applications which were being run on a multimillion-dollar large-mainframe computer. It turned out that somewhere near 50 percent of the central-processor machine cycles were absorbed just in sorting!

In smaller scale situations, putting a list in order is often part of a larger program, so it is still important to understand how to develop sort procedures that will work correctly and speedily. Happily, the technology of sorting is well understood, and many different algorithms exist. Therefore, comparative study of sorting algorithms gives useful information on and predictions of run-time performance.

In this chapter, we study various algorithms for sorting a list, the number of records of which is small enough for all of them to fit simultaneously into main memory. We call this *internal sorting*. For each of the algorithms we consider, we will study briefly how each one performs, in "big O" terms. Most algorithms are $O(N^2)$ or $O(N \log(N))$, but there are some exceptions.

9.2 SORTING TERMINOLOGY

Suppose we are given an array $A[1 . . N]$ of records, each with a key of some kind. (In the simplest case the record consists only of the key.) Then A is said to be *upward sorted* or *sorted in ascending order* if, for every index I from 1 to N, $A[I] \leq A[I + 1]$. If, on the other hand, for each I, $A[I] \geq A[I + 1]$, the array is said to be *downward sorted* or *sorted in descending order*.

A *sort algorithm* or *sort procedure* is one which, given A originally unsorted, will produce a sorted array. For simplicity, we shall use only ascending sorts, so "sorted" will mean "upward sorted."

An *internal sort* is a sort on an array of sufficiently small size that all records can fit into main memory at one time. An *external sort* is a sort of an array, the number of records of which is so large that some of them must reside on external storage (tape or disk) at any given instant.

Given an array A, we say that a record R1 *precedes* a record R2 if R1 is located at $A[I]$, R2 is located at $A[J]$, and $I < J$. A sort is said to be *stable* if, for any pair of records R1 and R2 such that R1 precedes R2 and key(R1) = key(R2) in the unsorted array, then R1 precedes R2 in the sorted array. In other words, a stable sort preserves the relative positions of records with identical keys.

An *in situ sort* is a sort in which the unsorted and sorted arrays occupy the same space, possibly with the use of a small, approximately constant amount of auxiliary working storage to carry out the sort. In other words, no copy of the array is needed for an *in situ* sort.

In the sections that follow, we present various internal sorting methods and consider their performance. Each of the algorithms presented is designed to operate on an array of arbitrary size whose contents are initially in arbitrary order. Accordingly, in predicting their performance, we can make no assumptions about the structure of the records they sort. We will, though, try to find the *best case*, *average case*, and *worst case* performance wherever possible. Years of work on sorting theory and practice have established that most internal sorts are of growth rate $O(N^2)$ or $O(N \log(N))$.

9.3 INTERNAL SORT ALGORITHMS WITH GROWTH RATE $O(N^2)$

The simplest and most straightforward internal sort methods are those with growth rate $O(N^2)$. These methods are easy to understand and require little memory beyond that necessary to store the array itself; they also have relatively small time-per-operation characteristics. Thus, for occasional sorting of reasonably small arrays, the payoff in simplicity and ease of debugging in using them is often worth their price of quadratic performance.

The five simple sorts to be considered are *simple selection, delayed selection, the bubble sort, linear insertion,* and *binary insertion.*

9.3.1 Simple Section

The simple selection sort is intuitively very easy to understand. Given the array $A[1..N]$ which we are to sort ascending, we *select* the smallest item in the array and place it in the first position, then the second smallest and place it in the second position, and so on. This is done for, say, the first position by comparing key($A[1]$) with key($A[2]$), and *exchanging* or *swapping* if key($A[2]$) is smaller. We then compare (the possibly new) key($A[1]$) with key($A[3]$), exchanging if necessary, and so on until key($A[1]$) and key($A[N]$) are compared. The procedure guarantees that the smallest item will end up in the first position.

This being the case, we then forget about $A[1]$, and do the same thing with $A[2]$ through $A[N]$, which will bring the second smallest item to the second position. If we call each scan of the partial array a *pass*, then we will finally execute a pass such that $A[N-1]$ and $A[N]$ find their proper places, and the array will be sorted. The process is illustrated in Figure 9–1, and a procedure appears in Figure 9–2. This procedure uses an auxiliary procedure

  Swap(VAR A: ARRAY OF ValueType; i,j: ValueType)

which interchanges two values. An invocation like Swap(T,lower,upper) interchanges the lowermost and uppermost values in the array T. The procedure Swap will be used by many of the algorithms in this chapter.

What is the performance of the simple selection algorithm? Its structure is a double loop with a decision inside. We accommodate the decision, as outlined in chapter 2, by assuming that the slower leg is always executed. So we assume that an exchange is done every time a comparison is done. The first pass requires $N-1$ operations, the second $N-2$ operations, and so on down to the $N-1$st pass, which requires one operation. Thus, the total number of operations is $(N-1) + (N-2) + \ldots + 1$, or $N \times (N-1)/2$. Multiplying across the parentheses, we get $(N \times N/2) - N/2$, and so the performance of the algorithm is $O(N^2)$, since the squared term will dominate the linear term for nontrivial N. (Even for $N = 10$, we are off by only 10 percent.)

The assumption that an exchange is done for each comparison corresponds to worst-case conditions where the original array is sorted downwards. If the array

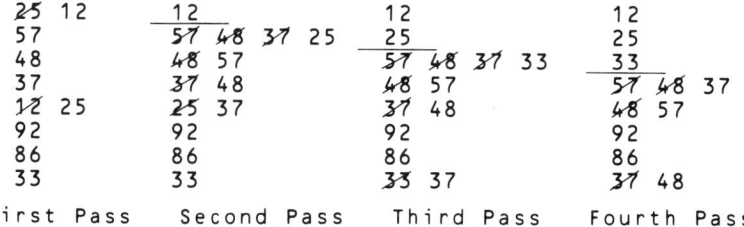

Figure 9–1 Simple selection sort (the reader can fill in the additional passes)

Sec. 9.3 Internal Sort Algorithms with Growth Rate $O(N^2)$

```
PROCEDURE Swap(VAR V: ARRAY OF ValueType; i,j: INTEGER);

   VAR Temp: ValueType;

   BEGIN

      Temp := V[i];
      V[i] := V[j];
      V[j] := Temp;

   END Swap;



PROCEDURE SelectSort(VAR V: ARRAY OF ValueType);

   VAR
      SlotToFill: INTEGER;
      Candidate:  INTEGER;

   BEGIN

      FOR SlotToFill := 0 TO HIGH(V)-1 DO
         FOR Candidate := SlotToFill+1 TO HIGH(V) DO

            IF V[Candidate] < V[SlotToFill] THEN
               Swap(V, SlotToFill, Candidate);
            END;

         END;
      END;

   END SelectSort;
```

Figure 9-2 Procedure for auxiliary selection sort

is sorted upwards, there will be no exchanges done at all. The actual execution time of this best-case scenario will then be faster, but the growth rate is still proportional to the square of the array size.

9.3.2 Delayed Selection

We can speed up the simple selection sort a bit if we try to reduce the number of exchanges that are made under less-than-best-case conditions. We do so by delaying any exchange until the end of the pass. For example, instead of exchanging $A[1]$ and $A[2]$ if $A[2]$ is smaller, we note in an auxiliary variable that $A[2]$ is the smallest key seen in the current pass (by setting the variable to 2, for instance), and then test that location's key against $A[3]$, keeping track of which is smaller. At the end of the first pass, the auxiliary variable will have the location of the smallest key; we then exchange that record with the one at $A[1]$.

Since in this improved algorithm we do only one exchange per pass, the overall running time will generally be faster than simple selection, even though the performance is still $O(N^2)$. Figure 9-3 shows the procedure for the delayed

```
PROCEDURE MinIndex(V: ARRAY OF ValueType; First: INTEGER): INTEGER;
VAR
    Min:     ValueType;
    Current: INTEGER;
    Index:   INTEGER;

BEGIN

    Min := V[First];
    Index := First;

    FOR Current := First TO HIGH(V) DO
      IF V[Current] < Min THEN
        Min := V[Current];
        Index := Current;
      END;
    END;

    RETURN Index;

END MinIndex;


PROCEDURE DelayedSelectSort(VAR V: ARRAY OF ValueType);

VAR
    Index:      INTEGER;
    SlotToFill: INTEGER;

BEGIN

    FOR SlotToFill := 0 TO HIGH(V)-1 DO
      Index := MinIndex(V,SlotToFill);
      Swap(V,SlotToFill,Index);
    END;

END DelayedSelectSort;
```

Figure 9-3 Procedure for delayed selection sort (and auxiliary routine to find minimum value)

selection algorithm. It uses an auxiliary function MinIndex which finds the index (location) of the smallest value in an array. Thus, an invocation like y:= MinIndex(A,i,N) will return to y the index of the smallest value in the section of the array A from the ith to the Nth locations, inclusive.

9.3.3 Bubble Sort

The bubble sort is another simple sort with $O(N^2)$ worst-case performance. According to its methodology, we compare the keys of *adjacent* items, exchanging when necessary. We begin with key(A[1]) and key(A[2]), then key(A[2]) and key(A[3]), and so on. At the end of the first pass, as shown in Figure 9-4, the "heaviest" item will have "sunk" to the bottom, one location at a time. We then start a second pass which runs through A[1] to A[N − 1], sinking the second-

Sec. 9.3   Internal Sort Algorithms with Growth Rate $O(N^2)$

```
   25              25              25
  5̶7̶ 48           4̶8̶ 37           3̶7̶ 12
  4̶8̶ 5̶7̶ 37        3̶7̶ 48 12        1̶2̶ 37
  3̶7̶ 5̶7̶ 12        1̶2̶ 48           48
  1̶2̶ 57           57              5̶7̶ 33
  9̶2̶ 86           8̶6̶ 33           3̶3̶ 57
  86 9̶2̶ 33        3̶3̶ 86           86
  3̶3̶ 92           92              92
  First Pass      Second Pass     Third Pass


  2̶5̶ 12           12              12
  1̶2̶ 25           25              25
  37              3̶7̶ 33           33
  4̶8̶ 33           3̶3̶ 37           37
  3̶3̶ 48           48              48
  57              57              57
  86              86              86
  92              92              92
  Fourth Pass     Fifth Pass      Sixth Pass (Sorted!)
```

Figure 9–4 Bubble sort

heaviest item down to the next-to-last position. As before, after $N - 1$ passes, we will have the array sorted.

At first glance, the bubble sort looks no better than the simple selection sort. But there is a way to improve it which can make a difference. Since only adjacent items are compared, if we ever make a complete pass in which no exchanges are necessary, we know that the array is sorted. Indeed, if the original array is *received* in sorted order, then only *one* pass is necessary to make that determination, and so the best-case performance is $O(N)$.

What we need to do, then, is maintain a Boolean, AnotherPassNeeded, which is initialized to false at the start of each pass, and then set to true whenever an exchange is made. If AnotherPassNeeded is false at the end of a pass, we can stop the sort. A procedure for this is given in Figure 9–5.

Accordingly, Bubble Sort has a running time of $O(N)$ in the best case and $O(N^2)$ in the worst case (when the array is originally in reverse order). In general, of course, it will lie somewhere in between.

You may be wondering why the algorithm is called "bubble" sort.

The reason is that a pass can just as easily be run upside down, comparing key($A[N]$) with key($A[N - 1]$), etc., and in that case, the "light" items "bubble up," instead of the "heavy" ones "sinking down." We chose the algorithm to proceed in the opposite direction in order to make it more intuitively comparable with the simple selection sort.

What factors determine how many passes will be required? It turns out that the most important factor is the fact that even though a "heavy" item can move all the way from top to bottom in one pass, a "light" one only moves up one position at a time. So the number of passes is determined by the number of

```
PROCEDURE BubbleSort(VAR V: ARRAY OF ValueType);

    VAR
        CurrentBottom:     INTEGER;
        AnotherPassNeeded: BOOLEAN;
        Top:               INTEGER;
        Current:           INTEGER;

    BEGIN
        Top := 0;
        CurrentBottom := HIGH(V);
        AnotherPassNeeded := TRUE;

        WHILE AnotherPassNeeded AND (CurrentBottom > 0) DO

            AnotherPassNeeded := FALSE;

            FOR Current := Top TO CurrentBottom-1 DO
                IF V[Current+1] < V[Current] THEN
                    Swap(V,Current+1,Current);
                    AnotherPassNeeded := TRUE;
                END;
            END;

            CurrentBottom := CurrentBottom - 1;

        END;

    END BubbleSort;
```

Figure 9-5 Procedure for bubble sort

positions in the longest upward trip. Because of this, the overall performance of the bubble sort can often be improved by running alternate passes in opposite directions, so that a "light" item which only moved one position in a given pass will get to move much farther in the next pass. An exercise requests you to write a program for this algorithm, which is sometimes called "shaker" sort, for obvious reasons.

9.3.4 Linear Insertion

Linear insertion is yet another simple sort with $O(N^2)$ running time. Its method is very similar to what one does in a game of cards where one receives cards one at a time and orders them in the hand. As each new card arrives, the player scans his or her hand, usually from left to right, searching for the correct place for the new arrival, and then inserts the arrival in that place.

In a programming context, suppose an N-element array A exists with $K < N$ elements in ascending order already in the first K locations. An algorithm to put a new arrival in its correct place is as follows:

TO PLACE A NEW ARRIVAL:

1. Search sequentially through the array until a key is found which is greater than that of the new arrival. Call its location J.

Sec. 9.3 Internal Sort Algorithms with Growth Rate $O(N^2)$

2. Make space for the new arrival by moving the contents of $A[J]$ through $A[K]$ to locations $A[J + 1]$ through $A[K + 1]$.
3. Insert the newly arrived element at $A[J]$.

To sort N new arrivals, then, we begin by inserting the first arrival in $A[1]$ and then looping $N - 1$ times through the preceding algorithm.

The foregoing discussion has assumed that there are "arrivals." Where do they arrive from? We can make this an *in situ* sort by having the new arrivals simply come from the array itself. Since the first K arrivals are sorted into the first K locations of the array, the $K + 1$st will fit in somewhere in the first $K + 1$ locations, and so the $K + 1$st location can be used to hold $A[K]$ as it is moved. In other words, the unsorted array "shrinks" from N elements down to none as the sorted one grows from no elements to N, and we can use the same physical space for both arrays, "back to back." This is shown in Figure 9–6.

Figure 9–7 gives a procedure in which the algorithm is in fact a bit simpler than discussed here: instead of moving a number of items after the new item's proper place has been found, we start the new item at the bottom of the sorted part of the array, moving it upward by exchanging, until it finds its proper place.

```
                25              25              25              25
             (57)  ← "new       57              48              37
                48    arrival"  (48)            57              48
                37              37              (37)            57
                12              12              12              (12)
                92              92              92              92
                86              86              86              86
                33              33              33              37
         Original array    57 inserted     48 inserted     37 inserted

                12              12              12              12
                25              25              25              25
                37              37              37              33
                48              48              48              37
                57              57              57              48
                (92)            92              86              57
                86              (86)            92              86
                33              33              (33)            92
           12 inserted     92 inserted     86 inserted     33 inserted
```

Figure 9–6 Linear insertion sort

```
PROCEDURE LinearInsertionSort(VAR V: ARRAY OF ValueType);

VAR
   Top:            INTEGER;
   Bottom:         INTEGER;
   CurrentBottom:  INTEGER;
   current:        INTEGER;

BEGIN

   Top := 0;
   Bottom := HIGH(V);

   (* TREAT EACH ITEM AS TEMPORARY BOTTOM *)
   FOR CurrentBottom := Top+1 TO Bottom DO

      (* MOVE CURRENT BOTTOM ITEM UP
         UNTIL IT FINDS ITS HOME *)
      FOR current := CurrentBottom TO Top+1 BY -1 DO
         IF V[current] < V[current-1] THEN
            Swap(V, current, current-1);
         END;
      END;

   END;

END LinearInsertionSort;
```

Figure 9-7 Procedure for linear insertion sort

To see that the linear insertion sort has performance $O(N^2)$, consider how many comparisons need to be made to place the $K + 1$st arrival. If we assume that all original orderings are equally probable, then, on the average, $K/2$ comparisons will be necessary to find the proper place for the $K + 1$st element. Furthermore, an average of $K/2$ exchanges will be required to make space. To sort the whole array, then, involves a number of operations characterized by a series that will sum once again to $(N - 1) \times N/2$, giving us a performance of $O(N^2)$.

9.3.5 Binary Insertion

Linear insertion can be speeded up by noticing that because the first K elements are already in order by the time the $K + 1$st arrives, we can find the proper place for the $K + 1$st item by using a binary rather than a linear search. Since the binary search has performance $O(\log(K))$, the searching part of the algorithm runs much faster.

On the other hand, the sorting part of the algorithm is not speeded up at all, since it still takes $O(K)$ time. Since for nontrivial N a squared term will dominate a logarithmic term, the overall algorithm still has performance $O(N^2)$, even though using the binary search reduces the actual running time somewhat.

9.4 INTERNAL SORT ALGORITHMS WITH GROWTH RATE $O(N \log(N))$

Three sort algorithms with performance $O(N \log(N))$ are the *merge sort, heap sort,* and *quick sort*. In each of them, a price is paid for the improved "big O" performance, either in extra space required or in increased complexity of the algorithm, or both. These sort algorithms show clearly that there are time/space and time/complexity tradeoffs that just cannot be avoided.

9.4.1 Merge Sort

In chapter 2 we gave a sketch of a recursive algorithm to sort a list by merging. Here, we shall develop a nonrecursive version of the merge sort, which sorts an array with performance $O(N \log(N))$. The price paid for the improved performance is that a second copy of the array is needed.

Consider the general algorithm for merging two *sorted* lists L1 and L2 to create a third list L3. (The lists are not necessarily linked; we are thinking abstractly here.) The sparse-vector addition algorithm of chapter 4 is a special case of such an algorithm.

The algorithm proceeds by comparing the key of the first item in L1 with the key of the first item in L2. The item with the smaller key is removed from its list and placed at the end of L3. (If the keys are equal, L1 is considered the smaller one.) At this stage, one of the lists has been shortened by one item.

Next, the two first items are compared again, and the one with the smaller key is removed and attached to L3. If the process is continued, eventually either L1 or L2 becomes empty. The remaining items in the nonempty list are then just removed and copied to L3. Each list is traversed exactly once, and every item is copied exactly once, so the performance of the merge is directly proportional to the total number of items in the two lists.

Several illustrations of the merge algorithm are given in Figure 9–8. We next need to consider how to use this merge to create a merge sort.

In any sort, we are given an unsorted array of N items. Let us think of this array as a collection of N *sorted* lists, each with one item in it. For simplicity, we assume that N is an exact power of 2; we shall remove this limitation later.

Now create a "blank" array to use as a result array. Then go through the original array, merging each pair of items into the result array. Items 1 and 2 will be merged into locations 1 and 2 of the result, and so on. When the merge is complete, the result array will contain $N/2$ sorted lists, each with *two* items.

Copy the result array back to the original, and then merge, from the original array, each *pair* of length-2 lists into the result. This will give $N/4$ lists of length 4. Again copy the result array back, and continue merging and copying longer and longer lists, until two lists of length $N/2$ are left in the original array. Finally,

L1	L2	L3=Merge(L1,L2)
3	4	3
5		4
9		5
		9

L1	L2	L3=Merge(L1,L2)
13	2	2
	5	5
	10	10
		13

L1	L2	L3=Merge(L1,L2)
2	1	1
4	3	2
8	6	3
10	11	4
13	15	6
		8
		10
		11
		13
		15

Figure 9-8 Several examples of merging

merge these into the result array, which is then sorted. The process is illustrated in Figure 9-9.

To calculate the performance of the merge sort algorithm, note that each merge pass does exactly $2N$ operations: each item is merged once and then copied back once. If N is a power of 2, there are $\log(N)$ passes, so the growth rate of the whole algorithm is $O(N \log(N))$. If N is not a power of 2, the number of passes is the logarithm of the next higher power of 2. (Prove this as an exercise!)

To turn the algorithm into a procedure, we show first in Figure 9-10 a fragment which merges two adjacent sections of an array into a result array. Figure 9-11 then shows the entire procedure for the merge sort.

The algorithm can be speeded up by avoiding the extra copying of the result array back to the original. This is done by alternating the original and result arrays, using a flag to keep track of which array is which. We leave the development of a program for this to an exercise.

Sec. 9.3 Internal Sort Algorithms with Growth Rate $O(N \log(N))$

Initially List Length=1	After 1st pass List Length=2	After 2nd Pass List Length=4	After 3rd Pass List Length=8	Finally List Length=16
23	14	-1	-1	-9
14	23	0	0	-3
0	-1	14	3	-1
-1	0	23	4	0
3	3	3	7	1
4	4	4	8	2
8	7	7	14	3
7	8	8	23	4
19	12	-3	-9	7
12	19	1	-3	8
1	-3	12	1	10
-3	1	19	2	12
10	-9	-9	10	14
-9	10	2	12	15
2	2	10	15	19
15	15	15	19	23

Figure 9–9 Merge sort (nonrecursive or "bottom up")

The merge sort is interesting in its own right as an internal sort, but its greatest utility is as a part of most external sort methods, where the lists to be merged reside on external files instead of in arrays. We consider this feature in chapter 10.

```
(* GO UNTIL ONE SUBARRY RUNS OUT *)

WHILE (Left < TopLeft) AND (Right < TopRight) DO
    IF TempArray[Left] <= TempArray[Right] THEN
        V[M] := TempArray[Left];
        INC(Left);
    ELSE
        V[M] := TempArray[Right];
        INC(Right);
    END;
    INC(M);
END;


(* NOW "COPY TAIL" OF WHICHEVER SUBARRAY REMAINS *)

WHILE Left < TopLeft DO
    V[M] := TempArray[Left];
    INC(Left);
    INC(M);
END;

WHILE Right < TopRight DO
    V[M] := TempArray[Right];
    INC(Right);
    INC(M);
END;
```

Figure 9–10 Fragment implementing a part of a merge sort

```
PROCEDURE MergeSort(VAR V: ARRAY OF ValueType);

VAR
    TempArray:          ARRAY OF ValueType;
    CurrentLength:      INTEGER;    (* LENGTH OF SUBARRAY *)
    Left, TopLeft:      INTEGER;    (* POSITION AND END OF LEFT *)
    Right, TopRight:    INTEGER;    (* POSITION AND LENGTH OF RIGHT *)
    M:                  INTEGER;    (* POSITION IN RESULT *)
    Max:                INTEGER;    (* LENGTH OF VECTOR *)

BEGIN
    Max := HIGH(V);
    CurrentLength := 1;
    WHILE CurrentLength < Max DO     (* NEW PHASE *)

        TempArray := V;
        Left := 0;
        M := 0;

        WHILE Left<= Max DO   (* FIND A PAIR OF SUBARRAYS *)
            Right := Left + CurrentLength;

            TopLeft := Right;
            IF TopLeft > Max THEN
                TopLeft := Max + 1;
            END;

            TopRight := Right + CurrentLength;
            IF TopRight > Max THEN
                TopRight := Max + 1;
            END;

            (* CODE TO MERGE SUBARRAYS GOES HERE *)

            Left := TopRight;
        END;

        (* DOUBLE SIZE OF SUBARRAYS, GO BACK FOR NEXT PHASE *)
        CurrentLength := 2 * CurrentLength;

    END;

END MergeSort;
```

Figure 9-11 Complete procedure for merge sort

9.4.2 Heap Sort

The heap sort is an important $N \log(N)$ algorithm for internal sorting. It is an unusual method in that no space penalty is exacted for its good performance: indeed, it is an *in situ* sort.

The algorithm uses the concept of a *heap*, which is a special kind of binary tree. Recall that an *almost complete binary tree* is a binary tree in which all leaves are at the same level, except for some which are one level higher and concentrated at the right side of the tree. We define a heap as an almost complete binary tree

Sec. 9.3 Internal Sort Algorithms with Growth Rate $O(N \text{ Log}(N))$

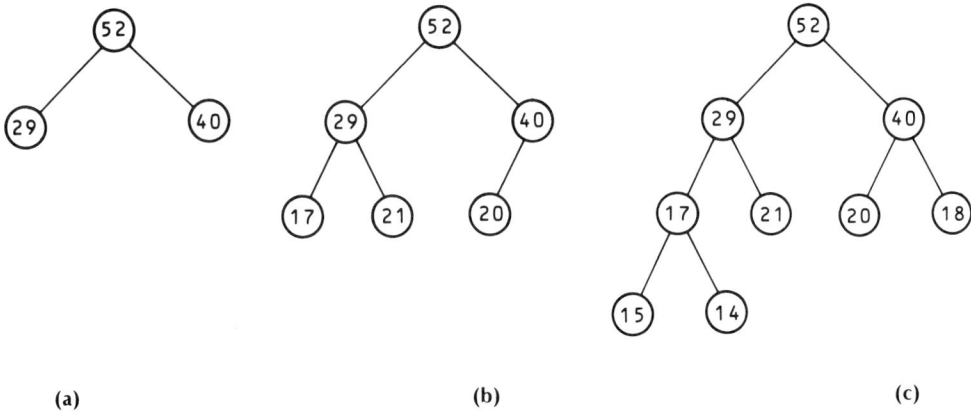

Figure 9-12 Some heaps

in which the key at every node is greater than or equal to the keys of its children. Note that a leaf is a heap by this definition.

One more definition will allow us to proceed: an *almost-heap* is an almost complete binary tree which fails to be a heap only because its *root* key may be smaller than one or both of its children's keys. Figure 9-12 shows some heaps, and Figure 9-13 shows some almost-heaps.

Creating a heap. We first show how to add a new value to an existing heap. Taking the heap from Figure 9-12(c) as an example, let us add a key 13 to it. We temporarily locate this new value in the next available leaf in the heap. (Note that because a heap is an almost complete tree, this position is always known.) Now, in order to maintain the heap property, the new arrival must be no larger than its parent. If it is, we are finished; otherwise, we exchange the new arrival

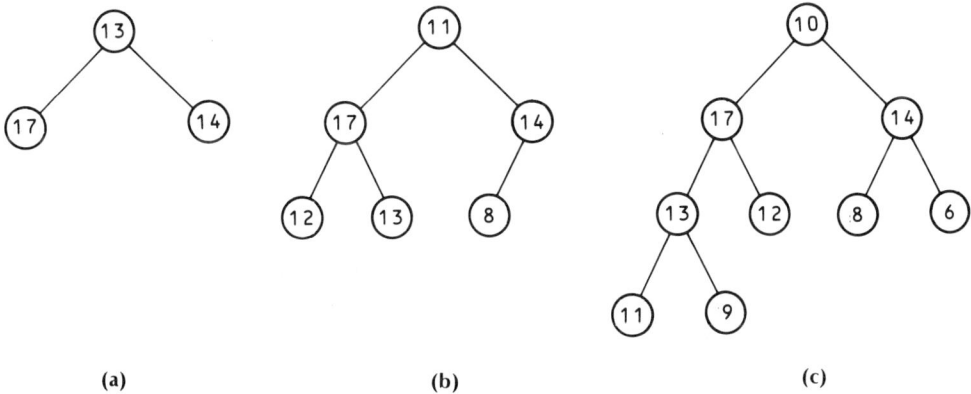

Figure 9-13 Some almost-heaps

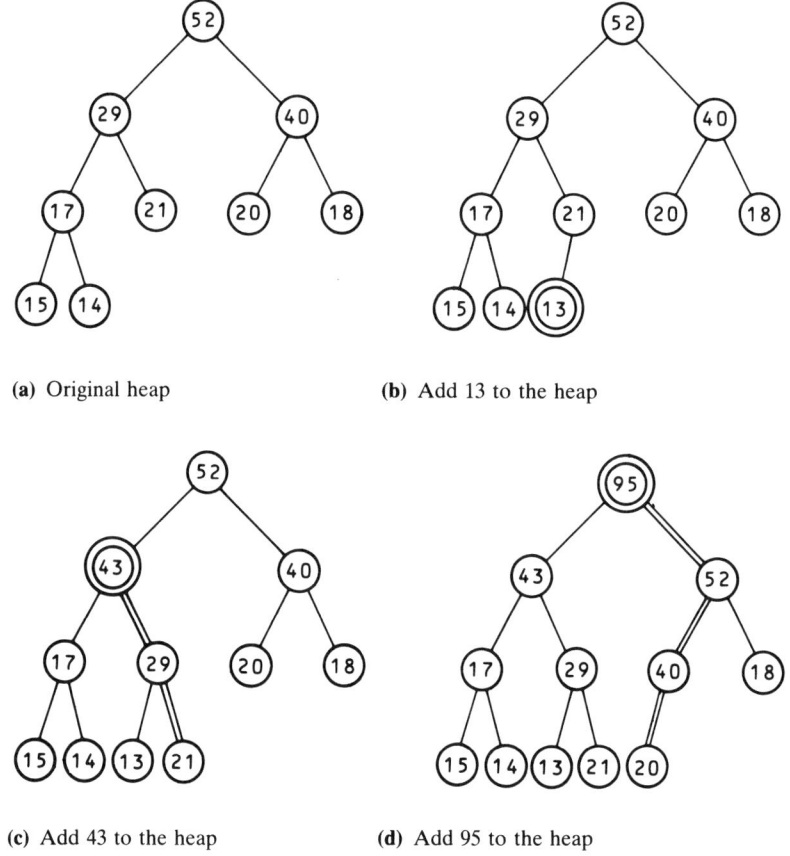

(a) Original heap

(b) Add 13 to the heap

(c) Add 43 to the heap

(d) Add 95 to the heap

Figure 9–14 Adding new keys to a heap

with its parent. The latter has the effect of moving the new arrival one level up in the heap. Notice that the subtree consisting of the new arrival and its children must still be a heap. But the new arrival may *still* be greater than its new parent. So we just continue the exchange process, moving the new arrival up in the heap until it is no greater than its parent. In so doing, we maintain the heap property throughout. Figure 9–14(a) shows our heap with 13 added, as it happens, as a leaf.

Let us now add the key 43 to the heap. Notice in Figure 9–14(b) how the other nodes are displaced in order to preserve the heap property. Similarly, adding 95 to the heap entails putting the 95 at the root, as shown in Figure 9–14(c).

Converting an almost-heap to a heap. Let us look at the almost-heap of Figure 9–13(c) and consider how to convert it into a heap. First, we need to exchange the root with the larger of its two children, which of course imposes the heap property with respect to the other branch. We now have the former root

Sec. 9.3 Internal Sort Algorithms with Growth Rate $O(N \log(N))$

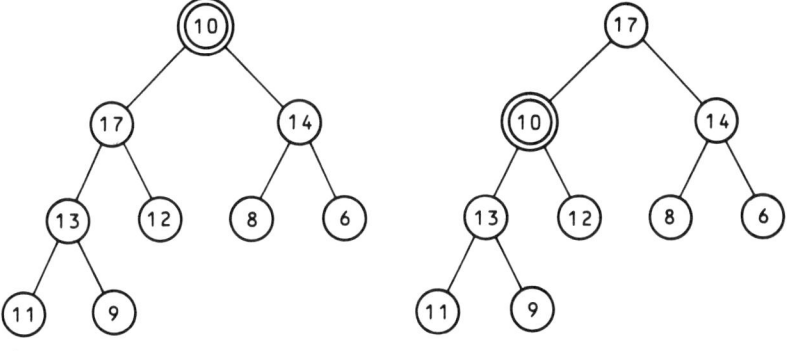

(a) Original almost-heap

(b) 17 is 10's larger child, so exchange 10,17

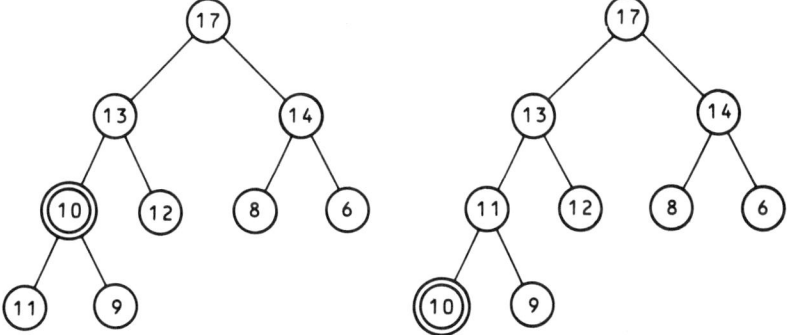

(c) Now 13 is 10's larger child, so exchange 10,13

(d) Now 11 is 10's larger child, so exchange 10,13; we have a heap!

Figure 9–15 Convert an almost-heap to a heap

located one level down, and possibly smaller than its children. So we exchange again with the larger child and continue the process until the former root key finds its proper place (i.e., no smaller than either of its children). Since only the root was out of place to begin with, the process leaves us with a heap. The steps in this exchange process are shown in Figure 9–15.

Sorting a list with a heap. Let us imagine that we have taken an unsorted list and built a heap one item at a time using the foregoing procedure. The largest item in the original list is now necessarily at the root of the heap.

Now take this largest key and exchange it with the key in the *rightmost* position of the *lowest level* of the heap. If we then (conceptually) cut this leaf off the tree, what are we left with? If the original heap had N nodes, we are left with an almost-heap of $N - 1$ nodes. (Remember, we are ignoring the rightmost lowest leaf.)

Now we convert the almost-heap to a heap of $N - 1$ nodes. It is clear that the second-largest key in the original list is now at the root. Accordingly, exchange it with the rightmost lowest leaf of the ($N - 1$-node) heap, and conceptually cut this leaf off, getting an $N - 2$-node almost-heap. Convert this to a heap, and then continue the process until all N keys have been removed from the heap. The procedure is demonstrated in Figure 9–16; the links to nodes that have been "cut off" are shown as dashed lines.

Looking at the resulting tree (no longer a heap, of course), we see that if we traverse it *level by level*, we visit the keys in ascending order.

The practicality of the heap sort. The practicality of the heap sort as an *in situ* sort can be understood by recalling that an almost complete binary tree can be represented easily in array form. Since all levels of the tree are complete except the lowest, we can store such a tree unambiguously, *level by level* in a linear array without using any pointers. In Figure 9–17 we show the heap of Figure 9–12(c) thus stored. All the "missing" nodes of the tree are concentrated at the right end of the array.

This implementation scheme is useful because the storage mapping function is straightforward and efficient to calculate, and level-by-level traversal of the tree is easy. If we call the array $A[1 .. N]$, then the ith node is at $A[i]$, its children (assuming it has children) are at $A[2i]$ and $A[2i + 1]$, respectively, and its parent is at $A[i\,\text{DIV}\,2]$, unless $i = 1$ (the root node). Furthermore, the ith level of the tree begins at $A[2^{i-1}]$, and all items at that level follow immediately. So, looking back at the tree resulting from the heap sort in the previous section, we see that in this implementation that tree would be a sorted array!

We can thus implement the heap sort for an arbitrary unsorted array as follows:

HEAP SORT:

1. $A[1]$ is trivially a heap.
2. Make a heap of the entire array by adding the values in $A[2]$ through $A[N]$ in turn. The heap "grows" in the left end of the array, the values yet to be added dwindle in the right end.
3. Remembering that the rightmost lowest key is now at $A[N]$, exchange it with $A[1]$ and convert the almost-heap $A[1] .. A[N - 1]$ into a heap.
4. Continue the process in (3) by exchanging $A[N - 1]$ with $A[1]$, converting to a heap, and so on.

The heap is now dwindling in the left end of the array, and the sorted list is growing in the right end.

Figure 9–18 shows procedures for (a) adding a new value to a heap, (b) converting an almost-heap to a heap, (c) building a heap from an array, and

Sec. 9.3 Internal Sort Algorithms with Growth Rate $O(N \, Log(N))$ **265**

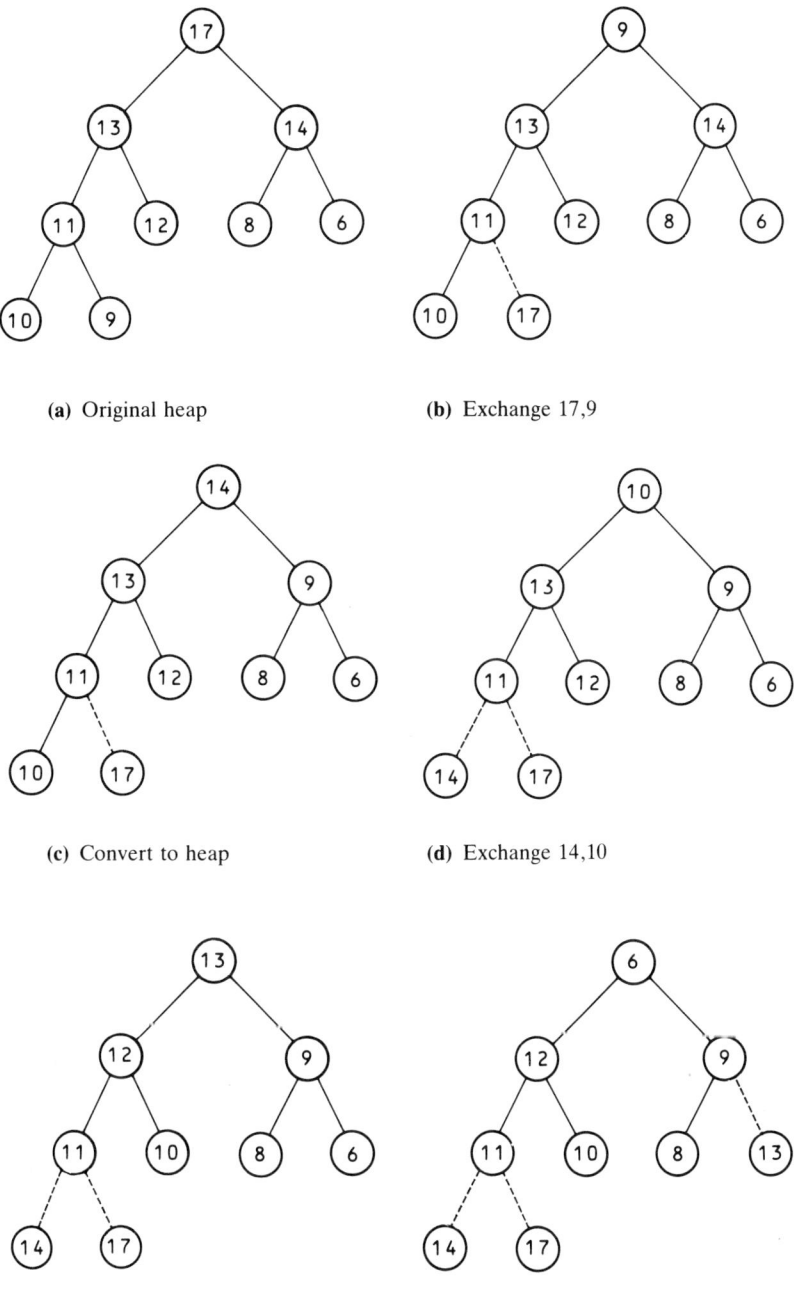

Figure 9–16 Sorting with a heap

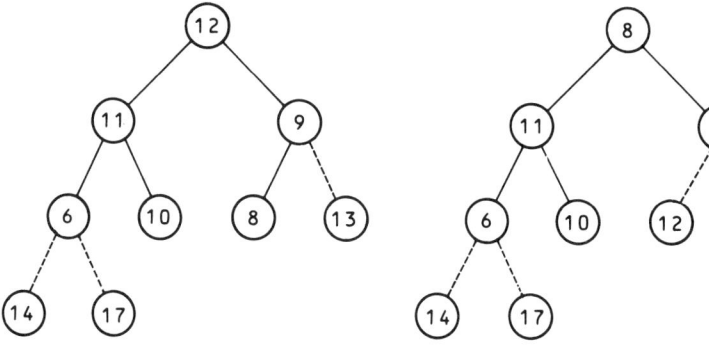

(g) Convert to heap

(h) Exchange 12,8

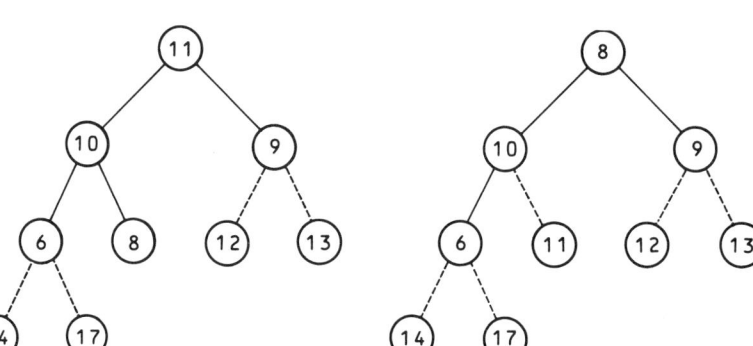

(i) Convert to heap

(j) Exchange 11,8

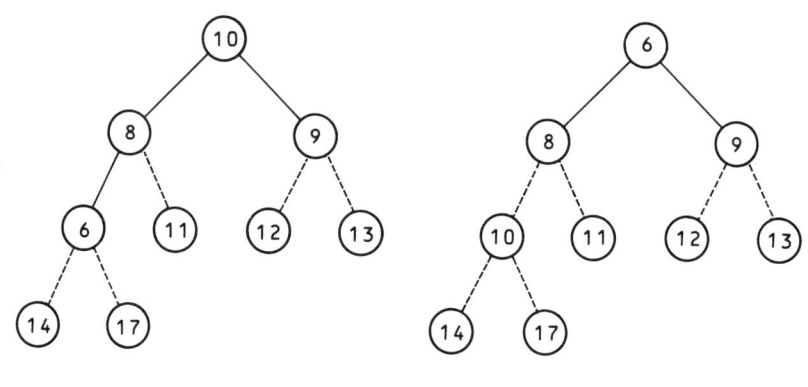

(k) Convert to heap

(l) Exchange 10,6

Figure 9–16 (*Continued*).

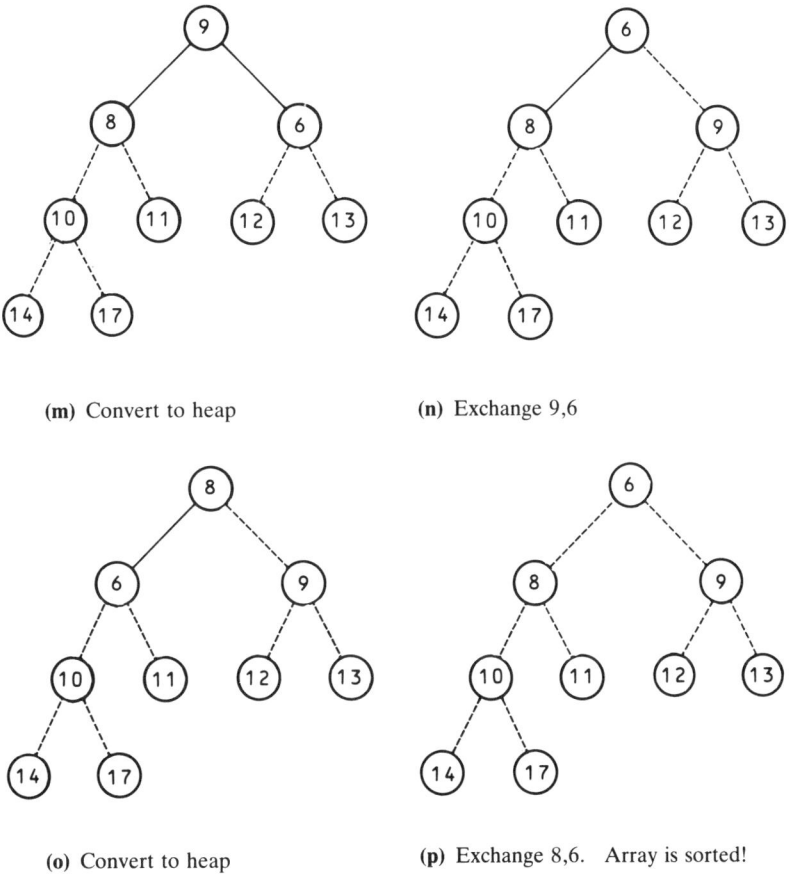

(m) Convert to heap
(n) Exchange 9,6
(o) Convert to heap
(p) Exchange 8,6. Array is sorted!

Figure 9–16 (*Continued*).

(d) sorting a heap. Figure 9–18(e) presents a driver program called HeapSort which runs the whole process.

What is the performance of the heap sort? We can estimate it conservatively by noting that since the tree we are using is balanced, its depth is equal to $\log_2(N) - 1$, where N is the number of nodes rounded up to the next higher power of 2. Now, in the procedure BuildHeap, items are moved upward in the tree by ExtendHeap. Since an item cannot move higher than the root, it cannot move more than $\log(N) - 1$ levels. Since there are N items (rounded upward), this gives performance $O(N \log(N))$ for BuildHeap. Similarly, in the procedure SortHeap, items are moved downward in the tree by AlmostHeapToHeap. Since no item can move down more than $\log(N) - 1$ levels, and since there are N items, we have performance $O(N \log(N))$ here as well. Thus, the overall performance is $O(N \log(N))$.

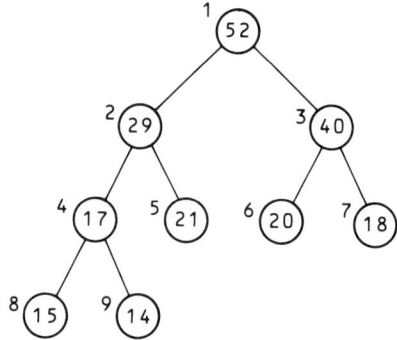

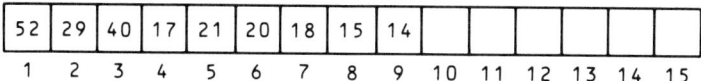

(a) Heap of Figure 9–12(c), with nodes numbered

52	29	40	17	21	20	18	15	14						
1	2	3	4	5	6	7	8	9	10	11	12	13	14	15

(b) Array view of the heap

Figure 9–17 Array and tree views of a heap

```
PROCEDURE ExtendHeap(VAR V:        ARRAY OF ValueType;
                         CurrentLast: INTEGER);

   (* ASSUMES V[0] .. V[CurrentLast-1] IS A HEAP;
      EXTENDS HEAP BY ADDING V[CurrentLast] TO IT. *)

   VAR
      child: INTEGER;
      parent: INTEGER;

   BEGIN

      child := CurrentLast;
      parent := child DIV 2;

      LOOP

         IF V[child] > V[parent] THEN
            Swap(V,child,parent);
            child := parent;
         END;

         IF parent = 0 THEN
            EXIT;
         ELSE
            parent := parent DIV 2;
         END;

      END;

   END ExtendHeap;
```

(a) Add new element to heap

Figure 9–18 Heap sort procedures

Sec. 9.3   Internal Sort Algorithms with Growth Rate $O(N \text{Log}(N))$ 269

```
PROCEDURE AlmostHeapToHeap(VAR V:       ARRAY OF ValueType;
                          CurrentLast: INTEGER);

(* ASSUMES V[0] .. V[CurrentLast] IS AN "ALMOST HEAP",
   THAT IS, IT WOULD BE A HEAP EXCEPT THAT V[0] MAY BE
   SMALLER THAN ONE OR BOTH OF ITS CHILDREN *)

   VAR
      parent: INTEGER;
      child:  INTEGER;
      placed: BOOLEAN;

   BEGIN
      parent := 0;
      child := 1;
      placed := FALSE;

      WHILE (child <= CurrentLast) AND NOT placed DO

         IF child+1 <= CurrentLast THEN
            (* PARENT HAS 2 CHILDREN *)

            IF (V[parent] >= V[child])    AND
               (V[parent] >= V[child+1]) THEN
                placed := TRUE;

            ELSE
               IF V[child] > V[child+1] THEN
                  (* LEFT CHILD IS LARGER *)
                  Swap(V,parent,child);
                  parent := child;
                  child := 2*parent;
               ELSE
                  (* RIGHT CHILD IS LARGER *)
                  Swap(V,parent,child+1);
                  parent := child+1;
                  child:= 2*parent;
               END;
            END;

         ELSE
            (* PARENT HAS 1 CHILD; IT MUST BE THE LEFT ONE *)
            IF V[parent] >= V[child] THEN
               Swap(V,parent,child);
            END;
            placed := TRUE;
         END;

      END;
   END AlmostHeapToHeap;
```

(b) Convert almost-heap to heap

Figure 9–18   (*Continued*)

```
PROCEDURE BuildHeap(VAR V: ARRAY OF ValueType);

   VAR
      CurrentLast: INTEGER;

   (* given an array V[0] .. V[HIGH(V)],
      build a heap in place *)

   BEGIN

      FOR CurrentLast := 1 TO HIGH(V) DO
         ExtendHeap(V,CurrentLast);
      END;

   END BuildHeap;
```

   (c) Build a heap in place

```
PROCEDURE HeapSort(VAR V: ARRAY OF ValueType);

      (* CODE FOR ExtendHeap       GOES HERE *)

      (* CODE FOR AlmostHeapToHeap GOES HERE *)

      (* CODE FOR BuildHeap        GOES HERE *)

      (* CODE FOR SortHeap         GOES HERE *)

   BEGIN

      BuildHeap(V);
      SortHeap(V);

   END HeapSort;
```

   (d) Sort the heap

```
PROCEDURE SortHeap(VAR V: ARRAY OF ValueType);

   VAR
      NewLargest: INTEGER;

   (* assuming V[0] .. V[HIGH(V)] is a heap,
      sort it in place.
      Also assume V has at least three elements. *)

   BEGIN

      FOR NewLargest := HIGH(V) TO 2 BY -1 DO
         Swap(V,0,NewLargest);
         AlmostHeapToHeap(V,NewLargest-1);
      END;

   END SortHeap;
```

   (e) Driver routine

Figure 9–18 (*Continued*).

Suppose the original array is already sorted? Since all the larger items are at the right-hand end of the array, they are at the bottom of the tree that is to be turned into a heap. Thus, these items will have further to move into their "heap" positions. Consequently, the heap sort's worst-case performance is for a sorted array. On the other hand, its best-case performance is achieved when the original array is sorted *downward*, since in that case it is a heap already!

The heap sort is interesting partly because it can be made to run with a relatively small time per operation: parents and children are calculated by dividing and multiplying by 2, respectively, and these operations can be implemented as single-bit shifts in either assembly language or in a high-level language which supports bitwise shifting.

9.4.3 Quicksort

The quicksort is one sorting method that has been shown by much experiment to perform well in the average case: on the average, a quicksort requires $O(N \log(N))$, even though its worst-case performance is $O(N^2)$.

The quicksort is often called a *partition sort*. In fact, it is a recursive method in which an unsorted array is first rearranged so that there is some record, somewhere in the middle of the array, whose key is greater than all keys to its left and less than or equal to all keys to its right. Then, once this middle record (which is probably not really exactly in the middle of the array) is found, the same method is again applied to sort the section of the array to its left, and then to sort the section of array to its right. The algorithm is thus another example of a "divide and conquer" method, where a structure is divided into two pieces by some criterion, and then the two pieces are attacked separately. Each piece is then subdivided, and so on, until the whole structure is processed. Philosophically, then, the method is in the same category with the binary search tree methods presented earlier.

The quicksort algorithm. The idea of the quicksort algorithm is to take a guess at a median or middle value in an array, i.e., an element in the array such that half the other elements in the array are less, and the other half greater, than the median. It would be a true median if exactly half the elements were greater, half were less, and we could partition the array into equally sized pieces. In general, however, we won't guess correctly, but whichever value we do guess will clearly let us partition the array into two unequally sized pieces such that one piece has all the smaller elements and the other piece all the larger ones.

How, then, shall we take a guess? Since we are not assuming any prior ordering in the array, any element has as good a chance of being the median as any other. So we might just as well take the first element in the array. In fact, we can be a little more clever than that: since the first few items in the array could all be the same value, we'll choose the leftmost *distinct* element. The procedure FindPivot presented in Figure 9–19 determines the location of the pivot element.

```
PROCEDURE FindPivot(VAR V:     ARRAY OF ValueType;
                    First:     INTEGER;
                    Last:      INTEGER;
                    VAR Pivot: INTEGER);

VAR
    Left:     INTEGER;
    Right:    INTEGER;
    Up:       INTEGER;
    FirstKey : ValueType;

BEGIN
    Left := First;
    Right := Last;
    FirstKey := V[Left];
    Pivot := -1;
    (* PIVOT WILL REMAIN -1 IF ALL KEYS EQUAL *)

    Up := Left+1;
    LOOP

        IF Up > Right THEN (* ALL KEYS IN SUBFILE ARE EQUAL *)
            EXIT;
        ELSIF V[Up] > FirstKey THEN
            Pivot := Up;
            EXIT;
        ELSIF V[Up] < FirstKey THEN
            Pivot := Left;
            EXIT;
        ELSE
            INC(Up);
        END;

    END;

END FindPivot;
```

Figure 9-19 Procedure to find pivot point

(Since our guessed "median" isn't likely to be a genuine median anyway, it is conventional to call it a pivot instead.)

Now, having found the location of the pivot, how do we execute the partition? The idea is to start two cursors moving, one rightmost from the left end of the array, the other leftward from the right end. The rightward-moving cursor (which we shall call "up") will keep moving as long as the elements it scans are less than the pivot; the leftward-moving one (which we shall call "down") will keep moving as long as the elements it scans are greater than the pivot. If the "up" cursor finds a value greater than the pivot and the "down" cursor finds one less than the pivot, those two values are exchanged. Then the cursors are started again from those points.

Eventually, the two cursors will meet. At the point where they meet, all values to the left are guaranteed to be less than the pivot, and all values to the right are guaranteed to be greater than the pivot. We call that meeting point the *partition point*, and the procedure Partition in Figure 9-20 computes it.

Now we can write a procedure Quick which first finds a pivot and then finds

```
PROCEDURE Partition(VAR V:            ARRAY OF ValueTyp
                    First:            INTEGER;
                    Last:             INTEGER;
                    Pivot:            ValueType;
                    VAR PartitionPoint: INTEGER);

  VAR
     Up:   INTEGER;
     Down: INTEGER;

  BEGIN

     Up := First;
     Down := Last;

     REPEAT

        Swap (V,Up,Down);

        WHILE V[Up] < Pivot DO
           INC(Up);
        END;

        WHILE V[Down] >= Pivot DO
           DEC(Down);
        END;

     UNTIL Up > Down;

     PartitionPoint:=Up;
  END Partition;
```

Figure 9-20 Procedure to partition array

```
PROCEDURE Quick(VAR V: ARRAY OF ValueType;
                First: INTEGER;
                Last:  INTEGER);

   VAR
      PivotPoint:     INTEGER;
      PartitionPoint: INTEGER;
      Left:           INTEGER;
      Right:          INTEGER;
      Pivot :         ValueType;

   BEGIN
      Left:=First;
      Right:=Last;
      FindPivot(V,Left,Right,PivotPoint);

      IF PivotPoint<>-1 THEN (* DISTINCT VALUES; SORT THEM *)
         Pivot:=V[PivotPoint];
         Partition(V,Left,Right,Pivot,PartitionPoint);
         Quick(V,Left,PartitionPoint-1);
         Quick(V,PartitionPoint,Right);
      END;

   END Quick;
```

Figure 9-21 The recursive procedure Quick

```
PROCEDURE QuickSort(VAR V: ARRAY OF ValueType);

  (* CODE FOR FindPivot HERE *)

  (* CODE FOR Partition HERE *)

  (* CODE FOR Quick     HERE *)

BEGIN
  Quick(V,0,HIGH(V));
END QuickSort;
```

Figure 9–22 Quicksort procedure

the partition point for that pivot. At that stage, the array is partitioned into a section with "smaller" values on the left and a section with "larger" values on the right. But the two sections are not yet sorted. On the other hand, we can sort them by calling Quick recursively, first to sort the left section and then to sort the right section. This recursive procedure appears in Figure 9–21.

Finally, we can write a driver called QuickSort which just calls Quick with the entire initial array as input. The entire procedure is illustrated in Figure 9–22; Figure 9–23 shows the various phases of the process as applied to a 10-element array.

For the average case, the quicksort performs in $O(N \log(N))$ time. Interestingly, its worst case, which approaches $O(N^2)$, occurs when the original array is already sorted. In that situation, every attempt to partition the array results in a left subarray of length 1 and all the rest of the items in the right subarray.

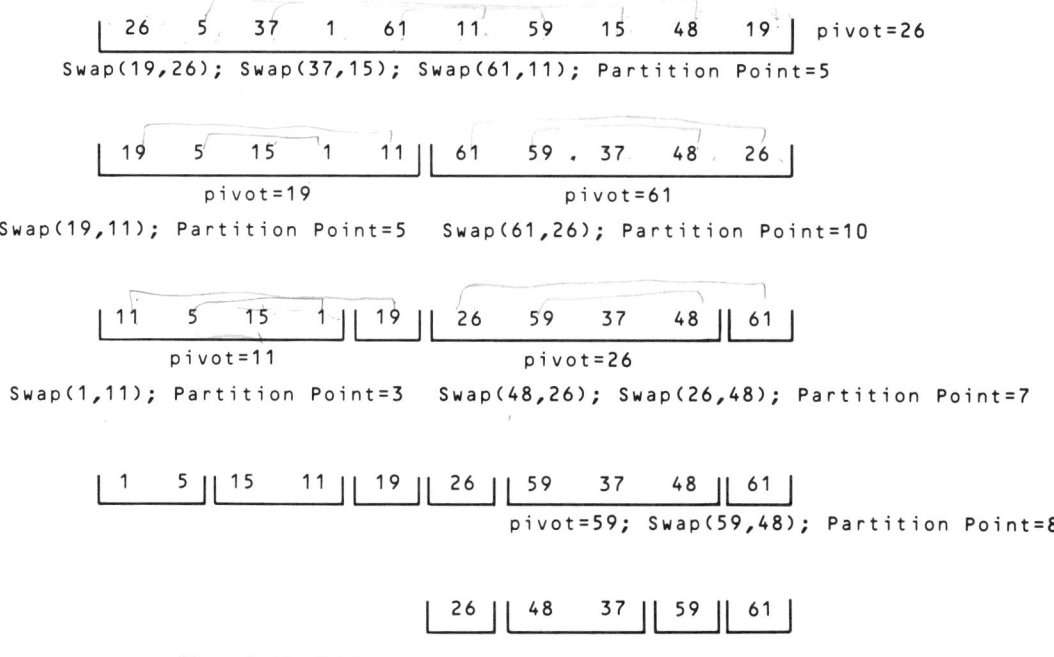

Figure 9–23 Quicksort applied to a 10-element array (the reader can complete the algorithm)

Sec. 9.5   Other Internal Sort Algorithms   275

9.5 OTHER INTERNAL SORT ALGORITHMS

In this section we introduce three sort algorithms whose performance does not fall neatly into either of the preceding categories.

9.5.1 Shell Sort

The quite popular Shell sort (named for its inventor, D. Shell) can be viewed as a modification of either the bubble sort or the linear insertion sort. In both of these sorts, an item moves toward its proper place only one "slot" at a time. The distance each item moves determines the overall running time; Shell sort tries to reduce this time by first putting the array in rough order. The algorithm does this by comparing items that are separated from each other rather than immediately adjacent ones. This is done by choosing a *distance* and sorting subfiles, each of which is made up of elements separated from each other by that distance. For instance, suppose the distance—call it d—is 3, and the total length of the array is 15. Then, as in the second column of Figure 9-24, we first sort the subarray {A[1], A[4], A[7], A[10], A[13]} (most authors use linear insertion to do this), then the subarray {A[2], A[5], A[8], A[11], A[14]}, followed by {A[3], A[6], A[9], A[12], A[15]}. This constitutes one phase.

Having put all these subarrays in mutual order—which moves small items nearer the top and large items nearer the bottom in larger steps than normal linear insertion—we reduce the distance. Each phase reduces the distance until, eventually, $d = 1$ and we sort the entire array by a final phase of linear insertion. By now each item needs to move only a very short distance to reach its home.

There is no general agreement in the literature on how best to choose the distances. Some authors advocate choosing a set of mutually prime distances. Others argue that letting $d_1 = N$ DIV 2 and $d_{k+1} = d_k$ DIV 2 is just as good. For simplicity we choose the latter method; a demonstration is given in Figure 9-24 and Figure 9-25 shows a procedure.

The performance of the Shell sort is in the neighborhood of $O(N\sqrt{N})$; the analysis is beyond the scope of this book.

9.5.2 Quadratic Selection

An interesting sort algorithm trades a penalty in space for a payoff in running time of $O(N\sqrt{N})$. Let $M = \sqrt{N}$, and divide up the unsorted array A into segments of size M (rounded up to the nearest integer, of course) and copy the segments into a square array A' of size $M \times M$. The algorithm then proceeds as follows:

QUADRATIC SELECTION SORT:

1. By comparing and swapping as in the delayed selection sort, get the smallest element in each row of A' into the first position of that row.

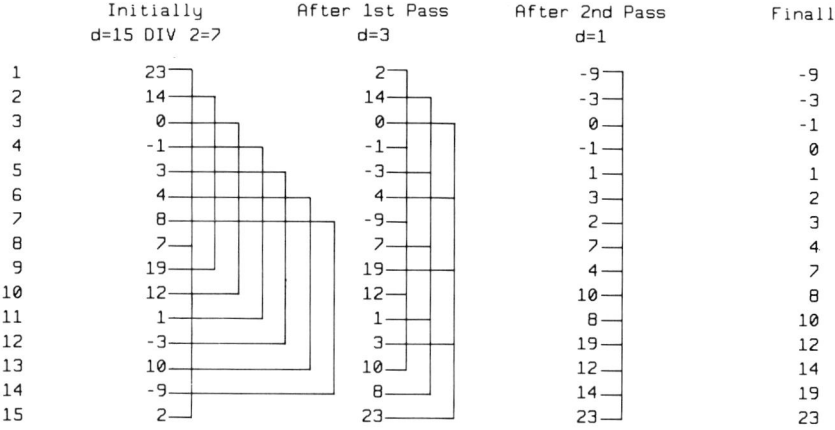

Figure 9-24 Shell sort applied to a 15-element array

```
PROCEDURE ShellSort(VAR V: ARRAY OF ValueType);

    VAR
        N:          INTEGER;
        Distance:   INTEGER;
        i,j:        INTEGER;

    BEGIN

        N := HIGH(V);
        Distance := N DIV 2;

        WHILE Distance > 0 DO

            FOR i := Distance TO N DO

                j := i - Distance;
                LOOP

                    IF (j >= 0) AND (V[j] > V[j+Distance]) THEN
                        Swap(V,j,j+Distance);
                        j := j - Distance;
                    ELSE
                        EXIT;
                    END;

                END;

            END;
            Distance := Distance DIV 2;

        END;

    END ShellSort;
```

Figure 9-25 Procedure for Shell sort

Sec. 9.5 Other Internal Sort Algorithms

2. Find the smallest element in the first column of A', and output it to the sorted array.
3. Replace that element in A' by the smallest element in the row from which it came, and "compress" the row by replacing the element just removed by the current last element in the row.
4. Continue the process until all the rows are empty. The original array is then sorted.

An example of the algorithm for quadratic selection is shown in Figure 9–26; writing a procedure for it is suggested as an exercise.

What is the running time of the quadratic selection? Since when we first create A' each row has at most M elements, and there are M rows, the initialization given in step 1 takes at most $M \times M = N$ operations. Step 2 then takes $M - 1$ operations, and step 3 takes a variable number, but surely no more than $M - 1$. However, we carry out steps 2 and 3 once for *each* item in the original array, or N times again. So we have the sum of an $O(N)$ term and an $O(N \times M)$ term, where, for nontrivial N, the second term dominates. Thus, the overall algorithm is $O(N \times M) = O(N\sqrt{N})$.

```
26  5  37  1  61  11  59  15  48  19  0  -3  7
```

(a) Original unsorted array

26	61	48	7
5	11	19	
37	59	0	
1	15	-3	

(b) Initial square array made from the array above

7	61	48	26
5	11	19	
0	59	37	
-3	15	1	

(c) Row minima located and placed at heads of rows

7	61	48	26
5	11	19	
0	59	37	
1	15		

Result Array -3

7	61	48	26
5	11	19	
37	59		
1	15		

-3 0

7	61	48	26
5	11	19	
37	59		
15			

-3 0 1

(d) A few steps of the algorithm (the reader can complete it)

Figure 9–26 Quadratic selection sort

278   Internal Sorting Methods   Chap. 9

9.5.3 Radix Sort

The radix sort is probably best explained in terms of electromechanical punched-card sorting machines. These machines were widely used during the 1930s, 1940s, and 1950s before computers became widespread; their popularity declined through the 1960s and 1970s, and they are hardly to be found anymore.

    A punched card has 80 data positions or columns, each with 12 rows. Ten of these rows are numbered 0 through 9, and each position can hold one character of data. If we assume for simplicity that all the data is numeric, then each character is one of the digits 0 through 9, and a digit in a given column is encoded by a single punch in the appropriate row of that column. A numerical value is a sequence of numeric digits and is encoded by a single punch in each of several consecutive columns. Figure 9–27 shows a section of a punched card with a six-digit number punched in it in positions 5 through 10.

    The card sorter has 13 bins or pockets, each capable of holding several hundred cards, and an equally large input hopper. The machine operates on one column at a time. The operator sets an indicator to the desired column, loads the input hopper with a face-up stack of cards, and then presses the start button. The machine, with much noise and furious movement of cards, places each card in the bin corresponding to the row in which a punch appears in the given column. The

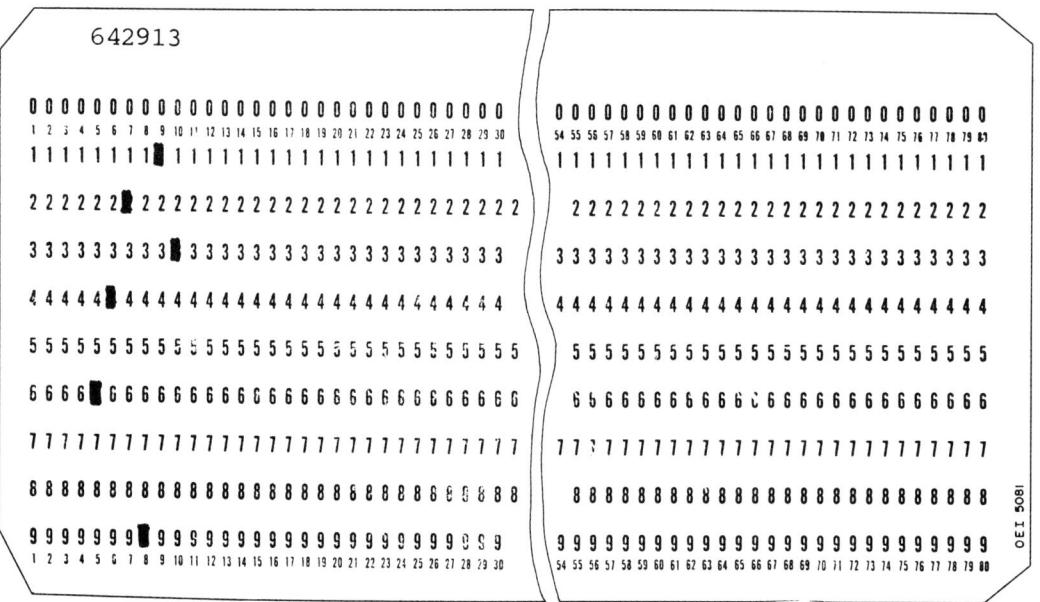

Figure 9–27   A punched card

Sec. 9.5   Other Internal Sort Algorithms   279

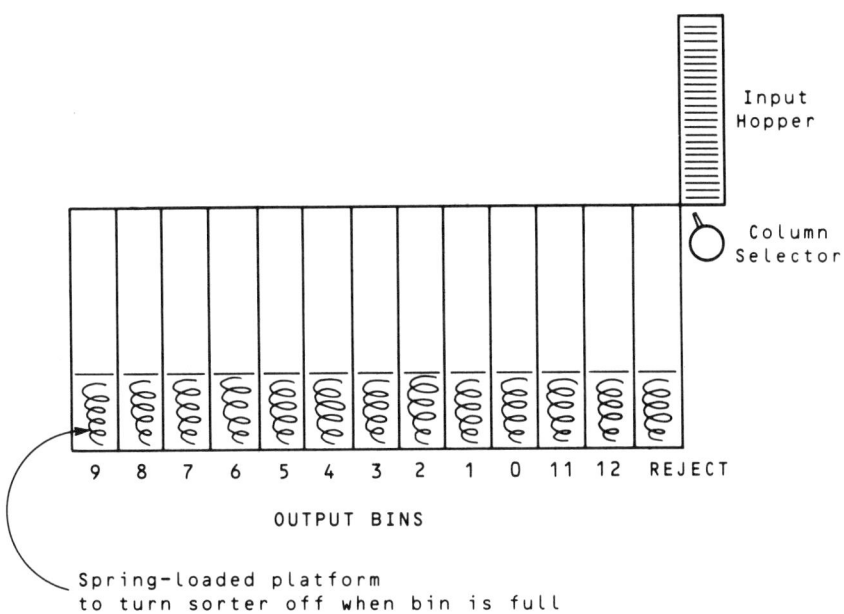

Figure 9-28 Electromechanical punched-card sorter

thirteenth bin collects those cards with no punch at all. Figure 9-28 shows a diagram of the machine.

How is such a machine used to sort a deck of cards on an entire key? The deck must be run through the card sorter once for each digit of the key, according to the following algorithm:

CARD SORTER ALGORITHM:

1. Set the column indicator to the rightmost (low-order) column of the key to be sorted.
2. Place the deck face up in the input hopper.
3. Start the sorter, and wait until input hopper is empty. (The machine stops by itself.)
4. Remove the decks of cards from each bin, and combine them into one deck with the contents of bin 0 on top and the contents of bin 9 on the bottom.
5. If not all the columns have been processed, move the column indicator one position to the left and repeat steps 2 through 5.

In Figure 9-29 we illustrate this kind of sort for an eight-card deck to be sorted on a three-digit key. As an exercise, determine why the sort must begin with the rightmost digit and move to the left, and not the other way around.

280   Internal Sorting Methods   Chap. 9

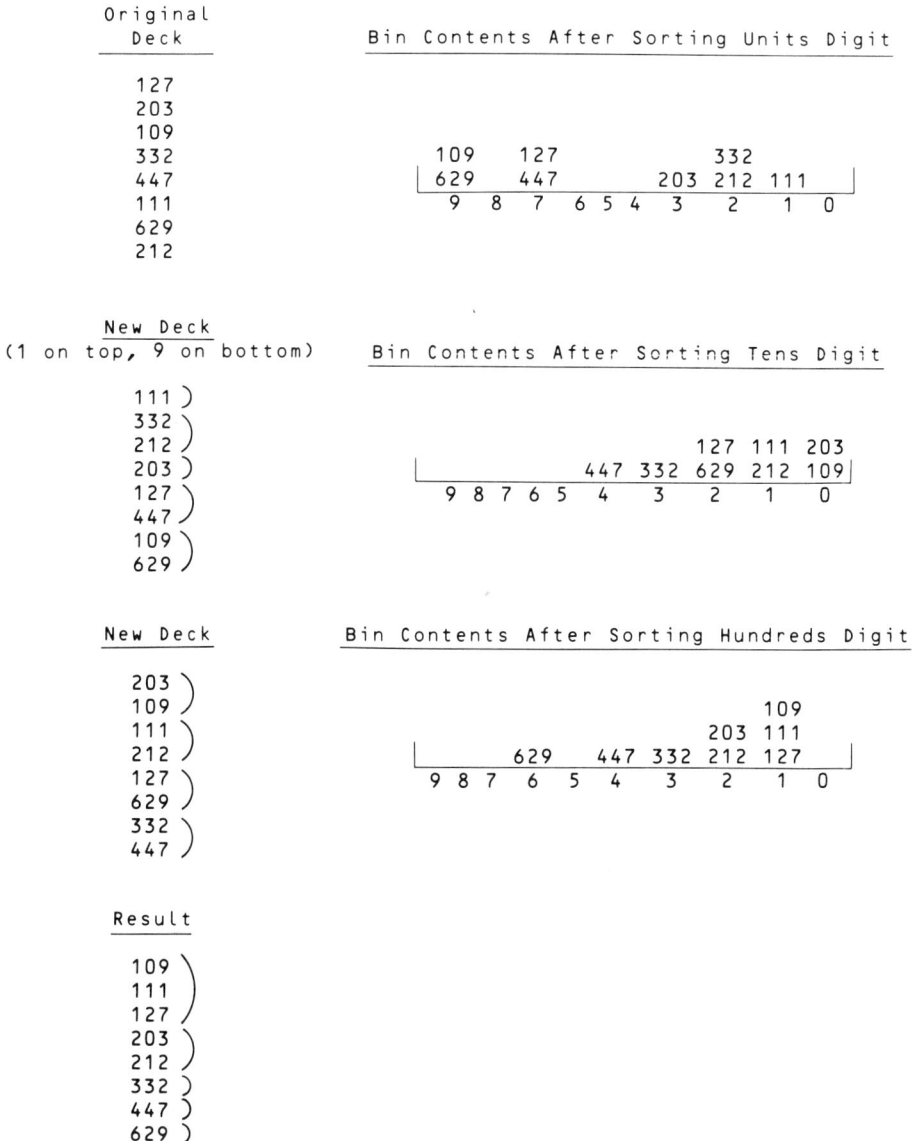

Figure 9–29 Radix sort, as used by a card sorter

    The algorithm can be adapted to operate on a computer, and works rather well if we realize that the keys are represented as binary sequences (like everything else on most computers). If the keys are unsigned binary integers or character strings, we can just treat them as bit sequences. We use just two "bins," generally arrays, and sort bit by bit, from the rightmost bit to the leftmost. The sort is called a *radix sort* because the number of bins is determined by the radix or base of the digits being sorted.

What is the performance of the radix sort? Clearly, the number of passes is determined by the number of bits in the key. Thus, for a fixed-size key, the number of passes is fixed. Since each pass examines each record exactly once, the total number of operations is a constant (the number of passes) times the number of records (N); in other words, the growth rate is $O(N)$. The radix sort is not very widely used because of the extra space required for the bins and because the (usually) large number of passes means that the $O(N)$ growth rate may well be dominated by the very large constant of proportionality.

9.6 DESIGN: PRIORITY QUEUES

A *priority queue* is a queue in which elements are enqueued as they arrive, as in the FIFO studied in chapter 5, but dequeued according to some priority scheme. Priority queues have many applications; a common one is the queueing system used in a multi-user operating system.

The discussion of priority queues appears in a sorting chapter because a common and very interesting implementation of priority queues is built on the heap-manipulation operations studied in section 9.4.2.

Assuming that each arriving element has a key field indicating its priority, we could imagine implementing a priority queue as a circular array with a difference. An element is inserted (enqueued) according to its priority; a dequeue operation is just like that of a FIFO queue—the head element is removed and the queue adjusted. The performance of an enqueue operation is clearly linear; that of a dequeue operation is clearly constant.

An implementation using a linked list would use as an enqueue operation the ordered-list insertion algorithm from chapter 4; a dequeue operation would simply remove the first element. Performance here is similar to the array implementation.

One particularly clever implementation of a priority queue uses a heap. Since a heap is just an array viewed differently, no more space is necessary than that required for a normal array queue. An enqueue operation is implemented as an ExtendHeap: the new element is moved up the heap until its priority (key) is greater than its children's priorities. A dequeue operation works just like one piece of the SortHeap algorithm: the first element of the array is removed (it is obviously the one with the largest key!), then the last element is moved to the first position. This leaves, precisely, an almost-heap. A call of AlmostHeapToHeap moves it into its proper place.

As we discovered in section 9.4.2, both of these operations are logarithmic. Comparing this implementation with the ones above, we have traded one constant-time operation and one linear operation for two logarithmic ones. In cases where the queue is likely to grow long, the tradeoff is advantageous. Implementing a priority queue in this fashion is left as an exercise.

9.7 SUMMARY

This chapter has presented a number of internal sorting methods, along with estimates of their performance. The discussion should enable anyone to make a sensible choice of a method for whatever sorting problem he or she is faced with. Clearly, if the list to be sorted is small, the best method is the one that's easiest to write, since in the case of a small list "people time" is more expensive than the computer time. On the other hand, if the list is large, particularly as part of an application that will be run frequently, it pays to think the problem through and choose wisely, because the computer time used in the sort will no longer be negligible.

CHAPTER 9 EXERCISES

1. Refer to the bubble sort procedure presented in section 9.3.3. Define "trip length" as the number of upward moves an element in the array must make on its way to its final position. Show that the number of passes required by the bubble-sort algorithm depends upon the maximum of all the trip lengths in the array.
2. In the bubble sort algorithm of section 9.3.3, we begin at the top of the array and move elements downward. Write a procedure for a bubble sort in which we start at the bottom of the array and move elements upward.
3. In the modified bubble sort often called "shaker sort," we run successive passes alternately in the upward and downward directions. Write a procedure for this sort. Why does this method sometimes offer improved performance?
4. Calculate the number of passes in the merge sort when the number of elements in the array is not an exact power of 2.
5. Write a modified merge sort procedure in which it is not necessary to re-copy the array after each pass. Hint: do this by "switching" alternately the input and output arrays.
6. Write a procedure implementing the quadratic selection sort of section 9.5.2.
7. Write a procedure implementing a decimal radix sort, assuming that the keys are all the same length and are represented as strings of digits.
8. Write a procedure implementing binary radix sort, as suggested in section 9.5.3. Note that since only two bins are required, only one additional array is needed because the array can be filled from both ends, all "0" elements inserted from the top of the array, all "1" elements from the bottom.
9. Design an experiment to do some measurements on a group of sort algorithms. Find out how to time a program on your computer, implement the algorithms as complete programs, and test them on files of length 4 to length 1024, doubling the file length each time (i.e., 4, 8, 16, 32, 64, . . . , 1024). Try best, worst, and random cases. Do the actual running times show that the "big O" predictions are correct?
10. Implement a module for handling priority queues according to the scheme discussed in section 9.6.

Chapter 10
SORTING EXTERNAL FILES

10.1 GOAL OF THE CHAPTER

Given the actual size of many files, and the size of main memory on most machines, the internal sorts discussed in the previous chapter are utterly inapplicable. To handle these sizable, complicated real-world situations requires that we direct our attention to external sorts instead.

External sorting algorithms are based on *merging*, as in the merge sort algorithm in the last chapter. The underlying principle of these algorithms is to break the file up into unsorted subfiles, sort the subfiles, and then merge the sorted subfiles into larger and larger sorted subfiles until the entire file is sorted. That is why these algorithms are often called *sort/merge* algorithms. Sort/merge algorithms generally begin with a *distribution* phase, which creates a number of sorted subfiles in external devices, followed by a series of *merge* phases, to produce one sorted file at the end.

10.2 THE TWO-WAY MERGE

The merge process for sequential files stored externally is so fundamental to the whole class of sort/merge algorithms that it makes sense to study it independently. The external merge is similar to the internal merge as applied to items in an array or linked list.

We start with two previously sorted sequential files F1 and F2, and merge them to produce a new sorted file F3. Assuming the two input files are sorted in ascending order on key *k*, we begin by reading one record from each file and writing the one with the smaller *k* to the output file. If the record came from F1, we read a new record from F1; if it came from F2, we read a new record from F2. Then we just continue the process.

Eventually, F1 or F2 becomes empty first. At this point, we copy the remaining records from the *other* file into F3. This is known as "copying the tail." When we are finished, F3 will be a completely sorted file.

A Modula-2 procedure for this process, called TwoWayMerge, is shown in Figure 10–1.

```
PROCEDURE TwoWayMerge(VAR F1, F2, F3: File);

VAR
   R1, R2: RecordType;

BEGIN
   ReadRecord(F1, R1);
   ReadRecord(F2, R2);

   LOOP                       (* merge until F1 or F2 runs out *)
      IF KeyPart(R1) < KeyPart(R2) THEN
         WriteRecord(F3, R1);
            ReadRecord(F1, R1);
         IF Eof(F1) THEN
            WriteRecord(F3,R2);
            EXIT;
         END;
      ELSE
         WriteRecord(F3, R2);
            ReadRecord(F2, R2);
         IF Eof(F2) THEN
            WriteRecord(F3,R1);
            EXIT;
         END;
      END;
   END;

   LOOP                       (* copy tail of F1 *)
      ReadRecord(F1, R1);
      IF Eof(F1) THEN
         EXIT;
      ELSE
         WriteRecord(F3, R1);
      END;
   END;

   LOOP                       (* copy tail of F2 *)
      ReadRecord(F2, R2);
      IF Eof(F2) THEN
         EXIT;
      ELSE
         WriteRecord(F3, R2);
      END;
   END;

END TwoWayMerge;
```

Figure 10–1 Procedure for two-way merge

Sec. 10.4   Simple Merge Sorting   285

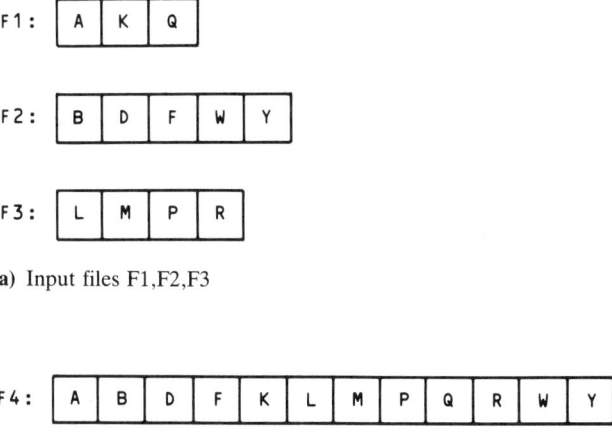

(a) Input files F1,F2,F3

(b) Output file F4, after merging

Figure 10-2   A three-way merge

10.3 THE K-WAY MERGE

A merge does not have to be limited to two files. If we have K sorted files, they can be merged simultaneously to produce a $K + 1$st sorted file. To do so, we read a record from each of the K files, write whichever record has the smallest key to the output file $F(K + 1)$, and then read a new record from whichever file the output record came from. When one of the K input files runs out of records, we continue the process with the remaining $K - 1$ files until only one file is left with records in it, at which point we copy its tail. A diagram of the process is shown in Figure 10-2; we leave it as an exercise to write the program.

Since in this merging process very little main memory is required (just the space for one record per file, plus the space for the program), the number of files we can merge is limited only by the number of sequential external files available to us.

10.4 SIMPLE MERGE SORTING

We can turn the K-way merge into a sort if we can produce the K sorted files required. In the ensuing discussion of how to do this, we shall use the term "tapes" to designate external devices, although these devices may actually be other media, e.g., disk files.

Suppose first that we are in the happy situation where, given an N-record unsorted file, we have enough tapes $K + 1$ and main storage R (records) such that $R = N/K$. The problem then becomes a simple one: use K tapes, read R records at a time from the $K + 1$st tape, sort them by an appropriate *internal* sort (perhaps a heap sort or quicksort), write this sorted file onto its own output tape, and then merge the K sorted files onto the $K + 1$st tape.

Unfortunately, however, this situation does not occur very often: typically, we are limited in both the amount of main storage and the number of tapes or other devices we have. An alternative solution, then, is as follows:

K-WAY DISTRIBUTION

1. Determine R, the number of records that can reasonably be sorted internally.
2. Determine T, the total number of tapes that are required and available, and let $K = T$ if T is an even number, or $K = T - 1$ if T is an odd number.
3. Sort R records at a time internally, writing the results in turn onto each of $K/2$ tapes, with file markers at their ends.
4. Continue the process in step 3, writing additional files onto the $K/2$ tapes.

At this point we have the situation depicted in Figure 10-3, where each of $K/2$ tapes has, say, M sorted subfiles on it. (Some of the tapes will probably have only $M - 1$ files, since we will run out of input records before we've put the Mth file on all $K/2$ tapes.) We have thus *distributed* the sorted subfiles onto the $K/2$ tapes, which can now be merged repeatedly onto the other $K/2$ tapes until a single sorted file is achieved:

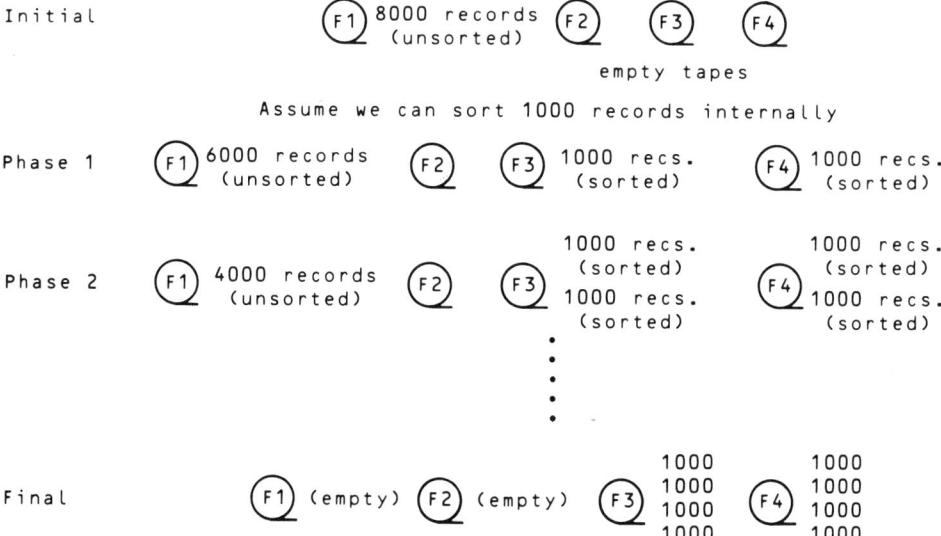

Figure 10-3 Distribution by internal sorting, k = 4

Sec. 10.5   The "Natural" Distribution   287

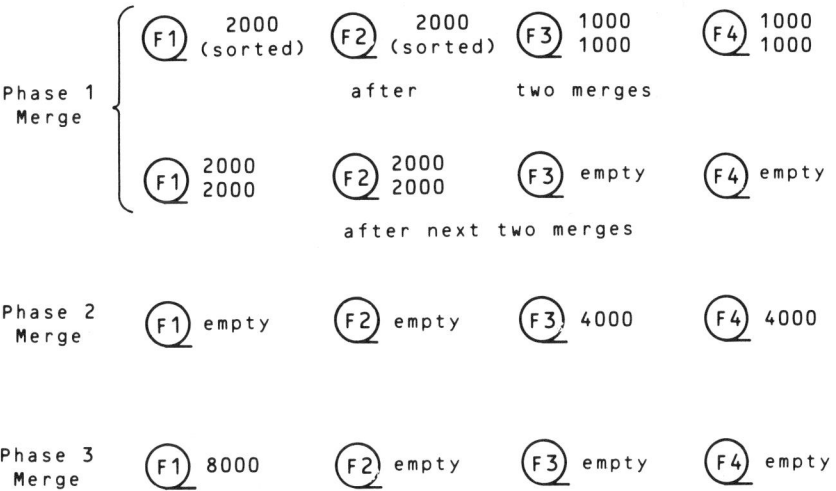

Figure 10-4 Merge process, starting with the distribution from Figure 10-3.

K-WAY SORT/MERGE, ONE PHASE:

1. Rewind all the tapes.
2. Do a $K/2$-way merge, using the first file from each tape and writing the output onto one of the $K/2$ empty tapes.
3. Repeat the process in step 2 for each of the rest of the $K/2$ empty tapes.
4. Repeat steps 1 through 3, merging in rotation onto the $K/2$ files until all the original $K/2$ files are empty.

We have now created files of $R \times (K/2)$ records on $K/2$ tapes. If we continue this process of "bouncing" between groups of $K/2$ tapes through additional phases, there will finally be at most one file on each of $K/2$ tapes. These can be merged in one final pass onto the output tape. The entire K-way merge process is illustrated in Figure 10-4.

10.5 THE "NATURAL" DISTRIBUTION

If we are severely limited in the main storage we can use, we cannot easily use the internal-sort method of distribution. Fortunately, another method of distributing sorted subfiles onto tapes arises out of the knowledge that most files arrive for sorting with their records in random order, which, oddly enough, means that there is already a certain amount of order in them. Using the term *run* to denote a series of records which already happen to be in the correct order, we can see that

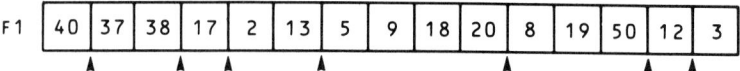

Figure 10-5 An unsorted file (note the runs, marked by arrowheads)

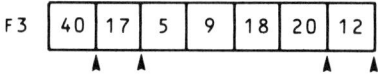

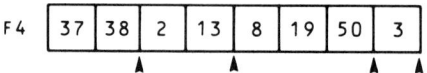

Figure 10-6 Distribution of the runs from Figure 10-5 onto two tapes (again note the runs)

any file consists of a sequence of runs. (In the worst case, the file is in *reverse* order, where no run is longer than one record.) Figure 10-5 shows a file divided into the runs which naturally occur in it.

To take advantage of the natural runs that occur in our file, we distribute the file onto $K/2$ tapes by simply reading a record at a time, comparing each key against the previous one to see if order is maintained, i.e., if the new record is part of the current run. If we are sorting from low to high, then the new record will fail to be part of the current run if its key is less than the previous key. We continue writing records onto a tape as long as they are part of a run. A record which is not part of the current run obviously begins a new run, in which case we switch to a new output tape.

We continue distributing runs onto tapes in rotation, just as we distributed sorted subfiles in the previous example. Finally, when all runs are exhausted, we are ready to merge as before. The distribution process for the file given in Figure 10-5 is shown in Figure 10-6.

The advantage of the "natural" distribution is that time and space are saved since we do not have to sort internally; the chief disadvantage of the method is that runs are of unequal length and we cannot predict how many of them there will be. It may turn out that many more merge phases than we would like will be required since runs may be very short—only one record long in the worst case, as mentioned.

10.6 POLYPHASE SORTING

In the K-way merge, it is not really necessary to empty all $K/2$ tapes before beginning a new phase: rather, we can use $K - 1$ tapes for input and *just one* for output, and then start a new phase whenever *any* tape is empty. All we need do is make the empty tape the output and the remaining $K - 1$ tapes (which still have runs

Sec. 10.6   Polyphase Sorting

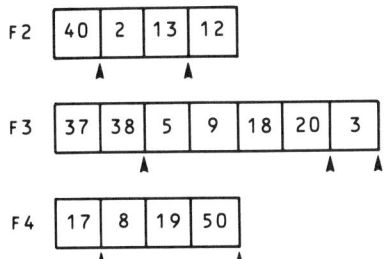

(a) Initial distribution of the file from Figure 10–5

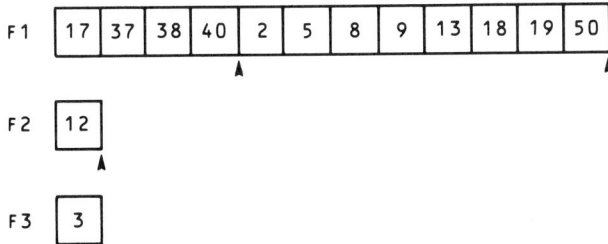

(b) First merge phase

F2 empty

F3 empty

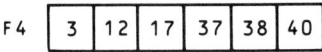

(c) Second merge phase

Figure 10–7   Polyphase merge

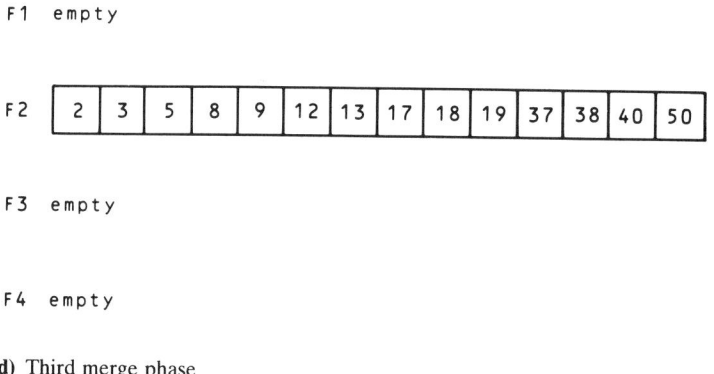

(d) Third merge phase

Figure 10-7 (*Continued*).

on them) the input. Such a procedure is known as *polyphase merging*. A polyphase merge using three tapes for input and one for output is shown in Figure 10-7, again using the file of Figure 10-5.

10.7 FIBONACCI DISTRIBUTION FOR POLYPHASE MERGING

The polyphase distribution just examined assigned runs to the $K - 1$ tapes in strict rotation, resulting in an (approximately) equal number of runs on each tape. It turns out that polyphase merging can be improved by a more optimal distribution of runs. This can be seen by working backwards from the ideal situation where the last phase consists of merging one run from each of $K - 1$ tapes onto the Kth tape, leaving $K - 1$ empty tapes and one sorted one. How could we have arrived at that situation?

To answer this question, let us look at an example where $K = 6$. Numbering the configurations such that the last one is denoted Last, we see that configuration Last $- 1$ is the one with one run on each of $K - 1$ tapes. What, then, should configuration Last $- 2$ be? It should be one with two runs on each of $K - 2$ tapes, one tape with one run, and one with none. And how about configuration Last $- 3$? We should have three tapes with four runs, one with three, one with two, and one with none. Several more configurations are given in Figure 10-8.

It should be clear by now that there is a systematic way to determine how a file with a given number of initial runs might be distributed to reach this ideal situation. Indeed, if we rewrite the configurations just to show the five tapes used for input and arrange the rows in descending numbers of runs, we can see that certain relationships exist among these numbers. The rearranged table is shown in Figure 10-9, and the relationships are given in Figure 10-10. They are closely related to the *Fibonacci sequences*, which are sequences of numbers well known in number theory.

Sec. 10.7 Fibonacci Distribution for Polyphase Merging

PHASE	F1	F2	TAPE FILE F3	F4	F5	F6	TOTAL NO. OF RUNS
Last-5	16	15	14	12	8	0	65
Last-4	8	7	6	4	0	8	33
Last-3	4	3	2	0	4	4	17
Last-2	2	1	0	2	2	2	9
Last-1	1	0	1	1	1	1	5
Last	0	1	0	0	0	0	1

Figure 10-8 Ideal last six phases of six-tape polyphase merge sort

The numbers of runs for "perfect" distribution can be gotten from tables and built into a program. The distribution is as follows:

FIBONACCI DISTRIBUTION:

1. Given $K - 1$ tapes, start with the second line of the table and try to match the distribution, if possible (one run per tape).
2. Repeat the next two steps as long as records remain on the input tape.
3. Increment the line of the table that needs to be matched.
4. Try to match the distribution required by this line.

Eventually, the input file will run out of runs. When this happens, we shall be part way through matching a line, since files do not ordinarily have *exactly* the right number of runs to match a line of the table. To get the merge process going, we then imagine that the remaining runs to match the table line have in fact been placed on the tapes; they are just empty runs. This gives us a fictitious "perfect" distribution; we just merge the empty or "dummy" runs with the real ones until we reach the next lower line of the table, and continue from there.

L=LEVEL	1	2	3	4	5	TOTAL NO. OF RUNS
0	1	0	0	0	0	1
1	1	1	1	1	1	5
2	2	2	2	2	1	9
3	4	4	4	3	2	17
4	8	8	7	6	4	33
5	16	15	14	12	8	65
6	31	30	28	24	16	129
7	61	59	55	47	31	253

Figure 10-9 "Perfect" distribution of runs onto five tapes

$$A_5^{L+1} = A_1^L$$
$$A_4^{L+1} = A_1^L + A_1^{L-1}$$
$$A_3^{L+1} = A_1^L + A_1^{L-1} + A_1^{L-2}$$
$$A_2^{L+1} = A_1^L + A_1^{L-1} + A_1^{L-2} + A_1^{L-3}$$
$$A_2^{L+1} = A_1^L + A_1^{L-1} + A_1^{L-2} + A_1^{L-3} + A_1^{L-4}$$

substituting f_L for A_1 gives

$$f_{L+1} = f_L + f_{L-1} + f_{L-2} + f_{L-3} + f_{L-4} \text{ for } L \geq 4$$
$$f_4 = 1$$
$$f_L = 0 \text{ for } L < 4$$

These are the Fibonacci numbers of order 4.

Figure 10–10 Relationship among entries in Figure 10–9

10.8 SUMMARY

In this chapter, we have examined some typical methods for sorting external files. As in so many other areas of computing, we have been able only to scratch the surface of this important application area; entire books can easily be (and are) devoted to the subject.

Indeed, on most large-mainframe systems, sort/merge subprograms are supplied as part of the standard libraries, and these have usually been thoroughly optimized to the underlying hardware and peripheral configuration. On the other hand, *someone* has to write these standard procedures, and it may turn out to be you at some point in your career. Besides, understanding the underlying algorithms of sort/merge programs will enable you better to assess the performance of the one available on your system, and to "tune" it if necessary.

CHAPTER 10 EXERCISES

1. Estimate the performance of the various merge-sort algorithms discussed in the chapter.
2. Write a program to carry out the simple merge sort discussed in section 10.4. If your computer does not allow (convenient) use of tapes, use disk files to simulate them.
3. Implement a polyphase sort as discussed in section 10.6. Use the "natural" distribution. Use disk files to simulate the tapes, if necessary.
4. Implement polyphase sort using the Fibonacci distribution, as presented in section 10.7.

BIBLIOGRAPHY

This bibliography is far from exhaustive and of course reflects the subjective judgments of its author. It is divided into three sections: other textbooks on data structures and algorithms, books on the Modula-2 language *per se*, and other relevant references. The Modula-2 list is annotated, because the listed books are very recent and quite varied in their approach to the language. The additional references are mainly a selection of papers with important original results, and recent surveys with extensive bibliographies of their own.

Two periodicals are of particular interest to Modula-2 enthusiasts *ACM SIGPLAN Notices* is the monthly publication of the Special Interest Group on Programming Languages of the Association for Computing Machinery, and often carries articles on Modula-2, especially discussions of shortcomings, standardization, possible extensions, and the like. *Journal of Pascal, Ada, and Modula-2* is published bi-monthly by John Wiley and Sons, and is entirely devoted to the three languages named in its title. Modula-2 and the other two languages receive about equal attention; applications and compiler comparisons appear frequently.

BOOKS ON DATA STRUCTURES AND ALGORITHMS

Aho, A.H., J.E. Hopcroft, and J.D. Ullman [1983]: *Data Structures and Algorithms.* Addison-Wesley, Reading, Massachusetts.

Amsbury, W. [1985]: *Data Structures from Arrays to Priority Queues.* Wadsworth Publishing Company, Belmont, California.

Baron, R.J., and L.G. Shapiro [1980]: *Data Structures and their Implementation.* PWS Publishers, Boston, Massachusetts.

Beidler, J. [1982]: *An Introduction to Data Structures.* Allyn and Bacon, Boston, Massachusetts.

Berztiss, A.T. [1975]: *Data Structures: Theory and Practice, 2d. ed.* Academic Press, New York, New York.

Coleman, D. [1979]: *A Structured Programming Approach to Data.* Springer Verlag, New York, New York.

Deo, N. [1975]: *Graph Theory with Applications to Engineering and Computer Science.* Prentice-Hall, Englewood Cliffs, New Jersey.

Feldman, M.B. [1985]: *Data Structures with Ada.* Reston Publishing., Inc., a Prentice-Hall Company, Englewood Cliffs, New Jersey.

Harary, F. [1969]: *Graph Theory.* Addison-Wesley, Reading, Massachusetts.

Horowitz, E., and S. Sahni [1978]: *Fundamentals of Computer Algorithms.* Computer Science Press, Rockville, Maryland.

Horowitz, E., and S. Sahni [1984]: *Fundamentals of Data Structures in Pascal.* Computer Science Press, Rockville, Maryland.

Knuth, D.E. [1973]: *Fundamental Algorithms, 2d. ed.* Addison-Wesley, Reading, Massachusetts.

Knuth, D.C. [1973]: *Sorting and Searching.* Addison-Wesley, Reading, Massachusetts.

Reingold, E.M., and W.J. Hansen [1983]: *Data Structures.* Little, Brown, Boston, Massachusetts.

Roberts, E.S. [1986]: *Thinking Recursively.* John Wiley and Sons, New York, New York.

Sincovec, R.F., and R.S. Wiener [1986]: *Data Structures Using Modula-2.* John Wiley and Sons, New York, New York.

Tenenbaum, A.M., and M.J. Augenstein [1986]: *Data Structures Using Pascal, 2nd ed.* Prentice-Hall, Englewood Cliffs, New Jersey.

Tremblay, J.P., and P.G. Sorenson [1984]: *An Introduction to Data Structures with Applications, 2d. ed.*, McGraw Hill, New York, New York.

Wirth, N. [1986]: *Algorithms and Data Structures.* [uses Modula-2] Prentice-Hall, Englewood Cliffs, New Jersey.

Zimmer, J.A. [1986]: *Abstraction for Programmers.* McGraw Hill, New York, New York.

BOOKS ON THE MODULA-2 PROGRAMMING LANGUAGE

Beidler, J., and P. Jackowitz [1986]: *Modula-2.* PWS Engineering and Computer Science, Boston, Massachusetts. Teaches Modula-2 as a first programming language; does not assume prior knowledge of Pascal.

Ford, G., and R.S. Wiener [1985]: *Modula-2: a Software Development Approach.* John Wiley and Sons, New York, New York. Assumes prior Pascal knowledge, covers all of Modula-2 and shows interesting applications, especially in the concurrent programming area.

Gleaves, R. [1984]: *Modula-2 for Pascal Programmers.* Springer Verlag, New York, New York. A rapid introduction focusing on the differences between Modula-2 and Pascal.

Knepley, E., and R. Platt [1985]: *Modula-2 Programming.* Can be used without prior Pascal experience, but not without programming experience.

Walker, Billy K. [1985]: *Modula-2 Programming with Data Structures.* Wadsworth Publishing Company, Belmont, California. Can be used without prior Pascal experience, but not without programming experience.

Weiner, R.S. [1986]: *Modula-2 Wizard.* John Wiley and Sons, New York, New York. Aimed at advanced programmmers with considerable Pascal experience. Covers the language concisely and thoroughly, but should not be mistaken for a text on the language.

Wirth, N. [1985]: *Programming in Modula-2, third corrected edition.* Written by the inventor of Modula-2, this is the "official" description of the language. Not intended to be read by those without a good bit of prior programming experience.

OTHER REFERENCES

Adelson-Velskii, G.M., and E.M. Landis [1962]: "An Algorithm for the Organization of Information," *Dokl. Akad. Nauk SSSE, Mathemat.,* **146**:2, pp 263–266. (balanced binary search trees).

Bauer, F.L., and K. Samelson [1960]: "Sequential Formula Translation," *Comm. ACM,* **2**:2, pp. 76–83. (translating with a stack).

Comer, D. [1979]: "The Ubiquitous B-Tree," *Comp. Surveys,* **11**:2, pp. 121–137.

Eppinger, J.L. [1983]: "An Empirical Study of Insertion and Deletion in Binary Search Trees," *Comm. ACM,* **26**:9, pp. 663–669.

Feldman, M.B. [1981]: "Information Hiding in Pascal: Packages and Pointers," *Byte,* **6**:11, pp. 493–498.

Feldman, M.B. [1984]: "Abstract Types, Ada Packages, and the Teaching of Data Structures," *Proc. Fifteenth SIGCSE Technical Symposium on Computer Science Education,* pp. 183–189.

Feldman, M.B. [1986]: "Modula-2 Projects for an Operating Systems Course," *Proc. Seventeenth SIGCSE Technical Symposium on Computer Science Education,* pp. 250–257.

Guttag, J.V., E. Horowitz, and D.R. Musser [1978]: "Abstract Data Types and Software Validation," *Comm. ACM,* **21**:12, pp. 1048–1064.

Hoare, C.A.R. [1962]: "Quicksort," *Comp. Journal,* **5**:1, pp. 10–5.

Liskov, B.H., and S.N. Zilles [1975]: "Specification Techniques for Data Abstractions," *IEEE Trans. on Software Engrng.,* **SE-1**:1, pp. 7–18.

Liskov, B.H., and S.N. Zilles [1977]: "Abstraction Mechanisms in CLU," *Comm. ACM,* **20**:8, pp. 564–576. (specification vs. implementation).

Lukasiewicz, J. [1951]: *Aristotle's Syllogistic from the Standpoint of Modern Formal Logic,* Clarendon Press, England.

Martin, W. [1971]: "Sorting," *Comp. Surveys,* **3** [4]: 147.

Maurer, W.D. [1968]: "An Improved Hash Code for Scatter Storage," *Comm. ACM,* **11**:1. (quadratic hashing).

Maurer, W.D., and T. Lewis [1975]: "Hash Table Methods," *Comp. Surveys,* **7**:1, pp. 5–19.

Nievergelt J. [1974]: "Binary Search Trees and File Organization," *Comp. Surveys,* **6**:3, pp. 195–207.

Perlis, A., and C. Thornton [1960]: "Symbol Manipulation by Threaded Lists," *Comm. ACM,* **3**:4, pp. 195–204. (threading binary trees).

Shaw, M. [1980]: "The Impact of Abstraction Concerns on Modern Programming Languages," *Proc. IEEE,* **68**:9, pp. 1119–1130.

Shell, D.L. [1959]: "A High Speed Sorting Procedure," *Comm. ACM,* **2**:7, pp. 30–32. (Shell sort).

Shell, D.L. [1971]: "Optimizing the Polyphase Sort," *Comm. ACM,* **14**:11, pp. 713–719.

Singleton, R.C. [1969]: "Algorithm 347: an Algorithm for Sorting with Minimal Storage," *Comm. ACM,* **12**:3 pp. 185–187 (Quicksort).

Stroustrup, B. [1982]: "Classes: an Abstract Data Type Facility for the C Language," *SIGPLAN Notices,* **17**:1, pp. 354–356.

Williams, J.W.J. [1964]: "Algorithm 232 (Heapsort)," *Comm. ACM,* **7**:6, pp. 347–348.

INDEX

Abstract data type (ADT), 6–8
  complexity reduction and, 7–8
  constructor operations of, 12, 22
  defined, 6–7
  fractions and, 9–17
    encapsulation of, 11, 19–20
  in older languages, 25–26
  for one-way linked lists, 99–102
  predicate operations of, 13
  selector operations of, 12–13, 22
  for text strings, 22–25
Abstraction, 1–5
Addition, matrix and vector,
    64–65, 68, 111–16
Address, hash, 236
Algorithm(s), 29–54
  card sorter, 279–81
  data structures and, 30
  defined, 30
  design, 30
  growth rates. See Growth rate of
    algorithms
  for maintaining dynamic table,
    49–53
  performance prediction and,
    39–49
    examples, 48–49
  recursive, 29, 31–39
    binary search, 36–38, 48
    merge/sort, 38–39, 49
    permutations of a set, 33–36,
      48
    reverse of a string, 32–33, 48
  sorting
    binary insertion, 256
    bubble, 252–54
    delayed selection, 251–52
    heap sort, 260–71
    linear insertion, 254–56
    merge, 257–60
    quicksort, 271–74
    simple, 250–51
  traversal
    breadth-first search, 163,
      173–77
    depth-first search, 163, 173,
      174, 175
    left-node-right (inorder), 193,
      194, 198, 200, 202
    left-right-node (postorder),
      193, 194, 199
    node-left-right (preorder),
      193, 194, 198
Allocate procedure, 95–99
Allocation of memory, 20–22
  for linked structures, 97–111
  nonstandard, 18

pointers and, 94–97
simulating, 119–27
See also Storage
Almost-heap, 261, 262–63, 281
Antibugging, 135
Antisymmetry, digraph, 166, 167
Arcs of digraph, 163–64
Arithmetic
  expressions. See Expression trees
  matrix, 68–70
  vector, 63–68, 111–16
  See also specific arithmetic
    operations
Array(s)
  of characters, strings as, 22
  column-major implementation,
    60
  declaration of size of, 56
  in Fortran, 56
  higher dimensional, 58–61
  indexed, 79–81
  key-ordered, 51–53
  one-dimensional, 56–58
  open array parameters, 63–68
  queue implementation using, 135
  circular, 135–41
  row-major implementation,
    60–61
  stack implementation using,
    142–43
  two-dimensional, 1, 3–4, 58–61
  unordered
    deleting items in, 51
    searching, 50–51
  See also Matrix (matrices);
    Vector(s)
Associative rule, 148

Backus, John, 56
Band matrices, 73–74
Bandwidth, 73
Benchmarks, 39
"Big O" notation, 29, 39–49
  See Also Growth rate of
    algorithms
Binary insertion sort, 256
Binary search
  performance of, 48
  recursive, 36–38, 48
Binary search trees (BSTs),
    200–208, 222
  balanced (AVL), 204
  delete operation in, 205–8
  search operation in, 204–5, 206
  traversing, 200
  update operation in, 201–4, 205

Binary tree (s), 188–95
  almost complete, 189–91,
    260–61
  balanced (height-balanced), 190,
    191
  complete, 188
  implementing, 191–93
  properties of, 188–91
  strictly, 188
  traversing, 193–95
Boolean (adjacency) matrix,
    169–70
Breadth-first search, 163, 173–77
B-trees, 182, 211–15
  of order K, 213–14
Bubble sort, 252–54
Bucket hashing, 240, 243–44
Buffer, 5

Card sorter, 279–81
Client program, 8–9
  use of library module in, 15–17
Closed hashing, 240
Clustering, 243
Code, hash, 236
Collisions in hash tables, 234, 237
  resolving, 240–45
    bucket hashing, 243–44
    linear probing, 240–43
    nonlinear probing, 243
    ordered hashing, 244–45
Column-major array
    implementation, 60
Comparison operation, 23
Compiler, 3
Complexity, abstract data type and
    reduction of, 7–8
Concatenation operation, 22
Concordance, 217
Connectivity, digraph, 167–68
  strong, 168
Constants, encapsulation of, 8
Constructor operation, 12, 22
Count, trip, 43
Cross (orthogonal) lists, 116–17
Cross-reference generator, 215–27
  scanner (parser), 221–27
  table handler, 218–21
Cursor pointer, 124
Cycles, digraph, 166–67

Data structures. See Structures,
    data
Deallocate procedure, 129–31
Decision (*if-then-elseif-else*)
    structure, 40, 41–43

297

Declaration, array size, 56
Defensive programming, 135
Delayed selection sort, 251–52
Depth-first search, 163, 173, 174, 175
Depth of tree, 183–85, 187
Dequeue constructor, 134, 135, 142, 281
Dereferencing operation, 18–19
Design language, 7
Digital search trees, 182, 208–11
Digraphs. *See* Graphs, directed (digraphs)
Division hash function, 237, 238
Divisor, greatest common (GCD), 10
Dot product, 65–67
Dynamic memory allocation. *See* Memory allocation

Edges of digraph, 163–64
Encapsulation, 7–8
 of fractions in ADT, 11, 19–20
 See also Module(s)
Enqueue operation, 134, 281
Equivalence relation, 180
Error handling, 66–67
Event-driven simulation, 155–56, 157–60
Exception handling and reporting, 66–67
Export, 8, 9, 16–17
 opaque, 19–20
Expression evaluation, stacks and, 144–50
Expression trees, 195–200
 constructing, 195–99, 215
 traversing, 199–200

Factorial, 31–32
 performance of, 48
Fibonacci distribution, 290–92
Fields, 104
FIFO (first in, first out) structure, 133–34
File(s)
 blocked, 1, 4–5
 sorting, 283–92
  distribution phase of, 283
  Fibonacci distribution for polyphase merging, 290–92
  K-way merge, 285
  "natural" distribution of subfiles, 287–88
  polyphase, 288–90
  simple merge sorting, 285–87
  two-way merge, 283–85
 symbol, 9
Finite-state machine, 178, 180, 221
 See also Cross-reference generator; Scanner, lexical

Forest, 209
Fortran, 3, 56
Fractions, 9–17
 classifying operations, 11–13
 encapsulating in ADT of, 11, 19–20
 implementation of, 11
 linked lists and, 127–31
 as mathematical entities, 10–11
 record structure for, 11
 using pointers, 18–19
Function, hashing, 234, 236, 237–40
 choosing, 237–40
Function/expression notation, 17–22

Garbage collector, 95, 127
Graphs, directed (digraphs), 163–81
 components, 163
 graph traversals, 173–77
 implementations of, 169–73
 lexical scanner, 163, 177–79
 properties of, 164–69
 state, 172–73, 177–79, 221, 223–24
 tree as, 163
 See also Tree(s)
Greatest common divisor (GCD), 10
Growth rate of algorithms, 39, 40–48
 constant, 40
 estimating, 40–48
  counting loops and, 43–45
  decision (*if-then-elseif-else*) structure and, 40, 41–43
  multiplicatively controlled loop and, 45–47
  sequence of simple statements and, 40, 41
  simple statements and, 41
  subprogram calls and, 47–48
 of internal sorts
  $O(N^2)$ rate, 249–56
  $O(N\log(N))$ rate, 257–74
 linear, 40
 logarithmic, 40
 $N \log N$, 40
 polynomial, 45
 quadratic, 40

Hashing, 234–47
 bucket, 240, 243–44
 closed, 240
 collisions (clashes) in, 234, 237
  resolving, 240–45
 division, 237, 238
 function, 234, 236, 237–40
  choosing, 237–40

hash address (code), 236
hybrid search strategies, 245
mid-square, 237, 238–39
ordered, 244–45
partitioning (folding), 237, 240
quadratic, 243
reporting in, 237
sequential and binary searches and, 235
synonyms, 234, 237
truncation, 237, 238
Head of linked lists, 98–101
Heap, 18, 94
 almost-, 261, 262–63, 281
 creating, 261–62
 garbage collection, 95, 127
 manager, 94
 sort, 260–71
  almost complete binary tree and, 189, 260–61
  converting almost-heap to heap, 262–63
  practicality of, 264–71
 See also Storage
Hiding information, 3

If-then-elseif-else structure, 40, 41–43
Implementation, 1–5
Import, 8, 9, 16–17
Increment function, 240
In-degree of vertex, 168–69
Indexed arrays, 79–81
Infix notation, 144
 converting to Polish notation from, 147–50
 algorithm for, 150–55
Information hiding, 3
Initializing loops, 45–46
Inner (scalar, dot) product, 65–67
Insertion sort
 binary, 256
 linear, 254–56
In situ sort, 249, 254, 255, 260
Internal sorting. *See* Sort(ing)
Irreflexivity, digraph, 165

Key-ordered arrays, 51–53
Key-to-address transformation, 234, 236
K-way merging, 285

Language(s), programming
 abstract data types in, 25–26
 design (pseudocode), 7
 high-level- (HLL), 2
 without record types, 86–87
Left-node-right traversal, 193, 194, 198, 200, 202
Left-right-node (postorder) traversal, 193, 194, 199

Lexical scanner, 163, 177–79, 221–26
Library module, 8–9
  in client program, 15–17
LIFO (last in, first out) structure, 133, 134, 151–52, 153
Linear growth rate, 40
Linear insertion sort, 254–56
Linear probing, 240–43
Linked lists, 82, 93–132
  dynamic allocation for, 97–111
    pointers and, 94–97
    simulating, 119–27
  fractions and, 127–31
  head of, 98–101
  node (cell) in, 97
  one-way, 97–99
    abstract data type for, 99–102
    with head and tail pointers, 102–4
    ordered, 104–9
  queue implementation using, 140, 141
  sparse matrices and, 116–17
  sparse vectors as, 111
  stack implementation using, 144
  text handling and, 117–19
  two-way, 109–11
  vector arithmetic and, 111–16
  See also Tree(s)
Lists
  adjacency, 170–71
  cross (orthogonal), 116–17
Logarithmic growth rate, 40
Loops
  counting, 43–45
  initialization of, 45–46
  modification step in, 46
  multiplicatively controlled, 45–47
  termination condition, 46
Lukasiewicz, Jan, 144

Machine, finite-state, 178, 180, 221
  See also Cross-reference generator; Scanner, lexical
Magnetic tape, 4–5
Magnitude, order of. *See* "Big O" notation
Manager, heap, 94
Mapping, storage, 57–58, 59, 84–86
Matrix (matrices)
  adjacency (Boolean), 169–70
    weighted, 172
  arithmetic, 68–70, 111–16
    addition (sum), 64–65, 68
    multiplication (product), 68, 69–70, 116
  band, 73–74
  classical representation of, 55–63
  dynamic storage scheme for, 84–85
  functional notation and, 83–86
  general module for, 85–86
  lower triangular, 71–73
  sparse, 55, 75–79, 82
    linked-list implementation of, 116–17
  symmetric, 73
  transpose of, 68–69
  two-dimensional arrays and, 58–61
  upper triangular, 73
  with "zero" elements, 71–74
Memory allocation, 20–22
  for linked structures, 97–111
  nonstandard, 18
  pointers and, 94–97
  simulating, 119–27
  See also Storage
Merge sort, 257–60
Merging, 38–39, 49, 283
  K-way, 285
  polyphase, 290–92
  recursive, 38–39
    performance of, 49
  two-way, 283–85
  See also Sort(ing)
Mid-square hash function, 237, 238–39
Module(s), 1
  definition, 8, 9
  implementation, 8, 9
  library, 8–9
    use in client programs, 15–17
  for matrices, 85–86
Multilist structure, 117
Multiplication
  matrix, 68, 69–70, 116
  vector, 65–67, 111–16
Multi-user operating system, 281

"Natural" distribution of subfiles, 287–88
NIL value, 97
Node (cell), in linked structure, 97
Node-left-right (preorder) traversal, 193, 194, 198
Nodes of trees, 191–93
Nonlinear probing, 243
Notation
  "big O," 29, 39–49
  function/expression, 17–22
  infix, 144, 147–55
  for matrices and vectors, 83–86
  Polish. *See* Polish (postfix) notation (RPN)
  for subscripts, 56
Numbers, 1, 2–3
  See also Fractions

Opaque export, 19–20
Opaque type vectors, 84
Open array parameters, 63–68
Operating system, multi-user, 281
Operations
  on binary search trees
    delete, 205–8
    search, 204–5, 206
    update, 201–4, 205
  classifying, 11–13
  comparison, 23
  concatenation, 22
  constructor, 12, 22
  dereferencing, 18–19
  matrix atithmetic, 68–70
  predicate, 13
  primitive, 142
  priorities (precedences) of, 149–50, 154, 155
    overriding with parentheses, 152, 155
  on queues, 134–35
  selector, 12–13, 22
  SET, 174
  stack, 141
  vector arithmetic, 63–68, 111–16
  See also specific data types
Ordered hashing, 244–45
Ordered pairs, 164
Orthogonal lists, 116–17
Out-degree of vertex, 168–69

Packaging, 7–8
Pairs, ordered, 164
Palindrome, 32
Parameters, open array, 63–68
Parsers, 221–27
Partial ordering relation, 181
Partitioning (folding) hash function, 237, 240
Partition point, 272–74
Partition sort, 271
Paths
  digraph, 166
  tree, 183
Performance prediction, 29, 30, 39–49
  examples, 48–49
  See also Growth rate of algorithms
Permutations of set, 33–36
  performance of, 48
Pointers, 17–22
  cursor, 124
  defined, 18
  dynamic memory management and, 94–97
  head and tail, 102–4
  See also specific data types
Points of digraph, 163–64

Polish (postfix) notation (RPN), 144–50
  converting from infix form to, 147–50
    alogrithm for, 150–55
    evaluating, 145–47
Polynomial growth rate, 45
Polyphase sorting/merging, 288–92
  Fibonacci distribution for, 290–92
Predicate operation, 13
Prediction, performance, 29, 30, 39–49
  examples, 48–49
  See also Growth rate of algorithms
Primitive operations, 142
Priorities of operations, 149–50, 154, 155
  overriding with parenthesis, 152, 155
Priority queues, 281
Probing
  linear, 240–43
  nonlinear, 243
Procedures, encapsulation of, 8
Product (multiplication)
  matrix, 68, 69–70, 116
  vector, 65–67, 111–16
Program, client. *See* Client program
Programming
  defensive, 135
  high-level-language (HLL), 2
  See also Cross-reference generator
Pseudocode, 7

Quadratic growth rate, 40
Quadratic hashing, 243
Quadratic selection sort, 275–77
QueueFirst selector, 134
QueueInit constructor, 134, 135
QueueIsEmpty predicate, 134
Queues, 133–41
  access to, 133
  array implementation of, 135
  circular, 135–41
  in breadth-first searches, 174
  in event-driven simulation, 155–56, 157–60
  linked-list implementation of, 140, 141
  operations, 134–35
  priority, 281
Quicksort, 271–74

Radix sort, 278–81
Rational numbers. *See* Fractions
Records, 6
  languages without, 86–87

structure of, for fractions, 11
  vectors as, 57
Recursion, 29, 30, 31–39
  binary search, 36–38, 48
  data structures and, 93
  data types and, 97
  defined, 29
  merge/sort, 38–39, 49
  permutations of a set, 33–36, 48
  reverse of a string, 32–33, 48
  See also Algorithm(s)
Reduction of complexity, 7–8
Reflexivity, digraph, 165
Relation
  equivalence, 180
  partial ordering, 181
Reporting
  exception, 66–67
  in hashing, 237
Resources, export and import of, 8, 9, 16–17
Reverse Polish notation. *See* Polish (postfix) notation (RPN)
Root of tree, 183
Row-major array implementation, 60–61
RPN. *See* Polish (postfix) notation (RPN)
Rule, associative, 148

Scalar product, 65–67
Scalar type, 6
Scanner, lexical, 163, 177–79, 221–26
Search(ing)
  binary, 36–38, 48, 235
    performance of, 48
    recursive, 36–38, 48
  breadth-first, 163, 173–77
  depth-first, 163, 173, 174, 175
  hybrid strategies, 245
  sequential, 235
  of unordered arrays, 50–51
  using binary search trees, 200–208, 222
  using digital search trees, 208–11
  See also Hashing
Selection sort
  delayed, 251–52
  quadratic, 275–77
  simple, 250–51
Selector operation, 12–13, 22
Self-referencing data type, 97
Sequential searching in hashing functions, 235
SET operations, 174
Sets, permutations of, 33–36, 48
Shaker sort, 254
Shell sort, 275–77
Simulation
  event-driven, 155–56, 157–60

of finite-state machine, 178, 180
of memory allocation, 119–27
Size, array, 56
Sort(ing), 38–39, 49
  card, 279–81
  external (files), 283–92
    distribution phase of, 283
    Fibonacci distribution for polyphase merging, 290–92
    K-way merge, 285
    "natural" distribution, 287–88
    polyphase, 288–90
    simple merge, 285–87
    two-way merge, 283–85
  in situ, 249, 254, 255, 260
  internal, 248–82
    binary insertion, 256
    bubble sort, 252–54
    delayed selection, 251–52
    with growth rate $O(N^2)$, 249–56
    with growth rate $O(N \log(N))$, 257–74
    heap sort, 189, 260–71
    linear insertion, 254–56
    merge sort, 256–60
    priority queues, 281
    quicksort, 271–74
    radix sort, 278–81
    shaker, 254
    shell sort, 275–77
    simple selection, 250–51
    terminology, 249
  partition, 271
  recursive, 38–39
    performance of, 49
  stable, 203
  See also Merging
Sparse matrices, 55, 75–79, 82
  linked-list implementation of, 116–17
Sparse vectors, 55, 75–79
  indexed, 76–78
  linked-list implementation of, 111
Spreadsheet programs, 74
Stable sort, 203
StackInit constructor, 141
StackIsEmpty predicate, 141
StackOverflow flag, 143
StackPop constructor, 141, 142
StackPush constructor, 141
Stacks, 133–34, 141–50
  access to, 133
  array implementation of, 142–43
  expression evaluation and, 144–50
  linked-list implementation of, 144
  operations, 141
  Polish notation and, 144–50
StackTop selector, 141, 142

StackUnderflow flag, 143
Statements, sequence of simple, 40, 41
State (transition) graph, 172–73, 177–79, 221, 223–24
Storage
  data type and, 57
  mapping, 57–58, 59, 84–86
  for matrices, 84–85
    band, 74
    symmetric, 73
    triangular, 72–73
  pool, 94
  scatter. *See* Hashing
  for vectors, 84–85
  See also Heap; Memory allocation
Strings
  abstract data type for, 22–25
  as array of characters, 22
  reverse of, 32–33, 48
    performance of, 48
  variable-length, 22
Structures, data, 6–7
  algorithms and, 30
  linked, 94
    See also Linked lists
  recursive, 93
  See also specific types of structures
Subfiles, "natural" distribution of, 287–88
Subprogram calls, estimating growth rate of, 47–48
Subroutines, 2
Subscripts, 56
  range of, 58
Subtrees, 183, 186
Sum (addition), matrix and vector, 64–65, 68, 111–16
Symbol file, 9
Symmetry, digraph, 165–66, 167
Synonyms in hash tables, 234, 237

Table(s)
  in cross-reference generator, 218–21
  dynamic, maintaining, 49–53
  state (state graph), 172–73, 177–79, 221, 223–24
  See also Array(s); Graphs, directed (digraphs); Matrix (matrices); Vector(s)
Tape, magnetic, 4–5
Termination, loop, 46
Text analysis. *See* Cross-reference generator
Text handling with linked lists, 117–19
Threading trees, 227–29
Trade-offs, 30

Transformation, key-to-address, 234, 236
Transitivity, digraph, 166
Transpose of matrix, 68–69
Traversal (s)
  algorithms
    breadth-first search, 163, 173–77
    depth-first search, 163, 173, 174, 175
    left-node-right (inorder), 193, 194, 198, 200, 202
    left-right-node (postorder), 193, 194, 199
    node-left-right (preorder), 193, 194, 198
  defined, 163
  of graph, 173–77
  threading, 227–29
  of trees, 182
    binary, 193–95
    binary search, 200
    expression, 199–200
Tree(s), 183–233
  binary, 188–95
    almost complete, 189–91, 260–61
    balanced (height-balanced), 190, 191
    complete, 188
    implementing, 191–93
    properties of, 188–91
    strictly, 188
    traversals of, 193–95
  binary search (BSTs), 200–208, 222
    balanced (AVL), 204
    delete operation in, 205–8
    search operation in, 204–5
    traversing, 200
    update operation in, 201–4
  cross-reference generator and, 215–27
    scanner (parser), 221–27
    table handler, 218–21
  defined, 182
  depth of, 183–85, 187
  as digraph, 163
  expression, 195–200
    constructing, 195–99, 215
    traversing, 199–200
  general, 208–15
    B-, 182, 211–15
    digitial search, 182, 208–11
  nodes of, 191–93
  paths of, 183
  root of, 183
  subtrees, 183, 186
  threading, 227–29
  traversals and, 182
  vertices, 183

Triangular matrices, 71–73
Trip count, 43
Truncation function, 237, 238
Types, data, 6–7
  encapsulation of, 8
  enumeration, 177, 179, 223
  opaque, 84
  recursive ("self-referencing"), 97
  storage requirements and, 57
  See also Abstract data type (ADT); specific types of data

Until structure, 45–46

Variables, 1, 2–3
  encapsulation of, 8
  pointer. *See* Pointers
  temporary, deallocation of, 129–30
Variations, performance. *See* "Big O" notation.
Vector (s)
  arithmetic, 111–16
    addition, 64–65, 68
    multiplication, 65–67
    open array parameters and, 63–68
  classical representation of, 55–63
  dynamic storage scheme for, 84–85
  functional notation and, 83–86
  one-dimensional arrays and, 56–58
  opaque type, 84
  as records, 57
  sparse, 55, 75–79
    indexed, 76–78
    linked-list implementation of, 111
  with "zero" elements, 71–74
Vertex (vertices), 163–64
  accepting and rejecting states, 177
  children of, 186
  current state, 177
  descendants of, 186
  in-degree, 168–69
  interior (nonterminal), 183
  leaf (terminal), 183
  level of, 185
  next state, 177
  out-degree of, 168–69, 183
  parent, 186
  siblings of, 186
  start state, 177
  visiting (touching), 173

Weights, 172, 177
While structure, 45–46

"Zero" elements, matrices and vectors with, 71–74